Chemical Sensors

Chemical Sensors

edited by
T.E. EDMONDS
Department of Chemistry
Loughborough University of Technology
Loughborough, UK

Blackie

Glasgow and London

Published in the USA by
Chapman and Hall
New York

1988

CHEMISTRY

Blackie and Son Ltd.
Bishopbriggs, Glasgow G64 2NZ
7 Leicester Place, London WC2H 7BP

Published in the USA by
Chapman and Hall
in association with Methuen, Inc.
29 West 35th Street, New York, NY 10001–2291

British Library Cataloguing in Publication Data

Chemical sensors.
 1. Chemical detectors
 I. Edmonds, T.E.
 543   TP159.C46

 ISBN 0–216–92255–0

Library of Congress Cataloging-in-Publication Data

Chemical sensors.

 Includes bibliographies and index.
 1. Chemical detectors. I. Edmonds, T.E.
TP159.C46C48 1987   660.2′8′00287   87–17841
ISBN 0–412–01601–X

Phototypeset at Thomson Press (India) Limited, New Delhi
Printed in Great Britain by Bell and Bain Ltd., Glasgow.

Preface

At the beginning of this book, and in the absence of guidance from IUPAC, it is appropriate to clarify the term 'chemical sensor'. A chemical sensor may be defined as a simple-to-use, robust device that is capable of reliable quantitative or qualitative recognition of atomic, molecular or ionic species. It is hard to imagine a field of applied chemistry in which a significant impact could not be made by such a device. Undoubtedly, it is this potential that has fuelled the contemporary preoccupation with chemical sensors. An unfortunate side-effect of this otherwise welcome interest is the use of the term 'chemical sensor' to add the chemical equivalent of a 'High-Tech gloss' to a rather ordinary device, publication, conference or research group. This loose usage of terminology is responsible in part for the ambiguity that surrounds many chemists' concepts of the form and function of chemical sensors. Further ambiguity arises from the extravagant claims that have been made for some sensors, and the impression that has been given of much 'verging-on-a-breakthrough' research. The research chemist engaged in sensor development should be mindful of the fact that the ultimate target for these devices is the real world, and that a successful laboratory device operating under well-defined conditions and careful calibration does not constitute a chemical sensor.

Research into chemical sensors is not a recent phenomenon; it has been under way for over 80 years. Indeed, two of the most successful devices, the glass pH electrode, and the non-dispersive infrared system, have been in routine use since the 1930s. Current research activity in sensors is more clearly identifiable, mainly because it is organized into coherent programmes, but the essential structure of a chemical sensor and the problems to be solved in developing one remain the same. A chemical sensor consists of two parts, a zone of selective chemistry, and a more or less non-specific transducer. The selective chemistry provides an interface between the transducer and a specific chemical parameter (usually concentration) of the target analyte. The function of this interface is twofold. Firstly, it must *selectively* interact with the target. Secondly, it must *transform* the desired chemical parameter into a chemical or physical signal to which the transducer responds. Clearly, the behaviour of the chemistry in this interface zone is of crucial importance to the overall performance of the sensor. Accordingly, the two major problems to be solved in sensor research concern the development of the interface chemistry, and the localization of this chemistry in or on an appropriate transducer.

The layout of this book reflects the nature of the problems mentioned in the previous paragraph. Species recognition lies at the heart of chemical sensing, and fundamental approaches to this process are discussed in the first two sections. Molecular and ionic recognition is a primary biological process, and not surprisingly there is a range of chemistries that can be exploited for sensing: Part 1 deals with these. Part 2 consists of a review of the efforts of synthetic organic chemists to produce structures which react in a highly specific way with target species. The material of Part 3 takes up the themes of the previous chapters, but with a more practical emphasis. The chapters in this section are written by authors experienced in the application of specific chemistry. Each chapter reflects the contributor's knowledge of the realities of implementing their specialist chemistry, often coupled with their vision for its future development. The final two parts of the book deal with transducers: an arbitrary division has been made between electrochemical and other transducers. Although some of the chapters in these sections contain appropriate reviews of current applications of the subject transducer in chemical sensing, this is by no means the primary aim of the contribution. In each case the authors have attempted to give fundamental information relating to the *modus operandi* of the transducer; information that is intended to help the newcomer in assessing the applicability of a transducer to a particular sensing problem.

This book is intended to be a handbook, giving practical information to its readers, as well as supplying ideas for future use. A broad range of chemistries are drawn together in this volume, the chapters of which are written by authors from industrial, academic, clinical and government laboratories. It is not an up-to-the-minute review of every piece of chemical sensor research; that is a job best left to the specialist journals. Well-trodden paths have been deliberately avoided, as has the tendency to dwell on 'Friday afternoon laboratory curiosities' that work at the end of the week but fail on Monday. Finally, this book is a team effort, and I am grateful to all the contributors for their work, and to the publishers for their forbearance during the long gestation period. From a personal standpoint I am happy to acknowledge the help and guidance I have had from Professor T.S. West under whose supervision I embarked upon chemical sensor research in 1973. In the early stages of this book discussions with Dr J.F. Alder helped to sharpen my ideas on the objectives. His clarity of thought and his candour were most welcome. My family have supported me admirably during this time, and since in many respects the greatest beneficiaries of sensor research should be people like them who will experience improved health care and a cleaner and safer environment, I dedicate this book to Diana, Thomas and Matthew.

TEE

Contributors

G.J. Bastiaans, Integrated Chemical Sensors, 44 Mechanics Street, Newton, MA 02164, USA

P.D. Beer, Department of Chemistry, The University of Birmingham, P.O. Box 363, Birmingham B15 2TT, UK

R.E. Belford, Department of Electrical Engineering, University of Edinburgh, King's Buildings, Edinburgh EH9 3JL, UK

B.J. Birch, Unilever Research, Colworth House, Sharnbrook, Bedford MK44 1LQ, UK

W.H. Dorn, Chemical Sensors Group, Department of Chemistry, University of Toronto, Toronto M5S 1A1, Canada

T.E. Edmonds, Department of Chemistry, Loughborough University of Technology, Loughborough LE11 3TU, UK

W.J. Feast, Department of Chemistry, University of Durham, Durham DH1 3LE, UK

S.J. Gentry, Health and Safety Executive, Steel City House, West Street, Sheffield S1 2GQ, UK

A.L. Harmer, Battelle-Europe, 1227 Carouge, Geneva, Switzerland

R.G. Kelly, Department of Electrical Engineering, University of Edinburgh, King's Buildings, Edinburgh EH9 3JL, UK

L.J. Kricka, Department of Clinical Chemistry, Wolfson Research Laboratories, Queen Elizabeth Medical Centre, Birmingham B15 2TH, UK

G.J. Moody, Department of Applied Chemistry, UWIST, P.O. Box 13, Cardiff CF1 3XF, UK

R. Narayanaswamy, Department of Instrumentation and Analytical Science, UMIST, P.O. Box 88, Manchester M60 1QD, UK

A.E. Owen, Department of Electrical Engineering, University of Edinburgh, King's Buildings, Edinburgh EH9 3JL, UK

I. Robins, Thorn EMI Central Research Laboratories, Dawley Road, Hayes, Middx UB3 1HH, UK

N.J. Seare, Department of Chemistry, Loughborough University of Technology, Loughborough LE11 3TU, UK

J.D.R. Thomas, Department of Applied Chemistry, UWIST, P.O. Box 13, Cardiff CF1 3XF, UK

M. Thompson, Chemical Sensors Group, Department of Chemistry, University of Toronto, Toronto M5S 1A1, Canada

G.G. Wallace, Department of Chemistry, University of Wollongong, P.O. Box 1144, Wollongong, NSW 2500, Australia

Contents

PART 5 NON-ELECTROCHEMICAL TRANSDUCTION

xivCONTENTS

1

MOLECULAR AND IONIC RECOGNITION BY BIOLOGICAL SYSTEMS

1 Molecular and ionic recognition by biological systems

L.J. KRICKA

1.1 Introduction

Recognition of a molecule by another molecule or group of molecules is a fundamental process of vital importance in all biological systems. Nature has evolved a vast array of biomolecules and biomolecular structures which exhibit an exquisite specificity in their molecular recognition properties.

Molecular recognition underlies many essential biological processes. For example, resistance to disease relies on the presence of antibody molecules which recognize and combine with specific molecules on the surface of an invading organism. Tissue is recognized by the immune system as compatible or incompatible (foreign) via tissue proteins coded by the genes of the Major Histocompatibility System. Olfaction and taste depend on the interaction of molecules with specific chemoreceptors and the subsequent neural coding of this information (1). Such biological systems provide the analyst with a rich source of molecules with specific binding properties, and some examples are listed in Table 1.1. By far the most important and versatile source are the immunoglobulins produced by the immune system, because it is possible to induce the production of specific immunoglobulins which will bind to a particular substance by immunizing animals with that substance. Many hundreds of different immunoglobulin molecules with diverse binding specificities have been produced in this way, and the technique has been refined through the development of monoclonal antibody technology (see section 1.3.1).

An effective sensor requires a molecular recognition component which is

Table 1.1 Range of molecules with molecular recognition properties and their specificities

Binder	Substance(s) bound
Immunoglobulins	Wide range of small and large molecules
Enzymes	Wide range of molecules
Lectins	Oligosaccharides
Receptors	Hormones
Avidin	Biotin
DNA	DNA, RNA
Protein A	IgG, IgM, IgA

capable of recognizing and binding one particular molecule amongst a mixture of molecules. The objective of this chapter is to survey the range of binding specificities of biomolecules and biomolecular structures (receptors) which may be useful as the molecular recognition component of a chemical sensor.

1.2 Characteristics of molecular recognition systems

1.2.1 *Specificity*

Ideally a molecular recognition system should exhibit specificity, i.e. it should only bind one particular molecule and not bind other types of molecules to any appreciable extent. This ideal is rarely achieved, and, in general, molecular recognition systems will bind a range of molecules, albeit to differing extents. For example, the enzyme glucose oxidase (EC 1.1.3.4) has specificity for beta-D-glucose which it binds and transforms to gluconolactone. However the enzyme will also bind other related substances such as 2-deoxy-D-glucose, 6-deoxy-6-fluoro-D-glucose, 6-methyl-D-glucose, and 4,6-dimethyl glucose (2). Characterization of the specificity of a particular binding system is an important prerequisite before it is used analytically. For the binding of antigens to immunoglobulins the degree of non-specificity in binding is measured by the 'cross-reactivity' of the immunoglobulin. Cross-reactivity is determined from results of competitive immunoassay standard curves obtained by incubating a limited amount of immunoglobulin and a fixed amount of labelled antigen with increasing amounts of the antigen or increasing amounts of the possible cross-reacting substance (3). Cross-reaction is then defined as the ratio of the weight of cross-reactant which reduces the binding of labelled antigen to 50% of the value in the absence of antigen (expressed as a percentage).

1.2.2 *Binding constants*

Interaction between a molecule (M) and a substance (B) with binding properties for that molecule can be described by the equation:

$$M + B \underset{k_2}{\overset{k_1}{\rightleftharpoons}} M{:}B$$

(k_1, rate of association; k_2, rate of dissociation). The affinity of the binder for the molecule is described by a binding constant (K) where:

$$K = [M{:}B]/[M][B] = k_2/k_1$$

For multivalent interactions (e.g. of antibodies with antigens) the term *avidity* has been introduced in order to emphasize the stabilization of complexes by the multiple binding interactions. Binding constants vary widely, from $10^3 \, \text{L M}^{-1}$ for lectins, up to $10^{15} \, \text{L M}^{-1}$ for the avidin–biotin system.

1.2.3 Chemical basis of binding

Hydrophilic, hydrophobic and hydrogen bond interactions are all involved to varying extents in biomolecular binding reactions. The majority of work in this area has centred on the interaction of enzymes with their substrates and the binding of protein antigens to antibodies. Catalytically important amino acid residues in the active site have been determined for several enzymes (e.g. thermolysin) and enzyme–substrate binding interactions have been studied using molecular modelling systems (4, 5). X-ray analysis of several enzymes has revealed the presence of a cleft, crevice or depression in the globular protein structure. This is the site of the active centre of the enzyme which binds the substrate and mediates the metabolic transformation. Figure 1.1 illustrates structural data for the zinc metallo-enzyme carboxy-peptidase A and shows the location of the key residuals Tyr 248, Glu 270, and Arg 145 in the active site.

The identification of which part of the surface of a protein (antigenic site, antigenic determinant or epitope) is in contact with the binding site of an antibody has received considerable attention. The antigenic site may be a continuous segment of polypeptide chain or may consist of two or more segments brought together in the tertiary structure of the protein. This latter type of antigenic site has been variously named as a neotope, topographical, conformation-dependent, or discontinuous determinant (6). Antigenic sites usually contain charged and polar amino acid residues (7, 8) although hydrophobic interactions also play a role in the binding reaction. Antigenically reactive regions of a protein are usually small, typically 6–7 amino acid residues. Binding to antibody is primarily via ionic reactions involving charged amino acids, with non-polar amino acids providing a stabilizing effect via hydrophobic interactions (7). Hydrophilic segments appear to be important in a wide range of protein macromolecule interactions (8); for example, complement binds to the most hydrophilic region of the Fc portion of an immunoglobulin (9), and DNA polymerase II binds to nucleic acids via its most hydrophilic segment (10).

1.3 Selected molecular and ionic recognition systems

1.3.1 Antibodies

The immune system is a highly developed and efficient molecular recognition system which has been extensively exploited for the production of antibodies with specific binding properties. *In vivo* a foreign molecule (an immunogen) induces B lymphocytes to proliferate and differentiate into plasma cells which

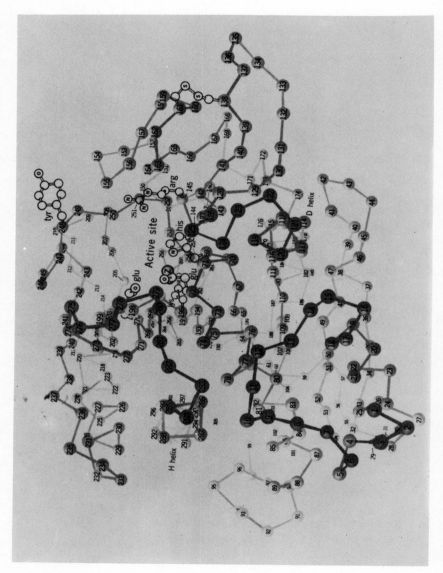

Figure 1.1 Structure of main chain of carboxypeptidase A showing location of the active site and the side chains of the active residues Tyr 248, Glu 270, and Arg 145. (Reproduced with permission from R.E. Dickerson and I. Geis (1970), *The Structure and Action of Proteins*, Harper and Row, New York.)

secrete antibodies with specific binding properties for the set of antigen determinants (epitopes) on the immunogen (11). An immunogen can induce, therefore, the production of a large number of antibodies (polyclonal response) with a range of binding constants, and it has been estimated that even a single antigenic determinant may induce as many as 10^4 different antibody molecules with a range of binding constants (12).

Several different classes of antibody (immunoglobulin isotypes) have been recognized; these have been classified into five major groups—IgA, IgD, IgE, IgG and IgM (Table 1.2). Most attention has focused on immunoglobulins of the IgG class. IgG is composed of two pairs of polypeptide subunits, designated heavy (H) and light (L) chains, linked via disulphide bridges as shown in Figure 1.2. The C-terminal part of each subunit has a region of constant amino acid sequence, whilst the N-terminal has a variable amino acid sequence. Binding to antigen occurs via the N-terminal part of the molecule. Controlled enzymatic digestion (e.g. by papain) produces three fragments—two identical Fab fragments and an Fc fragment. The Fab

Table 1.2 Immunoglobulins

Immunoglobulin	Molecular weight (kD)	Carbohydrate (%)	Valence for antigen binding
IgA1	160	7–11	2
IgA2	160	7–11	2
IgD	184	9–14	?
IgE	188	12	2
IgG1	146	2–3	2
IgG2	146	2–3	2
IgG3	170	2–3	2
IgG4	146	2–3	2
IgM	970	10–12	5

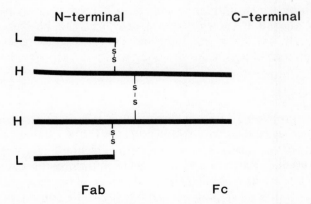

Figure 1.2 Structure of an immunoglobulin G molecule (H, heavy chain; L, light chain).

fragments retain the antigen binding properties of IgG but, unlike the intact molecule, are monovalent rather than divalent. In human sera IgG is present in four isotypic variations, IgG1, IgG2, IgG3, and IgG4. These differ in the structure of the constant region of the gamma chains.

Binding constants for the interaction of antibodies with antigens show a wide variation. Anti-carbohydrate antibodies have K values of the order of $10^5\,M^{-1}$, whilst polypeptide and certain hapten antibodies have K values approaching $10^{14}\,M^{-1}$ (e.g. anti-gastrin, anti-fluorescein) (13, 14).

The ability of a molecule to elicit an immune response is in part linked to its size. Few molecules with a molecular weight < 2000 are immunogenic. However, antibodies to small molecules (haptens) can be produced if the hapten is chemically linked to a larger carrier molecule (e.g. bovine serum albumin), and the hapten carrier molecule complex used as an immunogen.

Hybrid antibodies. In nature the two combining sites (Fab) of an antibody molecule have the same specificity. Antibody molecules can also be produced which have dual specificity. This is achieved chemically by preparing Fab fragments of two different antibodies and reforming the disulphide bridge between different Fab fragments (15).

Monoclonal antibodies. Antiserum produced by deliberate immunization of animals contains a diverse population of immunoglobulin molecules with various specificities arising from a multitude of stimulated lymphocyte clones. In addition, the immune response in animals, even of the same species, to the same antigen is variable, hence there is considerable inter-animal variability in the characteristics of antisera. For the use of antibodies as diagnostic tools, this rather unsatisfactory situation has been dramatically altered by the work of Kohler and Milstein (16) who showed that it was possible to immortalize a particular antibody-producing lymphocyte of a single specificity by fusing it with a myeloma cell. The resulting hybrid (hybridoma) cell line produces homogeneous antibody (monoclonal antibody) and this can be grown into large cultures or carried as antibody-secreting tumours in animals from which the ascitic fluid can be harvested. The hybridoma cells are stable cell lines and thus provide a permanent and limitless source of homogeneous antibody with a defined specificity (16).

1.3.2 Enzymes

Enzymes are protein molecules which catalyse chemical transformations of specific molecules (substrates). An integral step in the catalytic mechanism is the binding of the substrate to the enzyme (see section 1.2.3). Thus enzymes combine molecular recognition with molecular transformation. Amongst the nearly 2000 known enzymes there is a wide range of binding specificity and some representative examples of enzyme–substrate pairs are shown in Table 1.3. An advantage of enzymes in the context of biosensors is that

Table 1.3 Enzyme classes

Enzyme class	EC	Substrate
Oxidoreductases		
Alcohol dehydrogenase	1.1.1.1	ethanol
Transferases		
Aspartyltransferase	2.3.2.7	L-asparagine
Hydrolases		
Pectin esterase	3.1.1.11	pectin
Lyases		
Tryptophanase	4.1.99.1	L-tryptophan
Isomerases		
Mannose isomerase	5.3.1.7	D-mannose
Ligases (synthetases)		
D-Alanylalanine synthetase	6.3.2.4	D-alanine

molecular recognition of the substrate by the enzyme is accompanied by a chemical reaction and this can be readily detected by the appropriate sensor (e.g. thermistor for production of heat, electrode for production of ionic species).

1.3.3 Lectins

Lectins are proteins and glycoproteins which bind to oligosaccharides. Several hundred lectins are known, and some examples, together with their binding specificities, are listed in Table 1.4. The most extensively studied of the lectins is concanavalin A. This occurs as either a dimeric or a tetrameric molecule consisting of identical polypeptide subunits of molecular weight 26 kD, each of which carries a saccharide binding site (17, 18). Concanavalin A shows specificity for a wide range of oligosaccharides containing alpha-D-mannose of alpha-D-glucose residues (e.g. glycogen, amylopectin, dextrans, mannans). It also binds to glycoproteins such as carcinoembryonic antigen, ribonuclease B, and immunoglobulins. This property has been exploited in affinity purification. For example, jacalin, a lectin isolated from jackfruit seeds (*Artocarpus integrifolia*), binds to IgA but not IgG, and thus provides the basis of a facile method for preparing IgA which is free of IgG (19).

1.3.4 Receptors

Many substances exert their biological action via binding to cell-surface receptors. Receptors have been discovered for many types of molecule (Table 1.5), and generally receptors are proteins or protein aggregates

Table 1.4 Lectins

	Source	Molecular weight, kD (subunits)		$K_a(M^{-1})$	Specificity
Jacalin	*Artocarpus integrifolia* (jackfruit)	40	(4)	—	α-D-Galactose
Concanavalin A	*Conavalia einsformis* (jackbean)	55	(2)	2.06×10^4	α-D-Mannose α-D-Glucose
Soybean agglutinin	*Glycine max* (soybean)	110	(2)	3.0×10^4	α-D-Galactose N-Acetyl-D-galactosamine
Phytohaemagglutinin	*Phaseolus vulgaris* (red kidney bean)	140	(4)	—	N-Acetyl-D-glucosamine
Ricin (RCA$_1$)	*Ricinus communis* (caster bean)	120	(4)	1.5×10^4	β-D-Galactose
Wheatgerm agglutinin	*Triticum vulgaris* (wheatgerm)	23	(2)	1.3×10^3	(N-Acetyl-D-glucosamine)$_2$

Table 1.5 Receptors

Receptor	Molecular weight (kD)	
Asialoglycoproteins	26	
Epidermal growth factor	170	
Glucagon	188	
Insulin	460	(two subunits of 135 kD and two of 95 kD)
Low-density lipoprotein (LDL)	160	
Progesterone	110	
Transferrin	90	

(20, 21, 22). One of the most extensively studied receptors is the insulin receptor (23). This is a glycoprotein consisting of two alpha (MW 135 kD) and two beta subunits (MW 95 kD). The dissociation constant for insulin binding has been calculated to be 1.6×10^{-10} M (24). Similar high-affinity binding is encountered with other receptors; for example, the binding constant for the interaction of IgE with its receptor on the mast cell is in the range $1 \times 10^9 - 1 \times 10^{12}$ M^{-1} (25).

1.3.5 Intact chemoreceptors

Intact sensing structures (chemoreceptors) from natural sources provide an alternative to binding proteins for a range of small molecules. Crab antennules from the blue crab *Calinectes sapidus* are sensitive to amino acids and display some selectivity, e.g., the antennules contain a receptor for glutamate which is unresponsive to other amino acids such as glycine, alanine, proline, and taurine (26). Olfactory structures in other species offer selectivity for additional types of molecule which act as pheromones, e.g. the male moth *Bombyx mori* has specific receptors for the female sex attractant bombykol, and *Antherea polyphemus* has receptors for E6, Z11-hexadecadienyl acetate and E6, Z11-hexadecadienal.

1.3.6 Avidin

Avidin is a highly basic 66 kD glycoprotein (pI ~ 10) isolated from egg white. It binds four equivalents of the B group vitamin biotin. Binding is both selective and very strong ($K_D \sim 10^{-15}$ M) and is amongst the strongest non-covalent interactions known (27). More recently, a less basic avidin, called streptavidin (pI ~ 5, molecular weight 60 kD) isolated from *Streptomyces avidinii*, has been advocated because its less basic nature minimizes non-specific binding reactions which sometimes complicate certain applications of avidin (28).

1.3.7 *Protein A*

Protein A is a 42 kD polypeptide isolated from *Staphylococcus aureus* which binds to antibodies and to antibody–antigen complexes. It reacts with the IgG1, IgG2 and IgG4 subclasses of human IgG via the Fc part of the heavy chain (29). One molecule of protein A binds two molecules of IgG. Binding also occurs with IgA and IgM and with various animal immunoglobulins (30). An advantage of protein A is that it binds to immunoglobulins without interacting with the antigen binding (Fab) part of the molecule.

1.3.8 *DNA*

Deoxyribonucleic acid (DNA) is a polymer with a deoxyribose-phosphate backbone to which are joined purine and pyrimidine bases (Figure 1.3). A DNA molecule will bind to another DNA molecule to form a DNA duplex. The binding is very specific and is determined by the sequence of bases on the DNA molecule. Binding occurs via hydrogen bonding between adenine (A) and thymine (T) and between guanine (G) and cytosine (C). Thus a segment of DNA with a base sequence ATGC will only bind to a segment of another DNA molecule with the complementary sequence TACG. The stability of the DNA duplex is characterized by a melting temperature (T_m). This is the temperature at which the two DNA strands are half dissociated. Hydrogen bonding between complementary DNA molecules is very strong, as evidenced by the high melting temperatures which are typically greater than 60°C. Melting temperature, and hence strength of the DNA–DNA

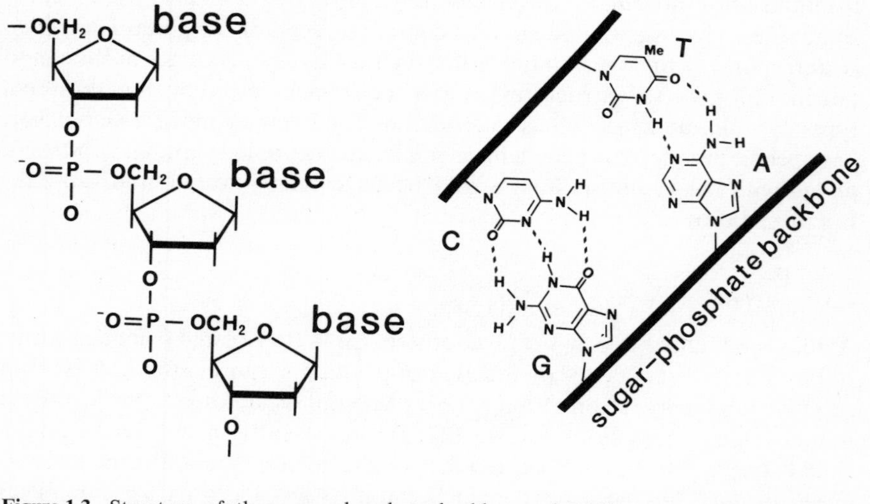

Figure 1.3 Structure of the sugar-phosphate backbone of DNA (*a*) and the purine and pyrimidine bases as hydrogen-bonded base pairs (*b*). (A, adenine; T, thymine; C, cytosine; G, guanine).

Table 1.6 Transport proteins

Transport protein	Substance bound
Albumin	Calcium ions
Apolipoproteins	Cholesterol, cholesterol esters, triglycerides, phospholipids
Caeruloplasmin	Copper ions
Coagulation factor VIII-related antigen	Coagulation factor VIII
Haemopexin	Haemin
Haptoglobin	Haemoglobin
Retinol binding protein	Retinol (vitamin A)
Steroid binding beta-globulin	17-Hydroxysteroids
Thyroxine binding globulin	Thyroxine
Transcortin	Cortisol
Transferrin	Ferric ions

binding interaction, increases as the length of the DNA molecule increases (31).

DNA will also bind to complementary molecules of ribonucleic acid. This is a polymer with a ribose-phosphate backbone carrying uracil, guanine, cytosine and adenine bases. DNA binds to RNA via hydrogen bonding between adenine and uracil and between guanine and cytosine.

1.3.9 Miscellaneous

Transport proteins provide a further group of molecules with molecular recognition properties. A large number of specificities are available and some examples are listed in Table 1.6.

1.4 Conclusion

Numerous binding specificities can be found amongst biomolecules and the most versatile are the enzymes and the immunoglobulins. A particular advantage of the latter is that it is possible to select and produce in large quantities individual immunoglobulins with a defined specificity. Fundamental studies may eventually facilitate the design and synthesis of biomolecules with a particular binding specificity, but currently this remains a remote possibility.

References

1. J. LeMagnen and P. MacLeod (eds.), *Olfaction and Taste VI*. Information Retrieval, London (1977).
2. M. Dixon, E.C. Webb, C.J.R. Thorne and K.F. Tipton, *Enzymes*, 3rd edn., Longman, London (1979).
3. B.A. Morris, in B.A. Morris and M.N. Clifford (eds.), *Immunoassays in Food Analysis*, Elsevier, London (1985) 21–51.
4. J. Greer, *Ann. N.Y. Acad. Sci. USA* **439** (1985) 44–63.

5. D.G. Hanauer, P. Gund, J.D. Andose, B.L. Bush, E.M. Fluder, E.F. McIntyre and G.M. Smith, *Ann. NY. Acad. Sci. USA* **439** (1985) 124–139.
6. R. Arnon, in F. Celada, V.N. Schumaker and E.E. Sercarz (eds.) *Protein Conformation as an Immunological Signal*, Plenum Press, New York (1983) 157–163.
7. M.Z. Atassi, *Immunochemistry* **12** (1975) 423–438.
8. T.P. Hopp, *J. Immunol. Methods* **88** (1986) 1–18.
9. M.B. Prystowsky, J.M. Kehoe and B.W. Erickson, *Biochemistry* **20** (1981) 6349–6356.
10. D.L. Ollis, P. Brick, R. Hamlin, N.G. Xuong and T.A. Steitz, *Nature* **313** (1985) 762–766.
11. L.E. Hood, I.L. Weissman, W.B. Wood and J.H. Wilson, *Immunology*, 2nd edn., Benjamin/Cummings Publishing Co., Menlo Park (1984).
12. G.W. Litman and R.A. Good (eds.), *Immunoglobulins*, Vol. 5, Plenum Press, New York (1978).
13. F. Karush, in G.W. Litman and R.A. Good (eds.), *Immunoglobulins*, Vol. 5, Plenum Press, New York (1978) 85–116.
14. R.M. Watt, J.M. Herron and E.W. Voss, *Mol. Immunol.* **17** (1980) 1237–1243.
15. U. Hammerling, T. Aoki, E. DeHarven, E.A. Boyse and L.J. Old, *J. Exp. Med.* **128** (1968) 1461–1469.
16. G. Kohler and C. Milstein, *Nature* **256** (1975) 495–497.
17. I.J. Goldstein and C.E. Hayes, *Adv. Carbohydrate Chem.* **35** (1978) 127–340.
18. G.L. Nicolson, *Int. Rev. Cytol.* **39** (1974) 89–190.
19. M.C. Roque-Barreira and A. Campos-Neto, *J. Immunol.* **134** (1985) 1740–1743.
20. M.S. Brown, R.G.W. Anderson and J.L. Goldstein, *Cell* **32** (1983) 663–667.
21. J. Kaplan, *Science* **212** (1981) 14–20.
22. G.S. Levey (ed.), *Hormone–Receptor Interaction*, Marcel Dekker, New York (1976).
23. R.A. Roth, D.O. Morgan, J. Beaudoin and V. Sara, *J. Biol. Chem.* **261** (1986) 3753–3757.
24. S. Jacobs and P. Cuatrecasas, in G.S. Levey (ed.), *Hormone–Receptor Interaction*, Marcel Dekker, New York (1976) 31–45.
25. H. Metzger, A. Goetze, J. Kanellopoulos, D. Holowka and C. Fewtrell, *Fed. Proc.* **41** (1982) 8–11.
26. S.L. Belli and G.A. Rechnitz, *Anal. Lett.* **19** (1986) 403–416.
27. N.M. Green, *Biochem. J.* **89** (1963) 585–591.
28. L. Chaiet and F. Wolf, *Arch. Biochem. Biophys.* **106** (1964) 1–5.
29. G. Kronvall and R.C. Williams, *J. Immunol.* **103** (1969) 828–833.
30. J. Goudswaard, J.A. van der Donk, A. Noordzij, R.H. van Dam and J.P. Vaerman, *Scand. J. Immunol.* **8** (1978) 21–28.
31. R.J. Britten and E.H. Davidson, in B.D. Hames and S.J. Higgins (eds.), *Nucleic Acid Hybridisation: A Practical Approach*, IRL Press, Oxford (1985) 3–15.

2

MOLECULAR AND IONIC RECOGNITION BY CHEMICAL METHODS

2 Molecular and ionic recognition by chemical methods

P.D. BEER

2.1 Introduction

Molecular recognition requires the ability of a molecular receptor (host) to discriminate and bind a certain substrate (guest) out of a plethora of structurally related molecules (Figure 2.1). The substrate may be a cation or anion (ionic recognition) or a neutral molecular species (covalent recognition). Successful selective receptor–substrate or host–guest complex formation results when the two species complement each other both in size and shape (geometry) and binding sites (energy). This complementarity is elegantly demonstrated by biological (biotic) systems through enzyme–substrate interactions, biosynthesis of proteins, and antigen–antibody reactions (see Chapter 1).

This chapter is primarily concerned with the recent syntheses and coordination chemistry of abiotic receptor systems, and hopes to provide through examples a series of broad principles for selective host–guest complexation applicable to the future design and development of new chemical sensors.

The chapter is split into two major sections, ionic recognition and molecular covalent recognition. Ionic recognition discusses synthetic receptor molecules designed to bind cations (metal, ammonium, bipyridinium ions) and anions (halide, azide, sulphate, phosphate, dicarboxylate ions etc.). The section on molecular covalent recognition describes the complexation of neutral (un-

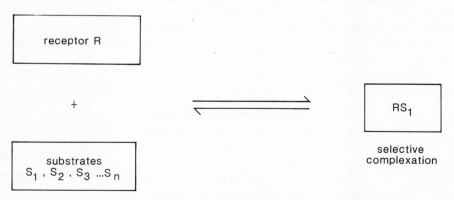

Figure 2.1 Molecular recognition by a receptor R (host) molecule for a certain substrate S_1.

charged) guest molecules (halogeno-methanes, aromatics) by abiotic cyclophane receptors.

Because of space limitations the references will not be exhaustive and the reader is encouraged to take note of the 'references cited therein'.

I IONIC RECOGNITION

In comparison to the formation of covalent bonds, a host–guest complex results from combinations of relatively weak intermolecular interactions, hydrogen bonding, van der Waals forces, electrostatic forces and so forth. Ion-dipole electrostatic attractions play a predominant role in the binding of *ionic* guest species, and this section reviews cationic and anionic host–guest complexation.

2.2 Cationic recognition 1: metal ion recognition

2.2.1 *Macrocyclic polyethers—crown ethers*

Dibenzo-18-crown-6 (1) was prepared by Pedersen (1) in 1967 quite by chance and the subsequent discovery (2) that (1) and other macrocyclic polyethers selectively complex biologically relevant alkali and alkaline earth metal cations has led to continued efforts to synthesize and investigate the coordination chemistry of related host compounds. The relationship between structure and cation selectivity has been intensely investigated by the systematic variation of the number of ether oxygen atoms, ring size, length of $(CH_2)_n$ bridge, substitution by other types of donor heteroatoms (N, S, P), and introduction of aromatic (benezene, biphenyl, naphthalene) and heteroaromatic systems (pyridine, furan, thiophene) in the ring (Figure 2.2).

The name 'crown' arises from both the appearance of models of these compounds and their ability to 'crown' cations by complexation. IUPAC nomenclature can be used to name these ligands; however, the result is complex and cumbersome. This has been resolved by using the more illustrative Pedersen crown notation. For example in (1) 'dibenzo' stands for both of the benzene substituents annexed to the ring, while -18- means the number of ring atoms.

The class name 'crown' is followed by the number of oxygen atoms in the polyether ring. A full account of the nomenclature can be found in reference 3.

Syntheses. The general method for preparing macrocyclic polyethers is the Williamson ether synthesis which involves nucleophilic displacement of halide ions from a dihaloalkane by the dianion derived from a diol (Scheme 1).

The product yields reported by Pedersen (2, 4) were generally low except for crown ethers containing five or six oxygen donor atoms. Dibenzo-18-crown-6 (1) for example was isolated in 45% yield under the conditions given in Scheme 1, and Pedersen (1) noted that 'the ring closing step either by a second

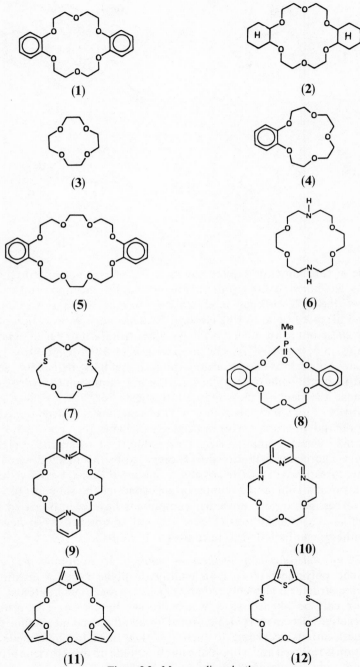

Figure 2.2   Macrocyclic polyethers.

B

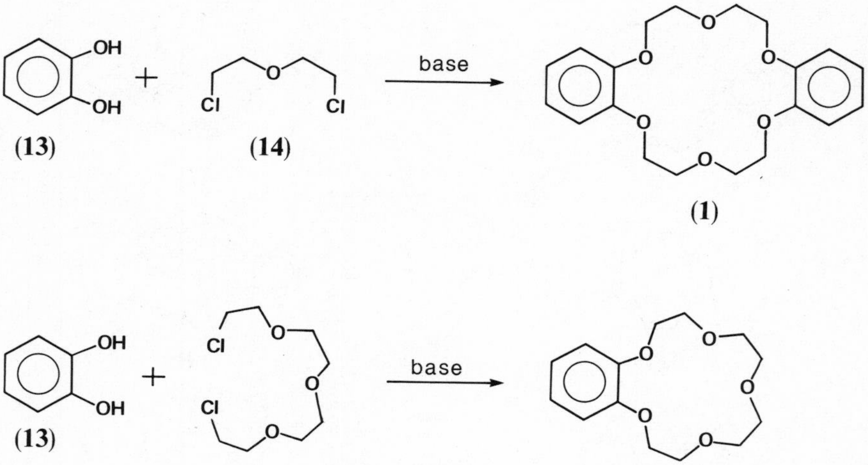

(13)   (14)   base   (1)

(13)   base   (4)

Scheme 1

molecule of catechol (**13**) or a second molecule of bis-(2-chloroethyl) ether (**14**), was facilitated by the sodium ion, which by ion-dipole interaction "wrapped" the three molecule intermediates around itself in a three-quarter circle and disposed them to ring closure'. Evidence for the operation of this *template effect* in crown ether synthesis comes from a number of synthetic procedures (5, 6) reported for the preparation of 18-crown-6 (**15**). Using potassium *t*-butoxide as base in tetrahydrofuran, yields as great as 60% can be obtained (Scheme 2); however, when tetra-*n*-butylammonium hydroxide is used instead the yield of (**15**) is reduced drastically (6).

Li$^+$ and Na$^+$ ions have been shown (7) to template the formation of 12-crown-4 (**3**) and 15-crown-5 (**16**) respectively (Scheme 3).

The benzo crown compounds can be reduced to make the cyclohexo derivatives. Dicyclohexo-18-crown-6 (**2**) prepared by this method was found to be a mixture of two of the five possible isomers—the *cis-syn-cis* and the *cis-anti-cis* forms (8). Differential complexation separates the isomers (9).

Many other macrocyclic polyether compounds have been prepared since Pederson's initial publications (Figure 2.2) and comprehensive reviews on their syntheses can be found in references 1, 10 and 11.

Properties of macrocyclic polyethers in solution. In lipophilic media the macrocyclic polyethers possess an endo-hydrophilic cavity consisting of hydrophilic structural elements (ether oxygens, nitrogen heteroatoms). This behaviour can be likened to a water droplet in oil (3). The resulting electronegative host cavity is ideally suited for alkali metal and alkaline earth metal guest cations according to their size (12). In hydrophilic media the polarization is reversed and a lipophilic interior made up of methylene groups, alkyl chains, etc., results (3).

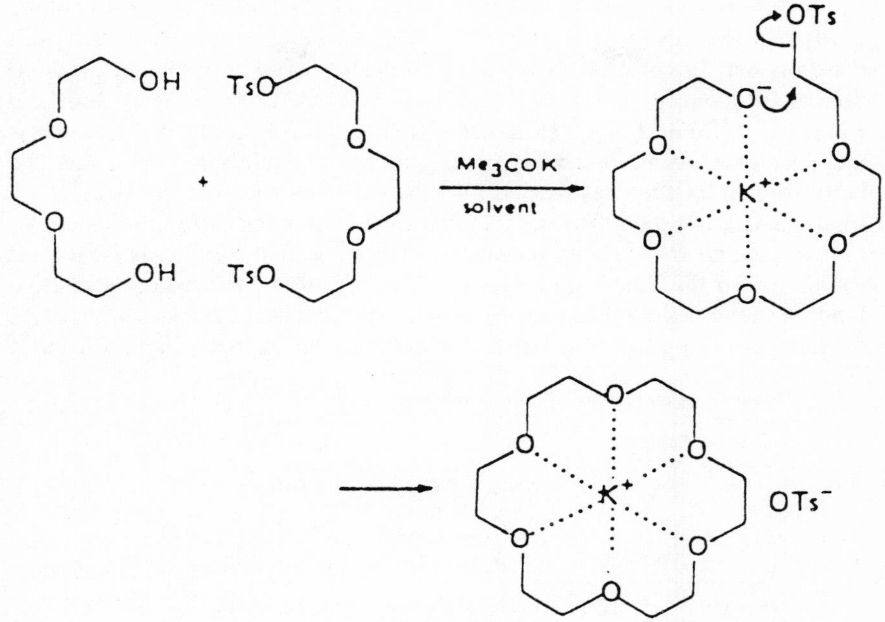

Scheme 2 (6) (15)

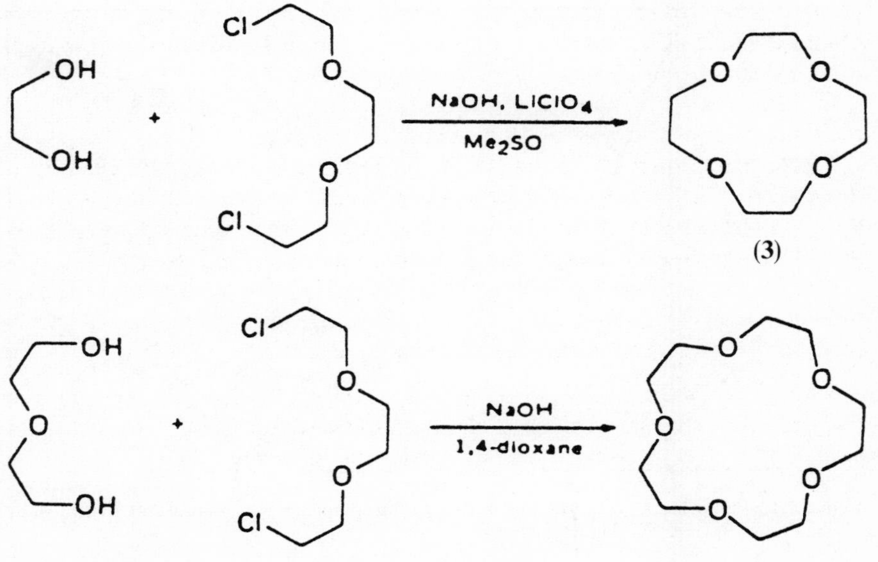

Scheme 3 (7)

Coordination chemistry: complex stability. The sizes of the hydrophilic cavities of the macrocyclic polyethers (13) can be compared with the radii of unsolvated ions of alkali metal ions (1, 4) (Table 2.1). The predicted 'Optimal Spatial Fit Concept' (circular recognition) between 12-crown-4 (**3**) and Li^+, 15-crown-5 (**16**) and Na^+, 18-crown-6 (**15**) and K^+ is confirmed by experimentally determined thermodynamic stability constants (15) (K_s), i.e. the better the 'fit' of the guest cation into the cavity of the host the larger K_s.[*] Figure 2.3 clearly demonstrates the relationship between cation radius/cavity radius size and the stability constant of the respective complexes (15).

The 'ion in the hole model' has limited usefulness for predicting relative binding capacities of metal cations with large macrocyclic polyether hosts. As the number of ring atoms increases, the macrocyclic flexibility increases and it

Table 2.1 Comparison of cation- and cavity radii

Cation	Radius[a] Å	Crown ether	Cavity radius[b] Å
Li^+	0.76	12-crown-4 (**3**)	0.6–0.75
Na^+	1.02	15-crown-5 (**16**)	0.86–0.92
K^+	1.38	18-crown-6 (**15**)	1.34–1.43
Cs^+	1.67	21-crown-7 (**17**)	1.7–2.1

[a] Based on x-ray structural analyses of oxides and fluorides (14, 15).
[b] Determined from Corey-Pauling-Kolton (CPK) molecular models (13).

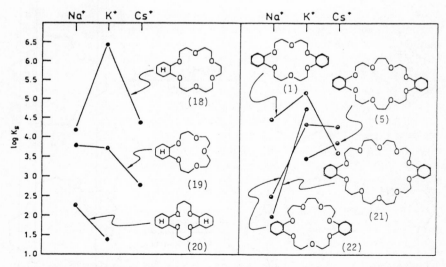

Figure 2.3 Plot of log K_s values of crown ether/cation complexes in methanol (15). Reprinted by permission of Springer Verlag, Heidelberg.

[*] A recent publication by Gokel *et al.* claims a higher K_s value for K^+ (**16**) complex over Na^+ (**16**) complex.[16]

becomes difficult to define the cavity diameter. For example, dibenzo-30-crown-10 (21) forms a very stable complex with K^+ (17). X-ray crystallographic data indicate that the host embraces itself around the cation to form a three-dimensional cavity with all ten oxygen donor atoms coordinated to the cation (18) (see Figure 2.9).

Numerous other factors, including charge on the cationic guest, the solvation enthalpies and entropies of the cation and ligand and the rigidity or flexibility of the host can influence K_s (15) and more detailed discussions on the thermodynamics and kinetics of cation–macrocycle interactions can be found in a recent extensive review (19).

Structures. Detailed structural studies of macrocyclic polyethers and their complexes can be found in references 13 and 20. Only a few pertinent examples will be given here. Dunitz *et al.* (21) determined the structure of 18-crown-6 (15) and the complexes of this host with Na^+, K^+ and Cs^+ (21, 22, 23). It is of interest to examine their results, bearing in mind the size relationships of cations and host cavities (Table 2.1). Table 2.1 predicts that Na^+ is too small, K^+ is just right and Cs^+ too large to fit into the cavity of an 18-crown-6 ligand. This is substantiated by the structural studies. Figure 2.4 shows Na^+ coordinated to all six oxygen atoms, with the ligand wrapping itself around the Na^+. In the K^+ complex the central ion is lying in the cavity of the crown ether surrounded by a nearly planar hexagon of oxygen donor atoms (Figure 2.5). With the Cs^+ complex, the cation is too large for the cavity and it lies above the plane of the oxygen donor atoms of the crown ether ligand (Figure 2.6). When the cation diameter is very much larger than the cavity diameter, sandwich complexes of 2:1 ligand:guest metal cation stoichiometry result. An example of such a complex is the K^+:2 benzo-15-crown-5 system (24) (Figure 2.7). The K^+ cation does not fit in the small cavity and so it is sandwiched between two crown ethers, coordinated to all ten oxygen atoms of the ligands. Conversely, when the cavity diameter is very much larger than the cation diameter, binuclear 1:2 ligand: cation stoichiometrical complexes or coiled structures still having 1:1 stoichiometry are found. An example of the former situation is the $2Na^+$-dibenzo-24-crown-8 (5) system (25) (Figure 2.8). Two Na^+ fit inside the host cavity; each has a coordination number of six and is coordinated to three oxygen donor atoms of the crown ether ligand leaving two oxygen atoms of the ligand not coordinated. The K^+-dibenzo-30-crown-10 (21) complex (18) (Figure 2.9) exemplifies the latter case, with the crown ether ligand folding and wrapping itself about a single K^+, resulting in all ten oxygen donor atoms coordinated to the guest cation.

Selectivity of crown ether complexation. The ability of a host macrocyclic polyether to recognize, discriminate and selectively bind a particular guest metal cation in the presence of other different cationic guests is of prime importance to the development of sensor devices. A measure for the selectivity of a particular crown ether towards two different cations M_1 and M_2 can be

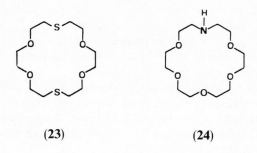

(23)   (24)

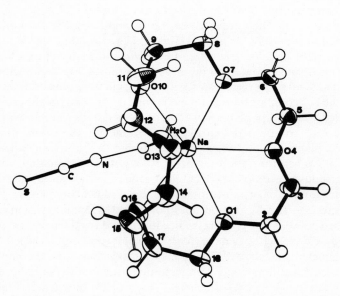

Figure 2.4 (Top) Aza and thia analogues of 18-crown-6. (Bottom) [NaSCN-H_2O-18-crown-6] with Na^+ complexed by the crown ether, a SCN^- and a water molecule (21). (By permission of International Union of Crystallography.)

expressed numerically by the ratio of the stability constants of the complexes LM_1 and LM_2 (L = crown ether ligand, M = metal cation), whereby the ion considered appears in the numerator and the one to be differentiated in the denominator (equation 2.1).

$$\text{selectivity} = \frac{K_s(LM_1)}{K_s(LM_2)} \tag{1}$$

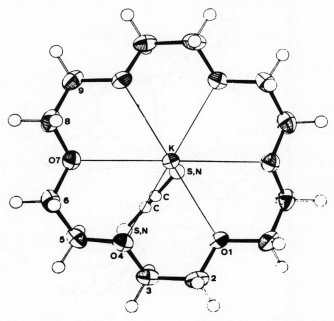

Figure 2.5   [KSCN-18-crown-6] (22). (By permission of International Union of Crystallography.)

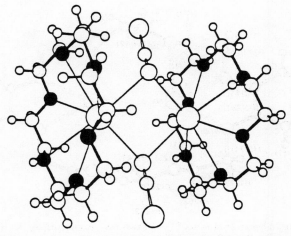

Figure 2.6   [CsSCN-18-crown-6] units bridged by SCN ligands. Cs^+ does not fit into the cavity of the crown ether ligands (23). (By permission of International Union of Crystallography.)

Numerically high factors correspond to high selectivity, low factors mean low discrimination. One can go on to distinguish two types of selectivity. A crown ether that exhibits high selectivity for several metal cations of similar size

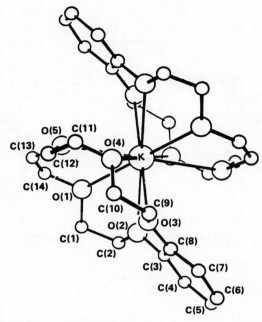

Figure 2.7 [K-2-benzo-15-crown-5]$^+$ with the K$^+$ sandwiched between two crown ether ligands (24). (By permission of The Royal Society of Chemistry.)

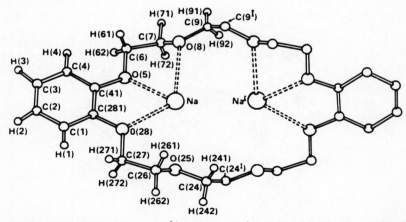

Figure 2.8 [Na$_2$-dibenzo-24-crown-8]$^{2+}$ with two Na$^+$ complexed by the crown ether (25). (By permission of The Royal Society of Chemistry.)

or charge but much lower selectivity for another group of ions, for example of smaller ionic radius, has what is known as *plateau selectivity* (12). For example Figure 2.3 shows (19) which exhibits low selectivity for Na$^+$/K$^+$ and high selectivity for K$^+$/Cs$^+$, and (22), vice versa. Another example is the antibiotic valinomycin which shows a high K$^+$/Na$^+$ selectivity but does not distinguish

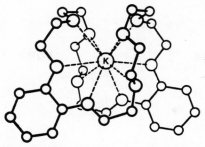

Figure 2.9 [K-dibenzo-30-crown-10]$^+$: the crown ether ligand embraces the cation (18). (By permission of The Royal Society of Chemistry.)

Table 2.2 Comparison of log K_s values for the complexation of 18-crown-6 and of some aza and thia analogues with K^+ and Ag^+

Crown ligand	(15)	(23)	(24)	(6)
K^{+a}	6.10	1.15	3.90	2.04
Ag^{+b}	1.60	4.34	3.30	7.80

[a] In methanol
[b] In water

at all well between K^+, Rb^+ and Cs^+. *Peak selectivity* refers to the situation where the crown ether clearly discriminates between all types of metal cations irrespective of their size and charge. The crown ether (18) exemplifies this (Figure 2.3) by showing distinct selectivity towards K^+ over Na^+ or Cs^+.

Effect of donor atom type on selectivity. Changing the type of donor atom at the binding site alters the nature of the ligand–cation interactions and can lead to quite subtle changes in complexation selectivity (15, 26). For example, the replacement of oxygen (hard base) as the donor atom by sulphur (soft base) in 18-crown-6 enhances the complexation of transition metal ions (soft acids), for example Ag^+, and reduces that of alkali metals (hard acids), for example K^+ (Table 2.2) (17, 27). With nitrogen as a donor atom, however, the complexation of transition metal ions (Ag^+) is also promoted without substantially diminishing that of alkali metals (K^+) (17).

Structurally related mixed oxygen and nitrogen donor and all nitrogen donor macrocyclic systems of variable ring size have recently been shown to display a discrimination mechanism termed 'dislocation discrimination' towards Zn^{2+} and Cd^{2+} guest cations (28, 29, 30). Figure 2.10 displays log K_s values for the formation of 1:1 complexes of Zn^{2+} and Cd^{2+} with (25)–(30) versus macrocyclic ring size (number of member atoms constituting the ring). Enhanced recognition for Zn^{2+} over Cd^{2+} is exhibited by the 15-membered macrocycle (26) relative to the 14- and 16-membered ring systems. Acyclic analogues of the 15-membered macrocycle, (29) and (30) showed no such discrimination.

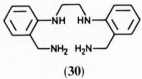

$n = 2$, R = H (**25**)   (**29**)
$n = 3$, R = H (**26**)
$n = 4$, R = H (**27**)
$n = 3$, R = Et (**28**)

(**30**)

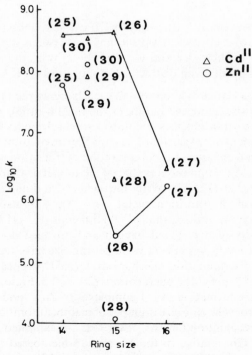

Figure 2.10 Log K_s values for formulation of the 1:1 complexes of Zn^{2+} ($\triangle$) and Cd^{2+} ($\bigcirc$) with (**25**)–(**30**) (30).

The incorporation and combination of sulphur, phosphorus (31, 32) or arsenic (33) donor atoms into macrocyclic rings of varying ring size provide future possibilities for transition metal ion recognition.

2.2.2 Bis crown ethers

The observation (24) that K^+ forms a 1:2 complex with benzo-15-crown-5 **(4)** (see Figure 2.7) has resulted in the synthesis of new type of macrocyclic polyether systems known as the bis crown ethers. These compounds contain two crown ether subunits linked together by a single bridge which can simply be a hydrocarbon chain (34, 35, 36), **(31)**–**(36)**, alkene (37), **(37)**–**(38)**, azo unit (38–41), **(41)**–**(42)**, or even a metallocene redox centre (42–44), **(39)**, **(40)**, **(45)**, **(46)** (Figure 2.11). Due to the cooperative effect of the two crown ether rings, bis crown ethers exhibit remarkable selectivity for alkali metal ions, forming 1:1 cation:bis crown ether intramolecular sandwich complexes (Figure 2.12). For example **(43)** is highly selective for Na^+ (36), **(39)** and **(40)** for K^+ (42) and **(33)** for Cs^+ (34b).

2.2.3 Macrobicyclic ligands—cryptands

The bridging of a two-dimensional monocyclic crown ether with an additional oligoether chain leads to a novel bicyclic three-dimensional type of host ligand (45). Lehn, their discoverer, called them cryptands, a name derived from the Greek word cryptos meaning cave. These macrobicyclic ligands through their special topology (12) completely encapsulate metal cations to form complexes termed cryptates. The first cryptands contain two bridgehead nitrogen donor atoms which are linked by three oligooxa chains of different length and number of oxygen donor atoms (46) (Figure 2.13).

Synthesis. The synthesis of the first cryptand **(50)** was successfully achieved in 1968 (45) by the synthetic sequence shown in Scheme 4.

The addition of alkali or alkaline earth metal cations to a solution of **(50)** in chloroform gave a new set of peaks in the ^1H nmr spectrum indicative of complexation. Further studies using ion-selective electrodes (47), pH-metric titration (47) and x-ray (48) analysis showed that the complexes formed were very stable and that the metal cation guest is located within the central cavity (48).

Synthetic pathways similar to the original one shown in Scheme 4 were used to prepare cryptands **(47)**–**(53)** (Figure 2.13) with varying host cavity sizes.

Nomenclature. The IUPAC nomenclature applied to compound **(50)** is 4, 7, 13, 16, 21, 24-hexaoxa-1, 10-diazabicyclo 8, 8, 8-hexacosane; this is quite a mouthful! Trivial names have been introduced and are generally used. The notation is very simple and is derived from three numbers, each designating the number of oxygen donor atoms in each bridge. The macrobicycle **(50)** in

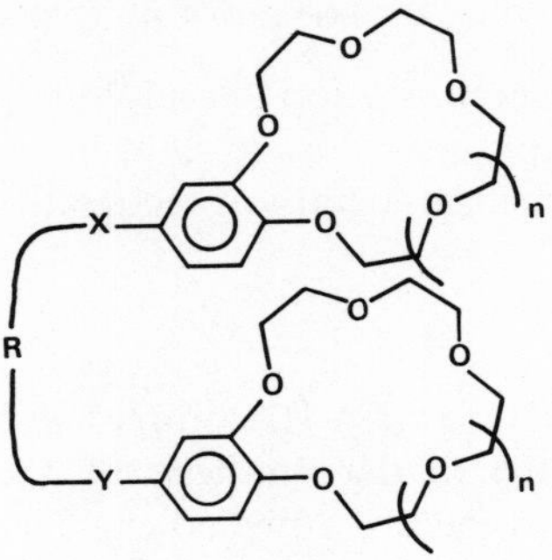

(31)–(33) X = Y = CONH, R = $(CH_2)_3$ $n = 1, 2, 3$ (34)
(34)–(36) X = CH_2, Y = 0, R = $(CH_2)_2$ $n = 1, 2, 3$ (35)
(37)–(38) X = Y = CH_2OCO, R = CH = CH $n = 1, 2$ (37)
(39)–(40) X = Y = CONH, R = $M(\eta^5\text{-}C_5H_4)_2$ M = Fe,
　　　　　　　　　　　　　　　　　　Ru $n = 1$ (42)

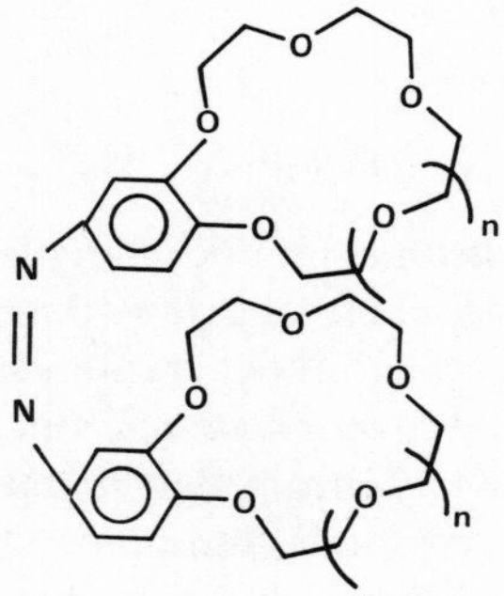

(41)–(42) $n = 1, 2$ (38–41)

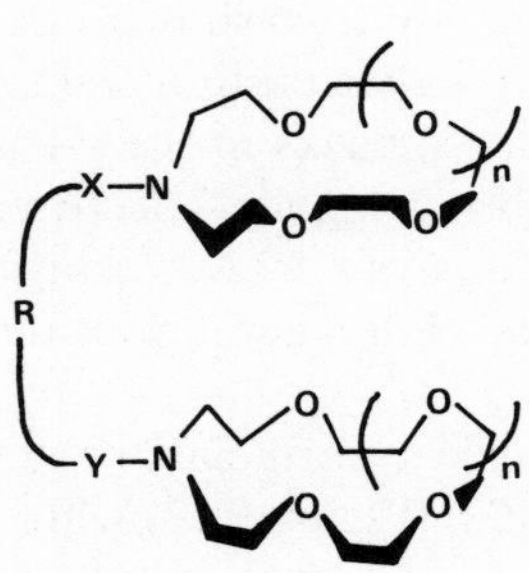

(43)–(44) X = Y = CH_2 R = $(CH_2)_n$ $n = 0, 1$ (36)
(45)–(46) X = Y = CONH R = $M(\eta^5\text{-}C_5H_4)_2$ M = Fe,
　　　　　　　　　　　　　　　　　　Ru $n = 1, 2$ (43, 44)

Figure 2.11 Example of bis crown ethers.

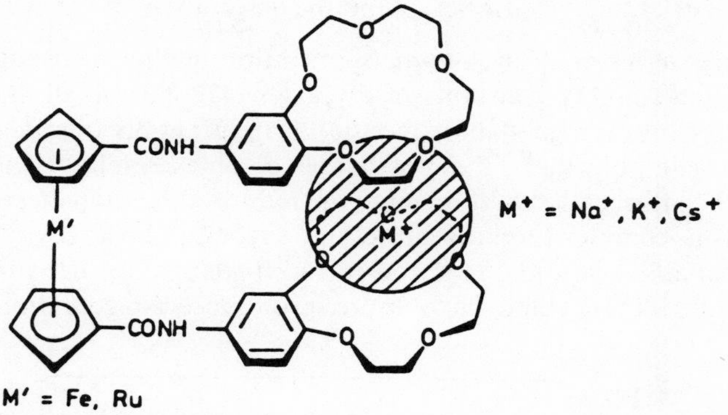

$M^+ = Na^+, K^+, Cs^+$

$M' = Fe, Ru$

Figure 2.12 Intramolecular sandwich complexes of the bis crown ethers (**39–40**) (42).

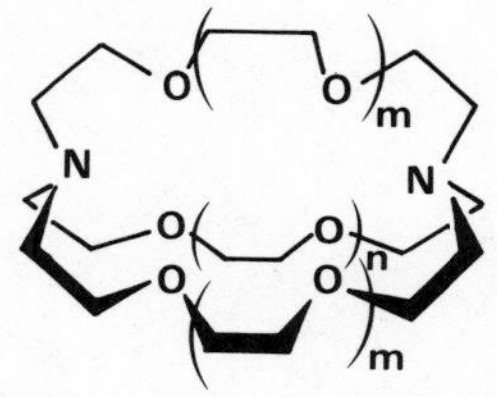

(**47**) [1.1.1] $m = n = 0$
(**48**) [2.1.1] $m = 0$ $n = 1$
(**49**) [2.2.1] $m = 1$ $n = 0$
(**50**) [2.2.2] $m = n = 1$
(**51**) [3.2.2] $m = 1$ $n = 2$
(**52**) [3.3.2] $m = 2$ $n = 1$
(**53**) [3.3.3] $m = 2$ $n = 2$

Figure 2.13 The diaza-polyoxa-macrobicyclic cryptands (46).

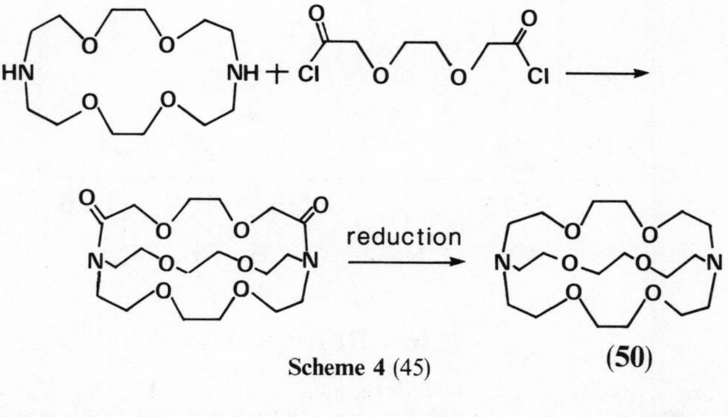

Scheme 4 (45)   (**50**)

this system is described simply as [2.2.2] cryptand (see Figure 2.13 for nomenclature of related compounds). To express the complexation of a metal cation (K^+ for example) with a cryptand the description $[K^+ \subset 2.2.2]$ is used.

Stability and selectivity. The pH-metric titration method is usually used to determine the stability constants of cryptates (47). For alkali and alkaline earth metal cations, high stability constants are generally observed. As with the monocyclic polyethers, the most stable complex results when the ionic radius of the metal cation best matches the radius of the cavity formed by the cryptand on complexation. Because the cryptand host cavity is three-dimensional and spheroidal in shape, it is well adapted for a 'ball-like' guest metal cation. Hence they have more pronounced recognition receptor

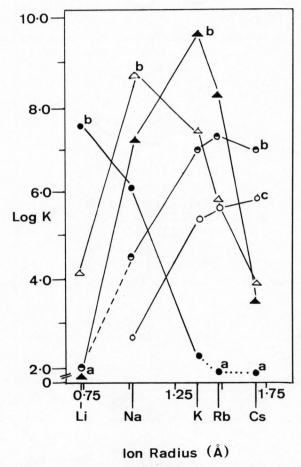

Figure 2.14 Selectivity of cryptands towards the alkali metal ions; ● = (**48**), △ = (**49**); ▲ = (**50**); ◓ = (**51**); ○ = (**53**); [a]value reported < 2.0; [b]solvent 90% methanol; [c]solvent methanol (47). Reprinted with permission from *J. Amer. Chem. Soc.* **97** (1975) 6700, © 1975 American Chemical Society.

Table 2.3 Stability constants (log K_s) of alkaline earth metal cryptates (47)

Ligand	Cavity radius (Å)	Ionic radius (Å)		
		Ca^{2+} (0.99)	Sr^{2+} (1.13)	Ba^{2+} (1.35)
[2.1.1](48)	0.8	2.5	< 2	< 2
[2.2.1](49)	1.1	6.95	7.35	6.30
[2.2.2](50)	1.4	4.4	8.0	9.5
[3.2.2](51)	1.8	2.0	3.4	6.0
[3.3.2](52)	2.1	2.0	2.0	3.65
[3.3.3](53)	2.4	< 2	< 2	< 2

properties for spherical metal cations than two-dimensional crown ethers (spherical recognition) (49). Figure 2.14 illustrates how, in proceeding through a series of these ligands of increasing size, each of the alkali metal ions is preferentially bound according to size, i.e the cryptands [2.1.1] (48), [2.2.1] (49) and [2.2.2] (50) preferentially complex Li^+, Na^+ and K^+ respectively. The correspondence between K_s values and the macrobicycle cavity and cation diameters is also found for alkaline earth cryptates (Table 2.3).

Structures. Weiss *et al.* (48) confirmed in 1970 from an x-ray structural study

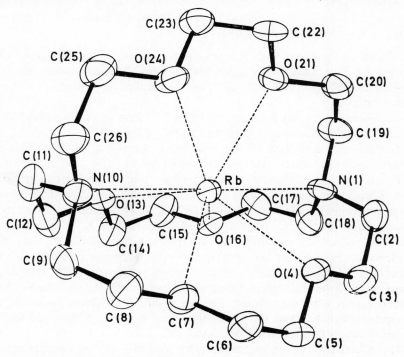

Figure 2.15 Structure of the RbSCN-[2.2.2] cryptate showing the Rb$^+$ completely enclosed by the cryptand host (48). (By permission of The Royal Society of Chemistry.)

of [2.2.2]–RbSCN complex that Rb$^+$ was located in the central cavity of the macrobicyclic system (Figure 2.15). In the cryptate [K$^+$ ⊂ 2.2.2], all eight donor heteroatoms are coordinated to the guest cation with the ligand in this complex adopting D$_3$ symmetry. Many other intensive structural studies on cryptates have now been reported and a compilation of these can be found in reference 50.

Cryptate effect. Cabbiness and Margenum (51) observed increased stability for the complexes of monocyclic ligands over those with an open chain analogue of similar composition and termed this extra stability the 'macrocyclic effect'. For example the monocyclic tetramine cyclam-Cu^{2+} complex is much more stable (by a factor of 10^4) than the open-chain tetramine Cu^{2+} complex. The same effect is found with the macrocyclic polyethers (17) (see Figure 2.16).

A similar effect occurs with the cryptates. Comparing the stability constants of [K$^+$ ⊂ 2.2.2] and its one-strand opened analogue (54) (known as a 'lariat ether', see later) the stability increases by a factor of 10^5 (Figure 2.16), and this is known as the cryptate effect or as the macrobicyclic cryptate effect (47). After comparing the thermodynamic quantities associated with formation of the [K$^+$ ⊂ 2.2.2] complex to those for the K$^+$-dicyclohexyl-18-crown-6 (2) complex, Kauffmann and co-workers concluded that the cryptate effect is of enthalpic origin (52).

Structural modifications relating to selectivity changes. Altering (i) the lipophilicity and (ii) the nature of the donor heteroatoms of these macrobicyclic host molecules can have profound effects on their selectivity behaviour towards metal cations.

The lipophilic cryptands shown in Figure 2.17 were prepared by analogous synthetic routes (53) to that described in Scheme 4.

The stability constants of K$^+$ and Ba^{2+} complexes of (55)–(57) were determined by pH-metric titrations and are shown in Table 2.4. The ionic radii of K$^+$ (1.33 Å) and Ba^{2+} (1.35 Å) are very similar and hence cavity/ionic radius size effects are minimized. Table 2.4 reveals that (50) [2.2.2] cryptand has a large preference for Ba^{2+} whereas (56) [2.2$_B$.2$_B$] forms complexes of identical stability and (57) [2.2.C$_8$] has a marked selectivity for K$^+$. This observed Ba^{2+}/K$^+$ selectivity variation can be explained in the following way. The presence of the two benzene substituents in (56) as compared to (50) has the effect of increasing the thickness of the organic layer separating the complexed guest cation from the solvent, hence reducing favourable interactions of the bound cation with the solvent. The net result is a destabilization of the complexes with K$^+$ and Ba^{2+}, but this effect is more pronounced for the divalent cation than for the monovalent one (12).

The selectivity differences between (50) and (57) can be rationalized by considering the number of oxygen donor atoms of the respective ligands, eight for (50) and only six for (57). Because K$^+$ and Ba^{2+} have respectively $n = 6$ and

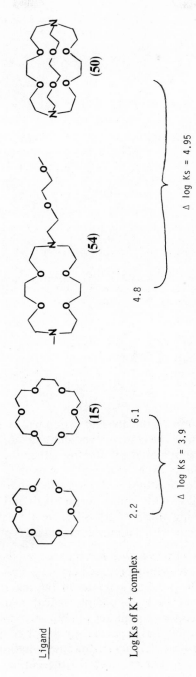

Figure 2.16 Increased stability constants observed for macrocyclic and macrobicyclic complexes.

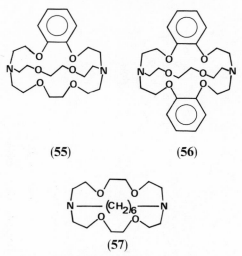

(55)   **(56)**

(57)

Figure 2.17   Lipophilic cryptands (53).

Table 2.4 Log K_s values and selectivity factors of monovalent cations versus divalent cations in methanol–water 95:5

Ligand	K^+	Ba^{2+}	K^+/Ba^{2+}
(50)[2.2.2]	9.45	11.5	1/110
(55)[2.2.2$_B$]	9.05	11.05	1/110
(56)[2.2$_B$.2$_B$]	8.6	8.5	1
(57)[2.2.C$_8$]	4.35	2	200

$n = 8$ for hydration numbers, the divalent cation will be more affected by the decrease in donor atoms. The control over alkali/alkaline earth metal cation selectivity can then be achieved by altering either the lipophilicity or the number of oxygen donors in the macrobicyclic ligands.

Replacing oxygen donor atoms with increasing numbers of nitrogens has the effect of decreasing stability constants for alkali and alkaline earth metal cations whilst increasing those of transition metal cations (54, 55). More pronounced effects occur with sulphur-containing cryptands (56, 57).

Macrobicyclic ligands incorporating heterocyclic subunits. Vogtle (58) and Newkome (59) have successfully prepared cryptands containing respectively one (58) and three (59) pyridine subunits in the macrobicyclic framework (Figure 2.18). The ligand (58) forms complexes with many metal cations, and binuclear cobalt and copper complexes of (59) were isolated.

More recently Lehn (60) has reported a number of 'photoactive' macrobicyclic ligands (60)–(64), containing anthracene, bipyridine and phenanthroline groups (Figure 2.19). Direct access to the Na^+Br^- complexes of the

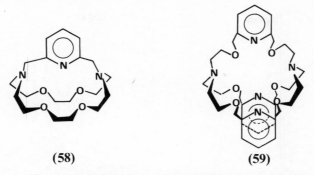

(58)   **(59)**

Figure 2.18 Cryptands containing pyridine subunits (58, 59).

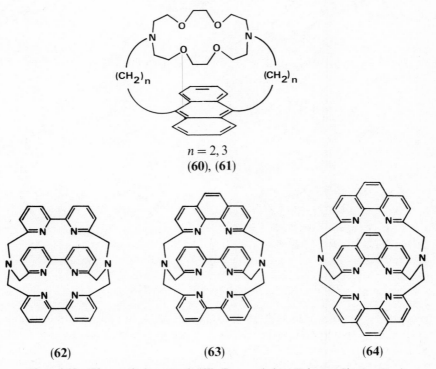

Figure 2.19 'Photoactive' cryptands (60). (By permission, *Helvetica Chimica Acta.*)

symmetrical cryptands (62) and (64) was achieved by a one-step macrobicyclization procedure. An x-ray crystal structure of the sodium cryptate of (64) shows the Na$^+$ cation bound inside the cavity coordinated to all eight nitrogen atoms (61). (Figure 2.20).

An analogous one-step synthetic procedure was used to prepare the

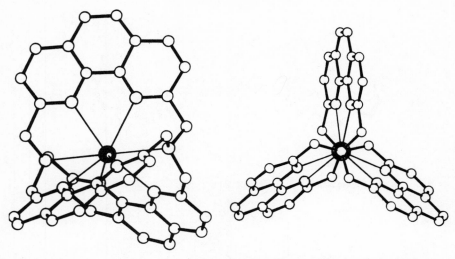

Figure 2.20 X-ray crystal structure of [Na$^+$ ⊂ (**64**)] complex (61). (By permission, *Helvetica Chimica Acta*.)

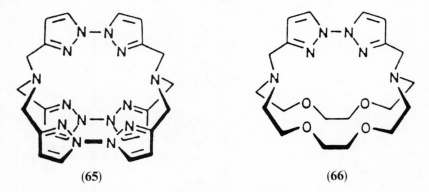

(65)   (66)

Figure 2.21   Pyrazolyl cryptands (62).

pyrazolyl cryptands (**65**) and (**66**) (62) (Figure 2.21). In contrast, the catechol cryptands (**67**)–(**69**) designed to act as iron sequestering agents require multi-step synthetic routes (63) (Figure 2.22).

Macrobicyclic ligands with carbon bridgeheads. The two carbon bridgehead macrobicyclic polyethers (**70**) and (**71**) synthesized by Parsons (Figure 2.23) are found to exhibit very high stability constants (64) (see Table 2.5) with (**70**) displaying the rare high selectivity of K$^+$/Rb$^+$.

The high stabilities are attributed to the large number of oxygen donor atoms included in the bicycles, and the fact that in (**70**) and (**71**) two successive oxygens are separated only by two carbon atoms even on the bridgehead. This

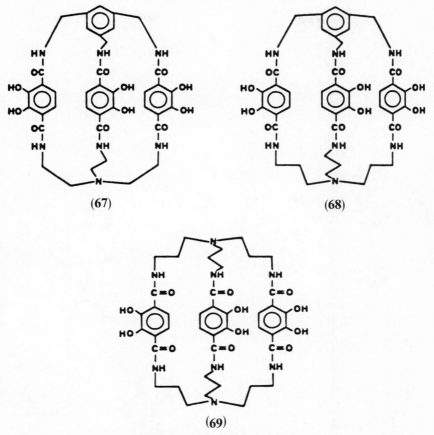

Figure 2.22   Catechol cryptands (63).

Table 2.5   Stability constants of carbon bridgehead macrobicycles in water (62)

	Na$^+$	K$^+$	Rb$^+$
(70)	5.4	5.7	3.8
(71)	3.5	4.3	4.4
(50)[2.2.2]	3.9	5.4	4.35
18-crown-6**(15)**	< 0.3	2.05	—

latter point is clearly demonstrated by the related macrobicycle **(72)** (Figure 2.23) exhibiting much lower stability constants, $\log K_s = 2.2$ for the K$^+$ complex in *methanol* (65).

Miscellaneous macrobicyclic ligands. Numerous other types of macrobicyclic cryptands have been synthesized in recent years, including ferrocene cryptands

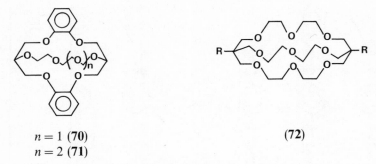

$n = 1$ **(70)**   **(72)**
$n = 2$ **(71)**

Figure 2.23   Carbon bridgehead macrobicyclic polyethers (64, 65).

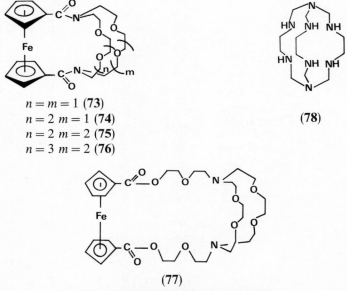

$n = m = 1$ **(73)**
$n = 2\ m = 1$ **(74)**   **(78)**
$n = 2\ m = 2$ **(75)**
$n = 3\ m = 2$ **(76)**

(77)

Figure 2.24   Miscellaneous cryptands (66).

(73)–**(77)** (66–68), sepulchrates **(78)** (69, 70) and chiral binaphthyl cryptands
(71) (Figure 2.24). Comprehensive reviews on cryptands can be found in
references 57 and 72.

2.2.4   *Lariat ethers*

At the borderline between the two-dimensional monocyclic polyethers and
three-dimensional cryptands is a relatively new type of host ligand, the 'lariat
ether' (73, 74).

   'Lariat ethers' is the trivial name given by Gokel (73) to macrocyclic

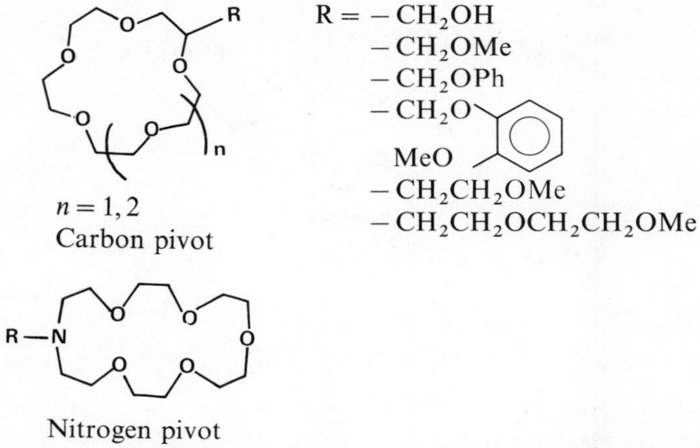

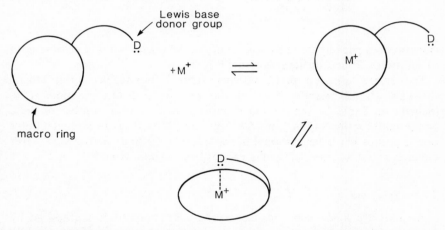

$R = -CH_2OH$
$-CH_2OMe$
$-CH_2OPh$
$-CH_2O$⟨benzene⟩

MeO ⟨benzene⟩
$-CH_2CH_2OMe$
$-CH_2CH_2OCH_2CH_2OMe$

$n = 1, 2$
Carbon pivot

$R-N$

Nitrogen pivot

Figure 2.25 Lariat ethers (73*b*). Reprinted with permission from *J. Amer. Chem. Soc.* **105** (1983) 586, © 1983 American Chemical Society.

Lewis base
donor group

macro ring

$+M^+$

M^+

Scheme 5 (73). Reprinted with permission from *J. Amer. Chem. Soc.* **104** (1982) 1672, © 1982 American Chemical Society.

polyethers having a pendant sidearm which bears one or more Lewis basic donor groups (Figure 2.25). These systems are designed so that the macro-ring can envelop the guest metal cation in the fashion normally associated with crown ether binding and the donor groups attached to the flexible sidearm further solvate the bound guest (Scheme 5).

Many lariat ether systems, having macro-rings of various sizes and sidearms bearing a variety of donor groups attached at either carbon (73*b*, 75) (carbon pivot) or nitrogen (76) (nitrogen pivot) (Figure 2.25), have been reported by Gokel and others (74, 75). As was hoped, alkali metal cation binding is generally enhanced by these ligands (73) and the donor-group-bearing

$$RNH_2 + ICH_2(CH_2OCH_2)_2CH_2I \xrightarrow[CH_3CN,83\ C]{Na_2CO_3}$$

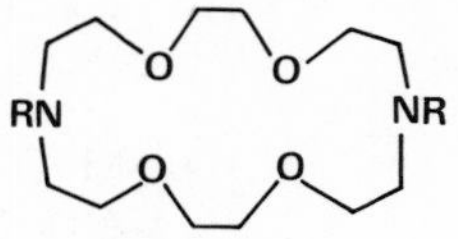

Scheme 6

Table 2.6 Stability constants for bibracchial lariat ethers

R	log K_s		
	Na$^+$	K$^+$	Ca^{2+}
(79) PhCH$_2$	2.72	3.38	2.79
(80) 2-MeO-C$_6$H$_4$CH$_2$	3.65	4.94	3.27
(6) H	1.5	1.8	—
(81) HOCH$_2$CH$_2$	4.87	5.08	6.02
(82) CH$_3$OCH$_2$CH$_2$	4.75	5.46	4.48
(83) EtOCOCH$_2$	5.51	5.78	6.78

sidearm does participate intramolecularly in the cation binding process both in solution (73) and in the solid state (77, 78).

Two-armed nitrogen pivot systems, termed 'bibracchial lariat ethers' (BiBLEs), have recently been synthesized in one-step reactions (79). (Scheme 6). Table 2.6 reports the stability constants of a number of these compounds for the Na$^+$, K$^+$ and Ca^{2+} guest cations. It can be seen that polar donor groups like ester carbonyl in the sidearms strongly favour Ca^{2+} over either Na$^+$ or K$^+$, but less polar groups like ethers favour K$^+$.

2.2.5 Spherands

Spherands are a new class of macrocyclic host compounds synthesized by Cram and his co-workers in which the ligating sites are conformationally organized during synthesis rather than during complexation (80). Whereas the crown ethers and, to a lesser extent, the cryptands, reorganize in order to bind a cationic metal ion guest, the spherands and the host parts of their complexes are conformationally the same (81). Spherands are thus composed of units that are much more rigid than those of crowns or cryptands (Figure 2.26).

These macrocycles show the highest stability constants known for certain alkali metal ions, and remarkable selectivity. For example, when the ionic diameter (1) for the guest metal cations becomes too large for the preformed cavity *no* complexation occurs. The three spherands (84)–(86) bind *only* Na$^+$ and Li$^+$; no binding to K$^+$ could be detected (81). This behaviour is in contrast to that of crowns and cryptands in which size-related peak selectivity is found.

Semi-rigid hemispherands (81, 82) have also been recently synthesized which

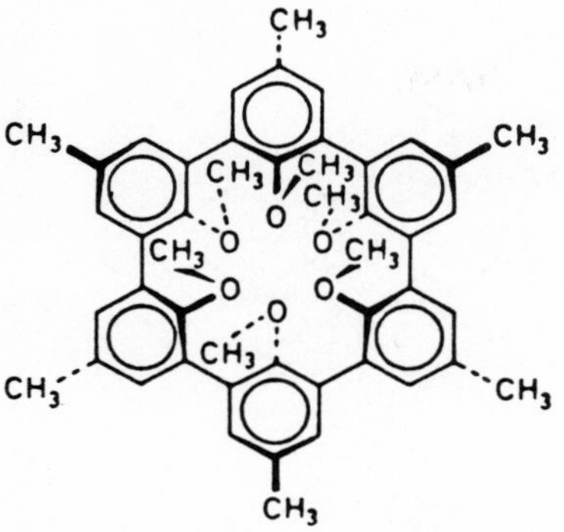

(84)

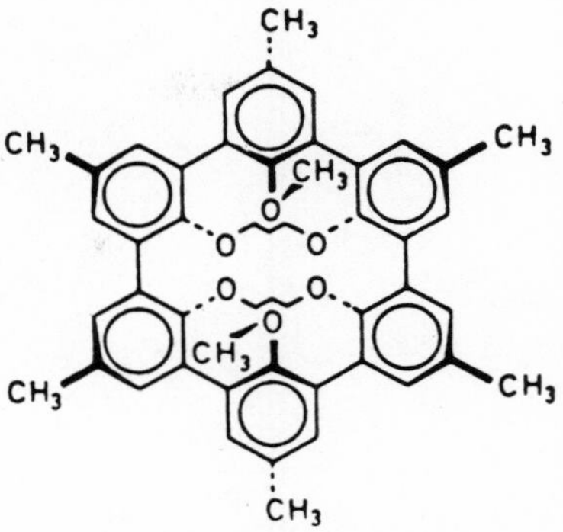

(85)

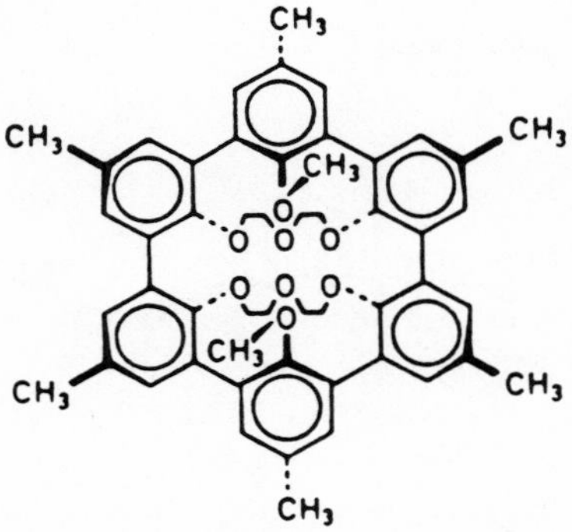

(86)

Figure 2.26 Spherands (80a, b).

consist of half spherand-half crown ether (**87**) (80) and half spherand-half cryptand (**88**) (83) (Figure 2.27).

These host molecules exhibit very high stability constants for alkali metal ions and are highly selective. For example, whereas the spherands (**84**) and (**85**) show Li^+ over Na^+ selectivity (factors of > 600 and 360 respectively), impressive selectivities of Na^+ over Li^+ are exhibited by cryptahemispherand (**89**) (420 000) and bridged hemispherand (**91**) (9500); (**90**) exhibits K^+ over Na^+ selectivity of 11 000.

2.3 Cationic recognition 2: molecular cation recognition

2.3.1 *Ammonium ion*

The simplest molecular cation is the ammonium ion which cannot be discriminated from K^+ by size very effectively. However, a clear difference in charge distribution exists, being spherical for K^+ and tetrahedral for NH_4^+. Selective binding of a tetrahedral guest requires the construction of a receptor

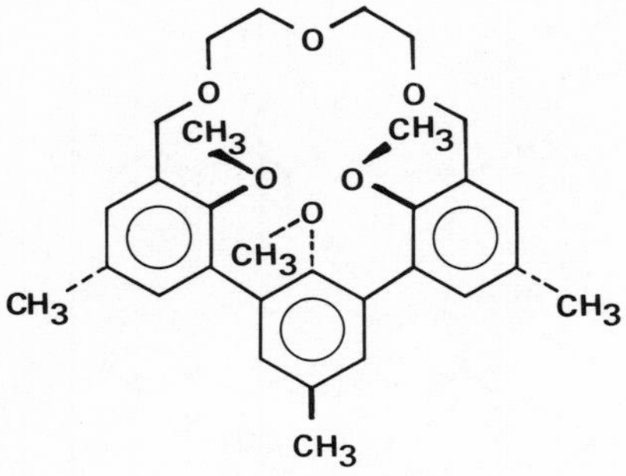

(87) 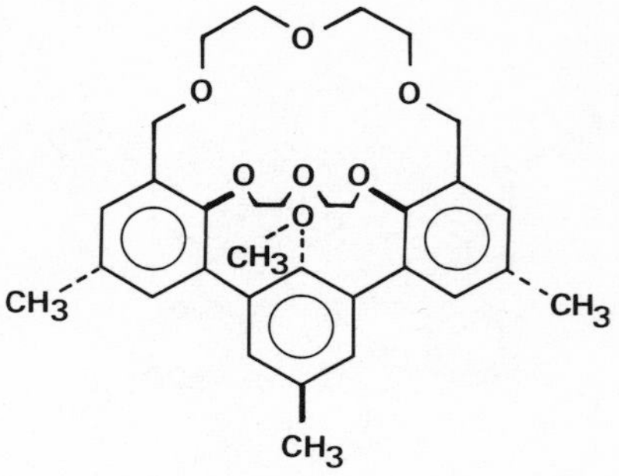   (91)

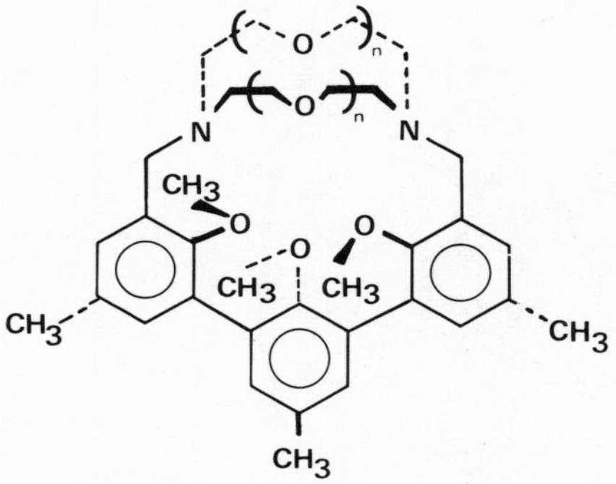

(88) $n = 1$ $m = 1$

(89) $n = 2$ $m = 1$

(90) $n = 2$ $m = 2$

Figure 2.27   Hemispherands (81–83)

molecule with a tetrahedral recognition site. This has been achieved by positioning four suitable binding sites at the corners of a tetrahedron and linking them with six bridges to form the spherical macrotricyclic cryptand (**92**) (84). This host binds the tetrahedral NH_4^+ exceptionally strongly and selectively as opposed to K^+, forming an ammonium cryptate (**93**) (85) (Figure 2.28).

The complex presents a high degree of structural and binding complementarity between the guest NH_4^+ and receptor (**92**). The NH_4^+ is held inside the cavity by a tetrahedral array of $^+$N-H---N hydrogen bonds and by electrostatic interactions with the six oxygen donor atoms. The strong binding in (**93**) results in a shift of the effective pK_a for the NH_4^+ by about six units higher than that of free NH_4^+. This illustrates how large may be the changes in substrate properties brought about by binding. Similar changes may take place when substrates bind to enzyme-active sites and biotic receptors.

Binding primary ammonium guest species. Macrocyclic polyethers and aza-polyethers selectively bind primary ammonium ions by anchoring the $-NH_3^+$ into the circular cavity with three $^+$N-H---X(X = O, N) hydrogen bonds (2, 86, 87) (Figure 2.29).

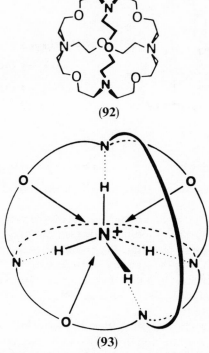

(92)

(93)

Figure 2.28 Spherical macrotricyclic cryptand and NH_4^+ cryptate (85).

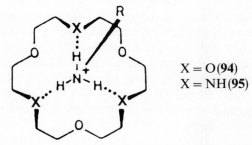

$$X = O\,(94)$$
$$X = NH\,(95)$$

Figure 2.29 Primary ammonium ion complexation by macrocyclic polyethers and aza-polyethers (87).

Comparative studies have demonstrated that the monocycle (95) forms much more stable primary ammonium complexes than (94) by the formation of three $^+$NH---N hydrogen bonds (88). The chiral tetrafunctional macrocycle (96a) forms the strongest primary ammonium complexes of any crown ether (89) (Figure 2.30). It is highly selective for binding primary ammonium ions against more highly substituted ones (central discrimination) (90). For example (96a) selectively binds the biologically active ions noradrenaline and

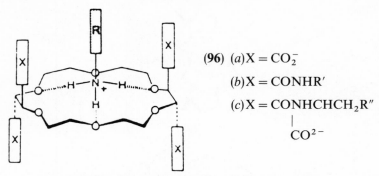

(96) (a) X = CO$_2^-$

(b) X = CONHR$'$

(c) X = CONHCHCH$_2$R$''$
$\qquad\qquad\qquad\quad|$
$\qquad\qquad\qquad\;\;$CO^{2-}

Figure 2.30   Chiral tetrafunctional macrocycle (89).

norephedrine in preference to their N-methylated derivatives, adrenaline and ephedrine.

The nature of the ligand side chain X in (96) can be varied and this affects the interactions (electrostatic, lipophilic, hydrogen bonding and charge transfer) between the side groups and the R group of the centrally bound guest Naturally this affects both the stability and selectivity of the complexes, which is called lateral discrimination (90). For example (96a) binds MeNH$_3^+$ more strongly than PhCH$_2$CH$_2$NH$_3^+$, whereas (96c) displays the same stability with both substrates (91).

Exotic macropolycyclic receptors have been synthesized by Lehn (92) and Sutherland (93), designed to bind bis-primary alkylammonium dications (100) (Figure 2.31). The cylindrical macropolycycles contain two macrocyclic subunits which are able to bind -NH$_3^+$ groups. The resulting supermolecules (101) (90) have the guest substrate located in the central molecular cavity of the receptor anchored by each terminal -NH$_3^+$ group to a di- or triaza-polyether macrocycle with three hydrogen bonds, as confirmed (92c) by the crystal structure of $^+$H$_3$N(CH$_2$)$_5$NH$_3^+$ $\subset$ (97b) (Figure 2.32).

Selectivity studies of these ligands towards $^+$H$_3$N(CH$_2$)$_n$NH$_3^+$ ($n=2$–9) substrates reveal (92, 93) that the substrate preferentially bound has a length complementary to the length of the molecular host cavity (linear molecular recognition) (Table 2.7).

2.3.2   Bipyridinium cations

Recently Stoddart and co-workers (94, 95) have elegantly demonstrated the successful binding of the bipyridinium herbicides diquat (102) and paraquat (103) by mono- and bicyclic polyether receptors. Dibenzo crown ethers are highly specific in their complexation of diquat, dibenzo-30-crown-10 (21) being the optimum receptor (94). X-ray structural analysis on the diquat (21)(PF$_6^-$)$_2$ complex (Figure 2.33) shows the flexible receptor molecule folding around the guest and achieving approximate parallel alignment of the two benzo rings in (21) with the bipyridinium ring of the diquat dication.

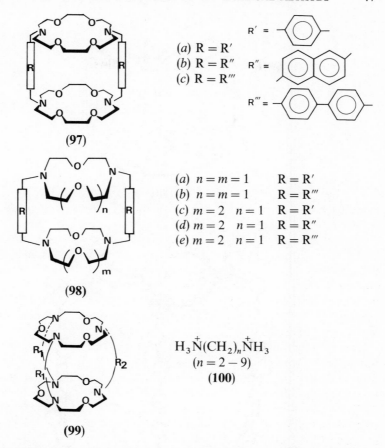

(97)

(a) R = R'
(b) R = R"
(c) R = R'''

R' =

R" =

R''' =

(98)

(a) $n = m = 1$ R = R'
(b) $n = m = 1$ R = R'''
(c) $m = 2$ $n = 1$ R = R'
(d) $m = 2$ $n = 1$ R = R"
(e) $m = 2$ $n = 1$ R = R'''

(99)

$$H_3 \overset{+}{N}(CH_2)_n \overset{+}{N}H_3$$
$$(n = 2 - 9)$$
$$(100)$$

Figure 2.31 Macropolycycles designed for complexation of diammonium dications (92b, 93a).

Table 2.7 Selectivity of the macropolycyclic hosts (97) and (98) towards bis-primary alkyl-ammonium dications $^{+}H_3N(CH_2)_n\overset{+}{N}H_3$, $n = 2$–9. Reproduced by permission from *Chem. Soc. Rev.* **15** (1986) 63.

Host	Selectivity (n)
98a	2 > 3
98b	4 < 5 6 > 7 > 8
98c	2 > 3 > 4
98d	3 < 4 > 5 > 6
98e	4 < 5 < 6 > 7
97a	3 > 4
97b	4 < 5 > 6 > 7 > 8
97c	7 > 8 > 9

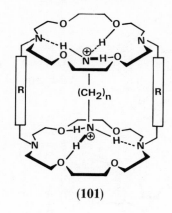

(101)

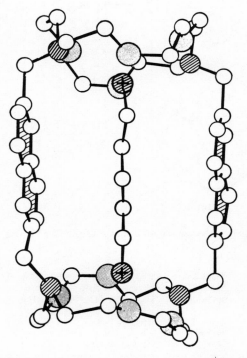

Figure 2.32 X-ray crystal structure (**92c**) of the complex $(H_3\overset{+}{N}(CH_2)_5\overset{+}{N}H_3 \subset (\mathbf{97}b))$]. (By permission of The Royal Society of Chemistry.)

The complex is held together by attractive electrostatic interactions which include charge transfer effects. Introducing a bridging unit between the two benzo rings of (**21**) produces a receptor (**104**) with a more rigidly defined host cavity, and this results in stronger binding of the diquat dication guest (95).

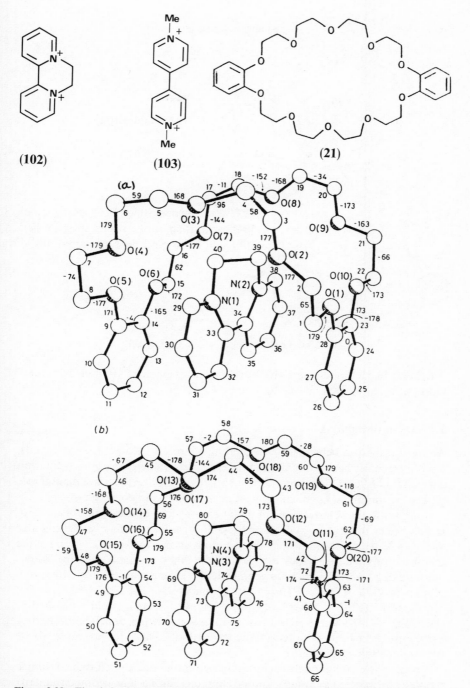

Figure 2.33 The skeletal representations (94) of the solid-state structures of two independent complexes (*a*) and (*b*) found in [diquat (**21**)]-[PF₆]₂. (By permission of The Royal Society of Chemistry.)

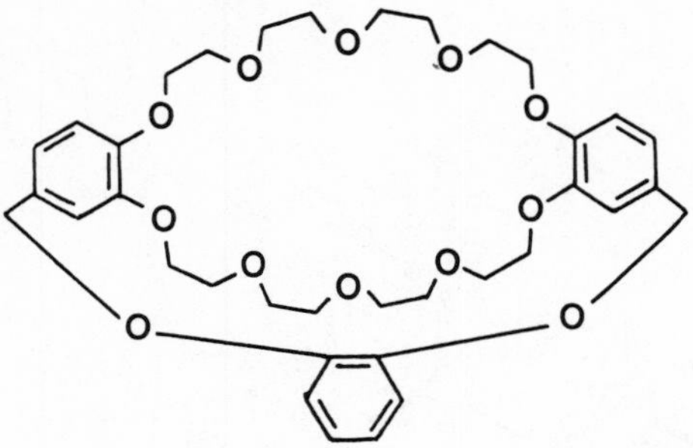

(104) binds diquat

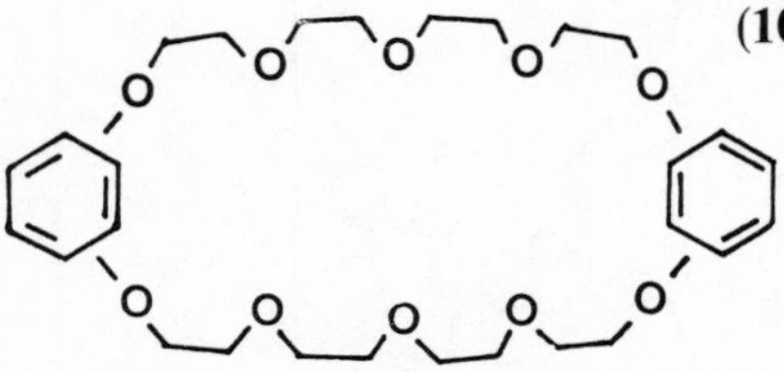

(105) *meta*: Bismetaphenylene crown ether
(106) *para*: Bisparaphenylene crown ether
(105) and **(106)** bind both paraquat and diquat

Figure 2.34 Receptors for diquat and paraquat (95).

Related bismetaphenylene **(105)** and bisparaphenylene **(106)** crown ethers bind paraquat as well as diquat (Figure 2.34).

2.4 Anion recognition

Anion coordination chemistry, the binding of anions by organic host ligands, has only recently been recognized and developed as a new area of coordination chemistry (49, 96). This is somewhat surprising in view of the fundamental role played by anions in chemical as well as in biological processes.

Comparing the relative sizes of metal cations and anions, anions are a lot larger. For example, the small anion F^- (1.36 Å) is approximately the same size as K^+ (1.33 Å). A list of selected anion radii is shown in Table 2.8 (97). Unlike metal cations whose shapes are likened to different-sized 'spherical balls', anions exhibit a wide variety of geometries (98): spherical (F^-, Cl^-, Br^-, I^-); planar (NO_3^-, CO_3^{2-}, RCO_2^-, etc.); linear (N_3^-, CN^-, SCN^-, etc.); tetrahedral (PO_4^{3-}, SO_4^{2-}, ClO_4^-, MnO_4^-, etc.); and octahedral ($Fe(CN)_6^{4-}$, $Co(CN)_6^{3-}$, etc.). Most anions are also pH-dependent, i.e. they exist only in a limited range of the normal pH scale, for example above pH 5–6 for the carboxylates.

A potential host ligand for a certain guest anion must therefore be designed to satisfy the particular anion's unique characteristics of size, geometry and pH dependence. In biotic systems the binding of anionic substrates by enzymes is achieved by ammonium ions of lysine residues (99) and guanidinium ions of

Table 2.8 Anion radii

Anion	Radius (Å)	Anion	Radius (Å)	Anion	Radius (Å)	Anion	Radius (Å)
F^-	1.36	OH^-	1.40	NO_2^-	1.55	CO_3^{2-}	1.85
Cl^-	1.81	CN^-	1.82	NO_3^-	1.89	SO_4^{2-}	2.30
Br^-	1.95	IO_3^-	1.82	$M_1O_4^-$	2.40	PO_4^{3-}	2.38
I^-	2.16						

arginine residues (100) within the enzyme structure. Thus initial research into this new exciting area of coordination chemistry has centred on the synthesis of macrocyclic and macropolycyclic host systems containing ammonium or guanidinium binding sites.

2.4.1   Anion receptors containing ammonium binding sites

Macropolycyclic systems: spherical anion recognition. In 1968 Simmons and Park (101) reported the first synthetic anion complexing agent, a diazamacro-bicyclicalkane host (Figure 2.35). Various halide ions (Cl^-, Br^-, I^-) depending on n are included into the diprotonated positive cavity. The stability of the resulting complex results from the electrostatic attraction between the positive cavity and the anion as well as from hydrogen bonds of the type $\overset{+}{N}$-H---X^----H-N$^+$ (Figure 2.36). X-ray structural analysis shows the Cl^- guest anion bound inside the cavity with a $\overset{+}{N}$(H)--Cl distance of 3.10 ± 0.01 Å (102).

In their tetraprotonated forms the spheroidal macrotricycles (**92**), (**108**), (**109**) form very stable complexes with Cl^- and Br^- (103) (Figure 2.37, Table 2.9).

The ligands (**92**)-4H$^+$ and (**108**)-4H$^+$ are selective for Cl^- over Br^- (selectivity > 1000), whereas (**109**)-4H$^+$ and (**107**)-2H$^+$ are less discriminating. The iodide ion, with an anionic radius of 2.16 Å, is too large for the respective cavities and is not enclosed. An x-ray structure of the complex $Cl^- \subset$ (**92**)-4H$^+$ confirmed the inclusion of the anion bound by a tetrahedral array of N-H--Cl$^-$ hydrogen bonds (104) (Figure 2.38).

Schmidtchen (105) has recently synthesized a series of macropolycylic compounds in which the binding sites are quaternary ammonium salts (Figure 2.39). They are able to coordinate Cl^-, Br^- and even, in the case of (**111**) and (**112**), I^- inside their respective host cavities. An x-ray structure (106) of the $4I^- \subset$ (**111**) shows one I^- encapsulated symmetrically into the spherical intramolecular cavity of (**111**) (Figure 2.40). Table 2.10 reports the stability constants determined by halogenide selective electrodes. Compared to Table 2.9, the stabilities are much lower and Br^- is more strongly bound than Cl^-. These values demonstrate that the combination of lower electrostatic interaction between $R\overset{+}{N}_4$---X^- due to steric hindrance and the lack of ionic hydrogen bonds in the quaternary ammonium ion results in the tertiary ammonium ion being a more efficient binder of anions. The larger intra-

C

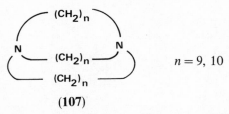

$n = 9, 10$

(107)

Figure 2.35 Diaza-macrobicyclicalkane host molecules for halide anion guests (101).

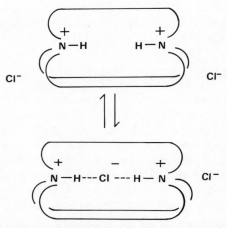

Figure 2.36 Schematic representation (101) of the binding of Cl⁻ in the protonated cavity of (107). Reprinted with permission from *J. Amer. Chem. Soc.* **90** (1968) 2431, © 1968 American Chemical Society.

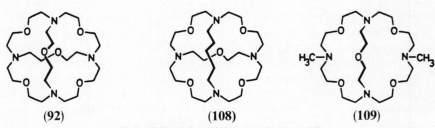

(92)   (108)   (109)

Figure 2.37 Spheroidal macrotricycles (103).

Table 2.9 Stability constants of halide complexes of tetraprotonated forms of (92) and (107)–(109)

Halide anion guest	(92) – 4H⁺	(108) – 4H⁺	(109) – 4H⁺	(107) – 2H⁺
Cl⁻	> 1.0	> 4.5	1.7	0.7
Br⁻	< 1.0	1.55	< 1.0	< 1.0

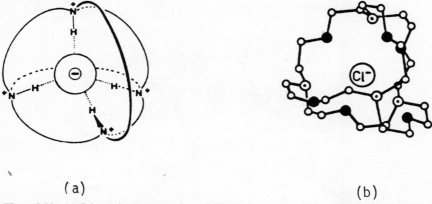

Figure 2.38   (*a*) Schematic representation of the enclosure of a Cl⁻ in (**92**)-4H⁺. (*b*) Crystal structure of the [Cl⁻ ⊂ **92**-4H⁺] complex (104). (By permission of The Royal Society of Chemistry.)

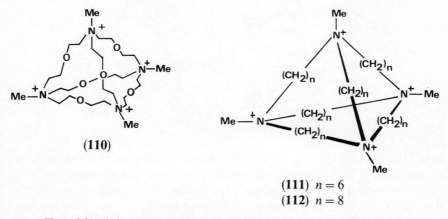

(110)

(111) $n = 6$
(112) $n = 8$

Figure 2.39   Anion receptors containing quaternary ammonium salts (105*a, b*).

Table 2.10   Stability constants of halide complexes of (**110**)–(**112**)

Halide anion guest	(110)	(111)	(112)
Cl⁻	1.0	1.3	<0.5
Br⁻	1.8	2.45	2.45
I⁻	—	2.2	2.4

molecular host cavity present in the quaternary ammonium compounds can explain the reverse selectivity of Br⁻ over Cl⁻.

Linear anion recognition. The hexaprotonated form of the ellipsoidal cryp-

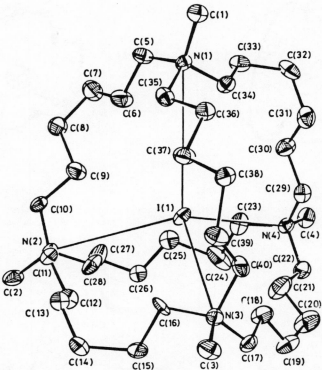

Figure 2.40 X-ray structure of the complex [4I⁻ ⊂ (111)] (106). (By permission of The Royal Society of Chemistry.)

tand Bis-Tren (113) synthesized by Lehn *et al.* (107) binds various monoatomic and polyatomic anions (107) and extends the recognition of anionic substrates beyond the spherical halides. This ligand is specifically designed to bind linear triatomic anions of the type XYZ, for example azide, N_3^-. Indeed ^{13}C nmr experiments (107) showed that the azide anion forms a strong 1:1 complex with [(113)-6H⁺]. X-ray structural analysis of [azide ⊂ (113)-6H⁺] anion complex has confirmed the complete inclusion of the anion, held in the host cavity by two pyramidal arrays of N-H--N hydrogen bonds, each of which binds one of the two terminal nitrogens of N_3^- (106) (Figure 2.41). The anion selectivity sequence of ClO_4^-, Cl^-, I, $MeCO_2^-$, Br^-, < HCO_3^- < NO_3^-, NO_2^-, ≪ N_3^- follows a trend of topological complementarity between host and guest rather than the physico-chemical properties of the anions. The non-complementarity between (113-6H)⁺ and the spherical halides results in much weaker binding and appreciable distortions of the ligand as seen in the crystal structures of the cryptates (114) (Figure 2.42) where the bound ion is F^-, Cl^- or Br^- (108).

The syntheses of related polyamine macrobicycles (Figure 2.43) with larger cavities have recently been reported in the literature by Lehn and co-workers

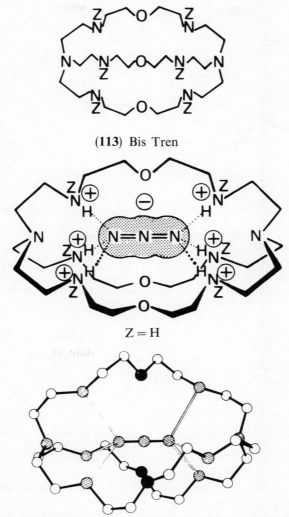

(**113**) Bis Tren

$Z = H$

Figure 2.41 X-ray structure of the azide [bis-tren-6H$^+$] anion complex (107, 108). C, unshaded; N, hatched; O, shaded symbols. Reprinted with permission from *J. Amer. Chem. Soc.* **100** (1978) 4914, © 1978 American Chemical Society.

(109). These molecules are designed so that when in their respective hexa- and nonaprotonated forms they can recognize and bind large anionic guest species.

Macrocyclic systems—polyazamacrocycles. Anion complexation can only occur when over a certain range of pH the anion exists and the host ligand is in its protonated form. With the macropolycyclic systems previously described, this prerequisite for binding is obeyed for halide and azide guest anions. However, for the important classes of carboxylate and adenosine

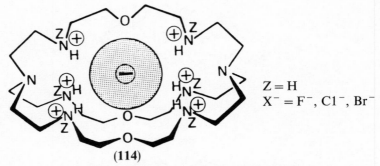

$Z = H$
$X^- = F^-, Cl^-, Br^-$

(114)

Figure 2.42   Schematic representation of **(114)** (108).

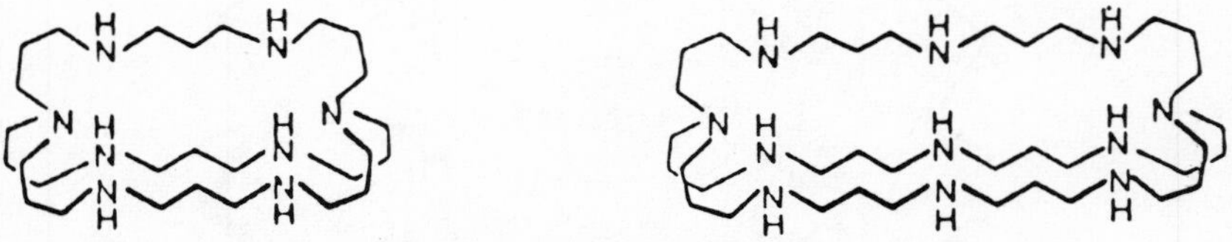

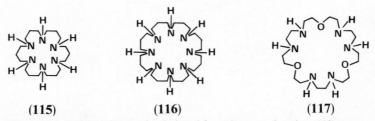

Figure 2.43   Polyamine macrobicyclic anion receptors (109).

(115)   **(116)**   **(117)**

Figure 2.44   Polyazamacrocycles designed for anion complexation (110).

phosphate anions this is not the case. Higher pH conditions are required for these anions to form.

Monocyclic polyazamacrocycles with propylene units separating the nitrogen atoms **(115)** and **(116)**, and **(117)** having two successive ethylene diamine units separated by five atoms, are found (110) to have respective pK_a values in the desired range (Figure 2.44).

Macrocycle **(117)**, based on ethylene diamine units, forms the most stable complexes with the highly charged anions oxalate, sulphate and to a lesser extent AMP, ADP and ATP. The receptor **(116)** has the largest cavity and exhibits the greater log K_s values for the large anions citrate, trimesate, $Co(CN)_6^{3-}$ and $Fe(CN)_6^{4-}$.

The binding of anions like $Co(CN)_6^{3-}$, $Fe(CN)_6^{4-}$ and $Ru(CN)_6^{4-}$ by **(115)**-6H$^+$ and **(116)**-8H$^+$ is an example of second sphere coordination (111) with the transition metal cation complexed by cyanide anions which are, in turn, bound to the receptors by multiple Ñ-H--NC-M$^-$ hydrogen bonds

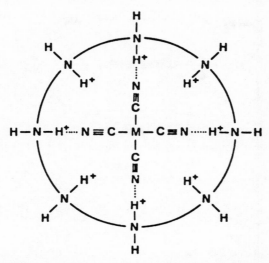

Figure 2.45 A possible structure for the $[Co(CN)_6 \ 116\text{-}8H^+]^{5+}$ complex. Two CN^- ligands which are perpendicular to the plane of the receptor are not shown.

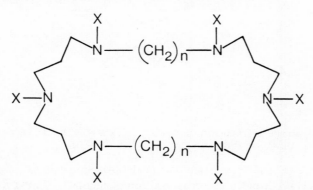

Figure 2.46 Ditopic receptors designed to bind dianionic guest substrates (114a). X = H; (118)n = 7; (119)n = 10.

(Figure 2.45). Electrochemical measurements (112) show that $Fe(CN)_6^{4-}$ and $Ru(CN)_6^{4-}$ both form complexes with (115)-6H$^+$ and (116)-8H$^+$ and that the anions are more difficult to oxidize in the presence of the receptors than in their absence. Also, recent photochemical studies on the $[Co(CN)_6^{3-}\text{-}116\text{-}8H^+]$ complex demonstrates that it is possible to protect, control and orient the ligand photosubstitution reactions of transition metal complexes by binding to appropriate receptor molecules (113).

'Ditopic receptors' designed to bind dianionic guest substrates have been successfully synthesized by Lehn and co-workers (114) (Figure 2.46). In their hexaprotonated forms they form strong complexes with various dicarboxylate salts with selectivity that depends on the chain length m of the

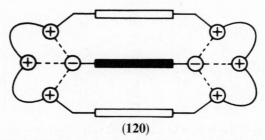

(120)

Figure 2.47 Structural complementarity between ditopic receptor and dicarboxylate dianion (114a).

(121)   **(122)**

Figure 2.48   Penta- and hexa-azamacrocycles (115).

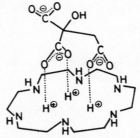

Figure 2.49   Schematic representation for the 1:1 complex of (**122**)-3H$^+$ with citrate (115b). (Reprinted by permission of Springer Verlag, Heidelberg.)

$\bar{O}_2C(CH_2)_mCO_2^-$ substrate. Selectivity peaks are found with $m = 2, 3$ for (**118**), $m = 5, 6$ for (**119**) as a result of structural complementarity (linear recognition) between the receptor and the dicarboxylate substrates (**120**) (Figure 2.47).

Kimura (115) has also described the binding of several polycarboxylate substrates by penta- and hexa-azamacrocycles (Figure 2.48). These investigations were carried out at neutral pH with (**121**) and (**122**) in their triprotonated forms. The citrate anion formed the most stable complexes with log K_s values of 3.00 and 2.38 respectively for (**121**)-3H$^+$ and (**122**)-3H$^+$ (Figure 2.49).

2.4.2   *Anion receptors containing guanidinium binding sites*

The guanidinium ion has a high pK_a value of 13.5 and thus it remains in its protonated form over a wide range of pH. This attractive feature has led to its incorporation into a variety of macrocyclic compounds (116) (Figure 2.50).

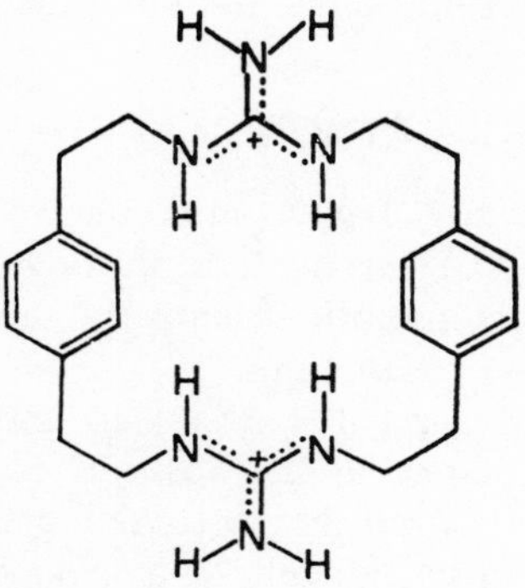

(123) 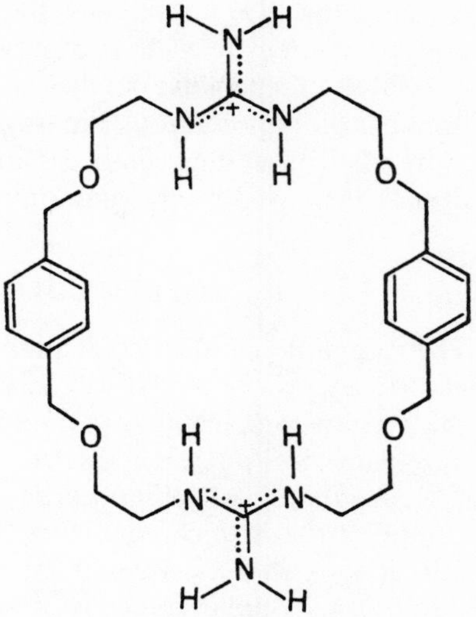   (124)

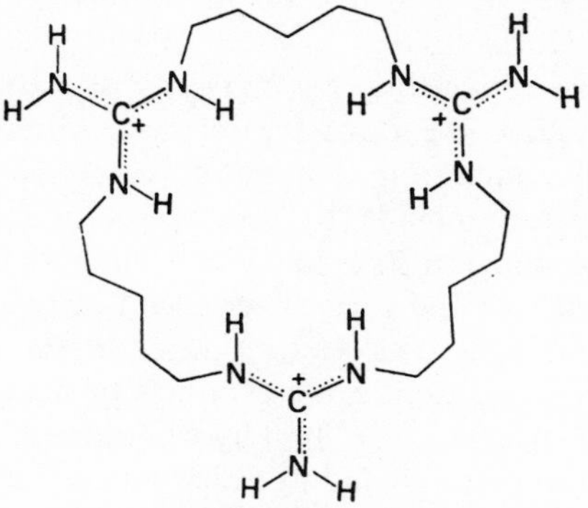

(125)

Figure 2.50 Macrocycles containing guanidinium binding sites (116).

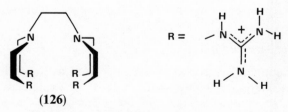

(126)

Figure 2.51 Acyclic guanidinium anion receptor (117).

Although these receptors were found to bind the phosphate anion PO_4^{3-}, their respective stability constants for complexation are quite small, indicating low stability. In fact the acyclic analogue (126) (Figure 2.51) has a stability constant for PO_4^{3-} binding only slightly lower than (123)–(125), implying that no macrocylic effect is observed.

Related comparative studies of acyclic ligands containing either guanidinium or ammonium as binding sites demonstrated that ammonium is a more efficient binding centre for anions (117). This is a result of higher charge density displayed by the ammonium ion compared to the guanidinium ion.

II MOLECULAR COVALENT RECOGNITION

The recognition and complexation of *neutral* (uncharged) guest molecules by abiotic receptors is a relatively recent area of chemical investigation, although this type of complexation plays a fundamental role in biotic systems e.g. the base-pairing of nucleic acids, or enzyme–substrate interactions.

Hydrophobic interactions are the binding forces largely responsible for the complexation of uncharged guest molecules and neutral apolar hosts. These interactions are very weak in solvents of low polarity but they become more significant in polar aqueous media. It is sometimes difficult to elucidate whether the guest molecule is enclosed in an *intramolecular* cavity of a single host or located in *intramolecular* cavities of the host lattice; only x-ray structural analysis can solve this. The majority of associations of synthetic neutral hosts with uncharged guests have only been observed in the crystalline state, mainly as lattice inclusion compounds, rarely as molecular inclusion complexes. This section intends to give a review of recently reported intramolecular inclusion complexes in which a simple integral stoichiometry exists between neutral host and guest species with emphasis towards selective recognition of neutral molecules. Hence lattice inclusion compounds such as the numerous inclusion cavities or channels of urea and thiourea (118), dianin (119), deoxycholeic acids (120), β-hydroquinone (119), etc. will not be described here. Cyclophanes, including calixarenes, the rigid cavitands and the inclusion complexes of water-soluble cyclophanes are considered.

The neutral molecular inclusion complexes of the naturally occurring cyclodextrins (121) and the synthetic macrocyclic polyethers have been comprehensively reviewed recently (122) and are not reported.

2.5 Cyclophane host molecules

The incorporation of hydrophobic subunits such as aryl, alkyl, napthyl, binaphthyl, etc., into a macrocyclic structural framework leads to synthetic host molecules containing large endolipophilic cavities which have the potential of accommodating neutral organic guest species.

As early as 1955, Stetter and Roos (123) recognized compounds of this type and synthesized bis(N, N′-tetramethylene benzidine (127) (Figure 2.52). They observed that (127) tenaciously retained stoichiometric amounts of benzene or dioxane (1:1), even after drying at 80° C for 5h, when crystallized from

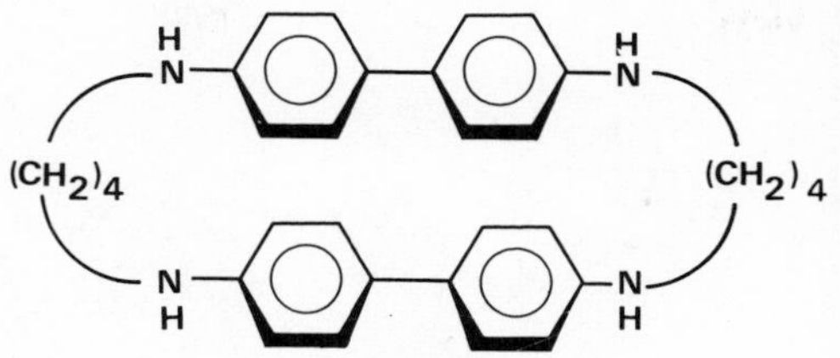

(127)

Figure 2.52 Bis(N, N′-tetramethylene benzidine) (123).

these solvents. Based on this, and in combination with molecular model considerations, Stetter proposed an intramolecular inclusion of benzene or dioxane in the cavity of the host molecule. The Stetter complexes were therefore considered classical examples of intramolecular inclusion compounds until 1982, when Saenger (124) showed by x-ray structural analysis of the benzene complex of (127) that the benzene molecules are *not* accommodated in the central cavities of the ligand molecules but fill interstices between them.

2.6 Calixarenes

The first examples reported of totally synthetic uncharged host–uncharged guest intramolecular inclusion complexes elucidated by x-ray crystal structures were the calixarenes (125, 126). This class of macrocyclic cyclophanes contains phenolic residues in a cyclic array linked by methylene groups all ortho to the hydroxy groups (Figure 2.53).

The calixarenes can be prepared by the condensation of *p*-tert-butylphenol and formaldehyde under basic conditions. Gutsche has recently reported the facile synthesis of the tetramer (128) known as *p*-tert-butyl calix[4]arene (125). The name calixarene comes from the similarity between the shape of a Greek vase known as a 'calix crater' and the shape of the cyclic tetramer (Figure 2.54). x-ray crystallographic determinations have established that in the solid state the calix[4]arene does exist in this 'cone' conformation (126, 128).

Many of the calixarenes retain the solvent from which they are crystallized. Crystal structures by Andreetti *et al.* (126) of (128)-toluene complex show the calix to be in the cone conformation and the toluene located in the centre of the calix, i.e. an 'endo'-calix complex (Figure 2.55). A recent publication by the same group (129) shows two molecules of (128) in the cone conformation encapsulating one molecule of anisole.

Selective complexation has been demonstrated (130) by crystallizing *p*-tert-butylcalix[4]arene (128) from 50:50 mixtures of two guest molecules such as benzene and *p*-xylene. Anisole and *p*-xylene are complexed in preference to most other simple aromatic hydrocarbons.

The isolation and characterization of solid state complexes does not necessarily indicate that similar complexes exist in solution. Little evidence in

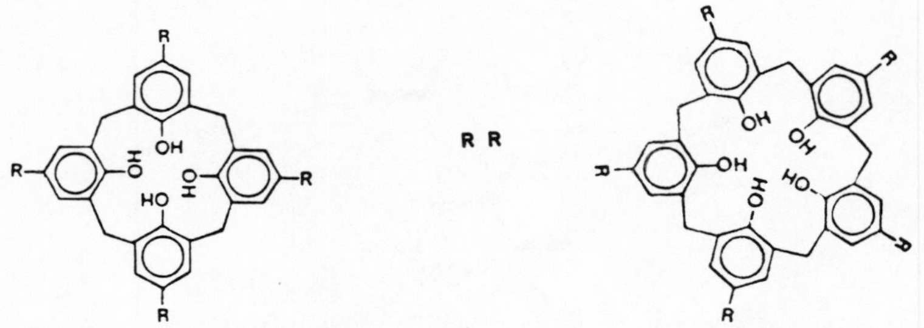

(128)

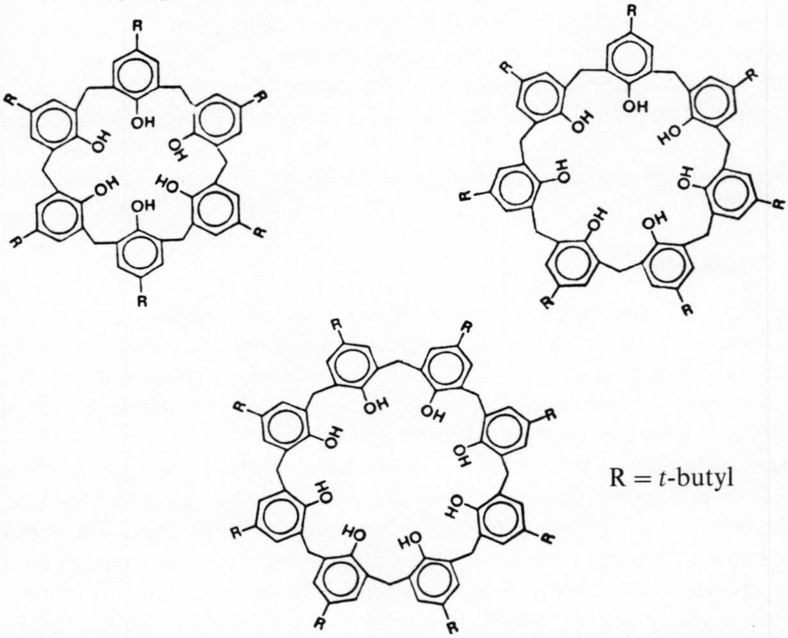

R = *t*-butyl

Figure 2.53 The calix [*n*] arenes (for the nomenclature of these compounds see reference 125). (By permission of Springer Verlag, Heidelberg.)

non-aqueous media for *solution* calixarene–neutral guest complexation has appeared in the chemical literature. The reason for this could stem from the conformational properties of calixarenes.

In solution, Cornforth (131) pointed out, four discrete forms of calix[4]arenes can exist, the 'cone', 'partial cone', '1, 2-alternate' and '1, 3-alternate' conformations (Figure 2.56). Kammerer (132) showed by dynamic ^{1}H nmr studies of *p*-alkylcalix[4]arenes that these conformations are rapidly interconverting on the nmr timescale. Hence no 'fixed' hydrophobic host cavity exists long enough to encapsulate a neutral organic guest molecule.

Modified calixarenes that are conformationally immobile, fixed in either the 'cone' or 'partial cone' conformation, have been successfully synthesized by

Figure 2.54 C.P.K. model of *p*-phenylcalix [4] arene (left) and calix crater (right). Reprinted with permission from *Acc. Chem. Res.* **16** (1983) 161, © 1983 American Chemical Society.

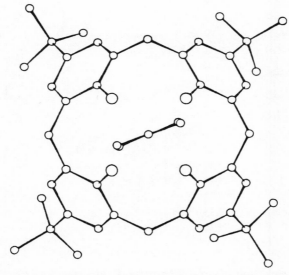

Figure 2.55 X-ray crystallographic structure of the 1:1 'endo'-calix complex of *p*-tert-butyl-calix [4] arene (**128**) and toluene (126). (By permission of The Royal Society of Chemistry.)

replacing the hydrogens of the OH functions with larger groups (131, 132) (Figure 2.57). Fixed 'cone' conformations are found with (**129**)–(**131**) and partial cone conformations with (**132**)–(**134**).

Another way of fixing the cone conformation has been achieved recently by the connection of two opposite *para*-positions by an aliphatic chain of appropriate length producing a 'bridged' calixarene (135) (**135**) (Figure 2.58).

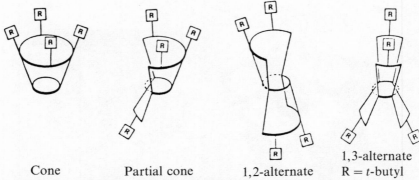

Cone   Partial cone   1,2-alternate   1,3-alternate
                                                            R = t-butyl

Figure 2.56 Conformations of the calix [4] arenes (131). Reprinted with permission from *Acc. Chem. Res.* **16** (1983) 181, © 1983 American Chemical Society.

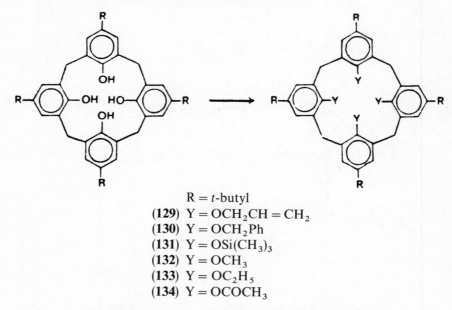

R = t-butyl
(129) Y = OCH₂CH = CH₂
(130) Y = OCH₂Ph
(131) Y = OSi(CH₃)₃
(132) Y = OCH₃
(133) Y = OC₂H₅
(134) Y = OCOCH₃

Figure 2.57 Conformationally immobile calixarenes (131, 132).

Solution complexation studies of these 'rigid' calixarenes with neutral organic guests have as yet not been reported.

2.7 Cavitands

Cram has proposed the class name 'cavitand' for synthetic organic host compounds that contain 'enforced' or 'rigid' cavities large enough to accommodate simple molecules or ions (136). The parent calixarenes (Figure 2.53) are not of this class, whereas the rigid modified ones (Figure 2.57)

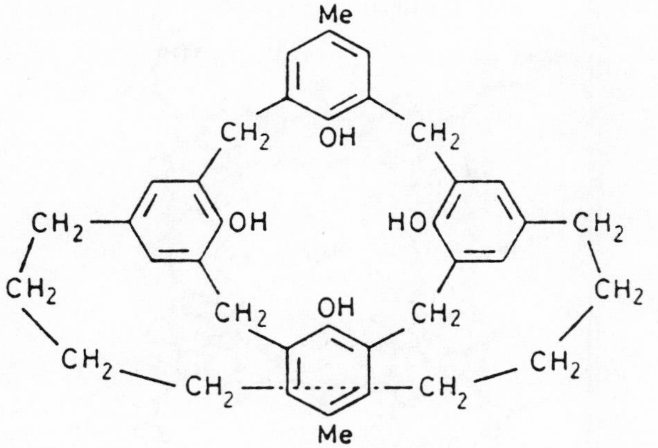

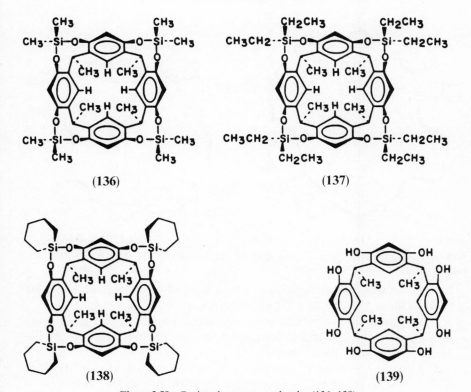

Figure 2.58 Bridged calixarene (135).

(138)

(136) (137)

(138) (139)

Figure 2.59 Cavitand receptor molecules (136–139).

are. Cram has prepared (137) the cavitands (136)–(138) from the conformationally mobile resorcinol–aldehyde condensation product cyclophane (139) (Figure 2.59). Treatment of (139) with appropriate dialkyldichlorosilanes in

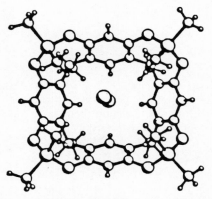

Figure 2.60 X-ray crystallographic structure of (**135**).2CS$_2$ complex (the CS$_2$ molecule which lies between molecules of complex is omitted) (137). Reprinted with permission from *J. Amer. Chem. Soc.* **107** (1985) 2574, © 1985 American Chemical Society.

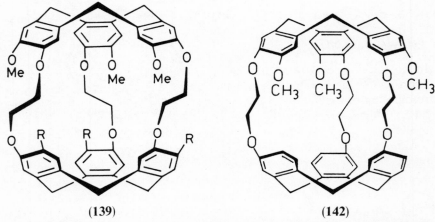

(**139**)   (**142**)

Figure 2.61 The cryptophanes (138*a*). (**139**), R = OMe; (**140**), R = H; (**141**), R = OCH$_2$CO$_2$H and Me is replaced by CH$_2$CO$_2$H.

tetrahydrofuran and triethylamine at high dilution gave the cavitands (**136**), (**137**), (**138**) in 37%, 9% and 7% yields respectively.

Evaporation of (**136**) in CS$_2$ gave crystalline (**136**)·2CS$_2$ of x-ray quality (Figure 2.60); one CS$_2$ molecule occupies the cavity of the host and one lies between molecules of the complex.

The results of ^{1}H nmr solution complexation studies of (**136**)–(**138**) in CDCl$_3$ and C$_6$D$_6$ with CS$_2$ were consistent with the solution structures of these complexes being similar to the crystal structure of (**136**)CS$_2$. These cavitands were also found to bind intramolecularly other linear molecules (CH$_3$C $\equiv$ CH and O$_2$) but binding of non-complementary guest candidates was not detectable by ^{1}H nmr.

2.8 Cryptophanes

Collet *et al.* (138–140) have recently reported the elegant syntheses of cavitands composed of two cyclotriveratrylene units connected by three di- or trimethylene bridges (Figure 2.61). These hollow host molecules have been named cryptophanes, and are suitable for inclusion of lipophilic guests in the interior of their three-dimensional enforced cavities. The cryptophanes **(139)**–**(142)** are chiral, and have been obtained in enantiomerically pure forms and their absolute configurations established.

Even in lipophilic solvents where hydrophobic effects and interactions are minimized, the cryptophanes strongly, reversibly and selectively complex neutral guests of complementary size such as the halogenomethanes. For example, enantio-selective complexation of CHFClBr by **(140)** has also been reported (138*a*). The water-soluble cryptophane **(141)** scavenges *trace* amounts of halogenomethanes from water.

2.9 Water-soluble cyclophanes

In biological systems the formation of intramolecular inclusion complexes by hydrophobic interactions in water plays a significant role in the binding and recognition of organic guests. Whereas in non-polar solvents hydrophobic binding forces are relatively insignificant, they are maximized in aqueous media.

Although various water-soluble cyclophanes have been successfully prepared (141), it is only recently that the first example of a crystalline complex of water-soluble cyclophane with a hydrophobic substrate has been described (142). Koga synthesized **(143)** by connecting two diaryl methane units and two bridging alkyl chains via four nitrogens. The free host molecule is water-soluble below pH 2, and its inclusion complex-forming ability was examined in acidic water by comparing the spectral changes of hydrophobic organic guests induced by **(143)** with those induced by an acyclic reference compound **(144)** (Figure 2.62). Fluorescence, ^{1}H and ^{13}C nmr spectra gave evidence for inclusion of a variety of aromatic guest species. Direct evidence of intramolecular inclusion came from an x-ray crystallographic study for the 1:1 complex with durene isolated from an aqueous solution (142). It was shown that, the guest molecule durene was fully enclosed in the cavity of the host molecule, located exactly in the middle of the hole.

Related macrocyclic cyclophanes synthesized by Tabushi (143, 144) utilize quaternary ammonium **(145)** (141) and sulphonium groups **(156)** (144) for their solubility as host molecules in water (Figure 2.63).

Shinkai and co-workers (145, 146) have incorporated sulphonic acid groups into a calix[6]arene skeleton to produce the first water-soluble calixarenes

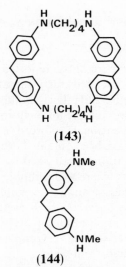

(143)

(144)

Figure 2.62 X-ray crystallographic structure of (**146**)-4H⁺ with durene (142). Reprinted with permission from *J. Org. Chem.* **50** (1985) 4478, © 1985 American Chemical Society.

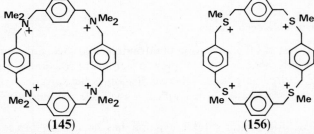

(145)   **(156)**

Figure 2.63 Cyclophane host molecules containing quaternary ammonium and sulphonium groups (141, 144).

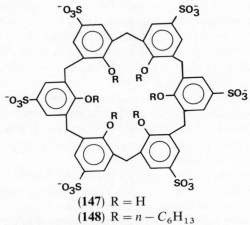

(147) R = H
(148) R = $n - C_6H_{13}$

Figure 2.64 Water-soluble calix[6] arenes (146).

(Figure 2.64). Fluorescence and absorption spectroscopic studies indicate that (147) and (148) are able to include naphthalene but not pyrene in the calix hydrophobic cavity.

2.10 Concluding remarks

This short review has attempted to summarize recent advances in molecular and ionic recognition. New developments in this exciting field are continuously being made. The possibilities of structural variations in improving the complexation selectivities of receptors toward specific guests will undoubtedly be accomplished in the future, benefiting the construction of new man-made chemical sensors.

References

1. (a) C.J. Pedersen, in R.M. Izatt and J.J. Christensen (eds.), *Synthetic Multidentate Macrocyclic Compounds*, Academic Press, New York, (1978), 1–51. (b) C.J. Pedersen, *Aldrichim. Acta* **4** (1971) 1.
2. C.J. Pedersen, *J. Amer. Chem. Soc.* **89** (1967) 7017.
3. E. Weber and F. Vogtle, *Top. Curr. Chem.* **98** (1981) 1.
4. C.J. Pedersen, *J. Org. Chem.* **36** (1971) 254.
5. J. Dale and P.O. Kristiansen, *J.C.S. Chem. Commun.* (1971) 670; *Acta Chem. Scand.* **26** (1972) 1471.
6. R.N. Greene, *Tetrahedron Lett.* 1972, 1793.
7. F.L. Cook, T.C. Caruso, M.P. Byrne, C.W. Bowers, D.H. Speck and C.L. Liotta, *Tetrahedron Lett.* 1974, 4029.
8. H.K. Frensdorff, *J. Amer. Chem. Soc.* **93** (1971) 4684.
9. R.M. Izatt, B.L. Haymore, J. S. Bradshaw and J. J. Christensen *Inorg. Chem.* **14** (1975) 3132.
10. D.A. Laidler and J.F. Stoddart, in S. Patai (ed.), *The Chemistry of Ethers, Crown Ethers, Hydroxyl Groups and their Sulphur Analogues*, Part 1, John Wiley, New York (1980) 1–53.
11. G.W. Gokel and S.H. Korzeniowski, *Macrocyclic Polyether Syntheses*, Springer Verlag, Berlin and Heidelberg (1982).
12. J.M. Lehn, *Structure and Bonding*, Springer Verlag, Berlin etc. (1973) 1.
13. N.K. Dalley, in R.M. Izatt and J.J. Christensen (eds.), *Synthetic Multidentate Macrocyclic Compounds*. Academic Press, New York (1978) 207.
14. R.D. Shannon, *Acta Crystallogr. Sect. A: Found. Crystallogr.* **32** (1976) 751.
15. R.M. Izatt, D.J. Eatough and J.J. Christensen, *Structure and Bonding*, Springer Verlag, Berlin etc. (1973) 161.
16. G.W. Gokel, D.M. Goli, C. Minganti and L. Echegoyn, *J. Amer. Chem. Soc.* **105** (1983) 6786.
17. H.K. Frensdorf, *J. Amer. Chem. Soc.* **93** (1971) 600.
18. M.A. Bush and M.R. Truter, *J. Chem. Soc. Perkin Trans.* **2** (1972) 345.
19. R.M. Izatt, J.S. Bradshaw, S.A. Nielsen, J.D. Lamb and J.J. Christensen, *Chem. Rev.* **85** (1985) 271.
20. M.R. Truter, *Structure and Bonding*, Springer Verlag, Berlin etc., (1973) 71.
21. P. Seiler, M. Dobler and J.D. Dunitz, *Acta Crystallogr.* **B30** (1974) 2741.
22. P. Seiler, M. Dobler and J.D. Dunitz, *Acta Crystallogr.* **B30** (1974) 2744.
23. P. Seiler, M. Dobler and J.D. Dunitz, *Acta Crystallogr.* **B30** (1974) 2733.
24. P.R. Mallinson and M.R. Truter, *J. Chem. Soc. Perkin Trans.* **2** (1972) 1818.
25. D.L. Hughes, *J.C.S. Dalton Trans.* (1975) 2374.
26. F. Vogtle and E. Weber, in S. Patai, (ed.), *The Chemistry of the Ether Linkage*, Supplement E, Part 1, John Wiley, London (1980) 59.
27. R.M. Izatt, R.E. Terry, A.G. Avondet, J.S. Bradshaw, N.K. Dalley, T.E. Jensen, J.J. Christensen and B.L. Haymore, *Inorg. Chim. Acta* **30** (1978) 1.
28. K.R. Adam, A.J. Leong, L.F. Lindoy, H.C. Lip, B.W. Skelton and A.H. White, *J. Amer. Chem. Soc.* **105** (1983) 4645.

70   CHEMICAL SENSORS

29. K.R. Adam, K.P. Dancey, B.A. Harrison, A.J. Leong, L.F. Lindoy, M. McPartlin and P.A. Tasker *J.C.S. Chem. Commun.* (1983) 1351.
30. K.R. Adam, C.W.G. Ansell, K.P. Dancey, L.A. Drummond, A.J. Leong, L.F. Lindoy and P.A. Tasker, *J.C.S. Chem. Commun.* (1986) 1011.
31. M. Ciampolini, N. Nardi, P. Dapporto, P. Innocenti and F. Zanobini, *J.C.S. Dalton Trans.* (1984) 575.
32. M. Ciampolini, N. Nardi, P. Drapporto and F. Zanobini, *J.C.S. Dalton Trans.* (1984) 995.
33. E.P. Kyba, R.E. Davis, S. Liu, K.A. Hassett and S.B. Larson, *Inorg. Chem.* 24 (1985) 4629.
34. (a) K. Kimura, T. Maeda and T. Shono, *Talanta* 26 (1979) 945; (b) K. Kimura, S. Kitazawa, T. Maeda and T. Shono, *Fresenius Z. Anal. Chem.* 313 (1982) 132.
35. M.J. Calverley and J. Dale, *J.C.S. Chem. Commun.* (1981) 684.
36. T. Ikeda, A. Abe, K. Kikukawa and T. Matsuda, *Chem. Letts.* (1983) 369.
37. K. Kimura, T. Tsuchida, T. Maeda and T. Shono, *Talanta* 27 (1980) 801.
38. S. Shinkai et al., *J.C.S. Perkin Trans.* 1 (1981) 3279.
39. S. Shinkai, T. Ogawa, Y. Kusano and O. Manabe, *Chem. Lett.* (1980) 283.
40. S. Shinkai et al., *J. Amer. Chem. Soc.* 103 (1981) 111.
41. S. Shinkai, *J. Amer. Chem. Soc.* 104 (1982) 1960.
42. P.D. Beer, *J. Chem. Soc. Chem. Commun.* (1985) 1115.
43. P.D. Beer, *J. Organometall. Chem.* 297 (1985) 313.
44. P.D. Beer and A.D. Keefe, *J. Organometall. Chem.* 306 (1986) C10.
45. B. Dietrich, J.M. Lehn and J.P. Sauvage, *Tetrahedron Lett.* (1969) 2885.
46. B. Dietrich, J.M. Lehn, J.P. Sauvage and J. Blanzat, *Tetrahedron* 29 (1973) 1629.
47. J.M. Lehn and J.P. Sauvage, *J. Amer Chem. Soc.* 97 (1975) 6700.
48. B. Metz, D. Moras and R. Weiss, *J. Chem. Soc. Chem. Commun.* (1970) 217.
49. (a) J.M. Lehn, *Acc. Chem. Res.* 11 (1978) 49; (b) J.M. Lehn, *Pure Appl. Chem.* 49 (1977) 587.
50. M. Dobler, *Ionophores and Their Structures*, John Wiley, New York (1981) 177.
51. D.K. Cabbiness and D.W. Margerum, *J. Amer. Chem. Soc.* 91 (1969) 6540.
52. E. Kauffmann, J.M. Lehn and J.P. Sauvage, *Helv. Chim. Acta* 59 (1976) 1099.
53. B. Dietrich, J.M. Lehn and J.P. Sauvage, *J. Chem. Soc. Chem. Commun.* (1973) 15.
54. J.M. Lehn and F. Montavon, *Helv. Chim. Acta* 59 (1976) 1566.
55. J.M. Lehn and F. Montavon, *Helv. Chim. Acta* 61 (1978) 67.
56. B. Dietrich, J.M. Lehn and J.P. Sauvage, *J.C.S. Chem. Commun.* (1970) 1055.
57. B. Dietrich, in J.L. Atwood, J.E. Davies, D.D. MacNicol (eds.), *Inclusion Compounds*, Vol. 2, Academic Press, London (1984) 337.
58. W. Wehner and F. Vogtle, *Tetrahedron Lett.* (1976) 2603.
59. (a) G.R. Newkome, V.K. Majestic and F.R. Fronczek, *Tetrahedron Lett.* (1981) 3035; (b) G.R. Newkome, V.K. Majestic and F.R. Fronczek, *Tetrahedron Lett.* (1981) 3039.
60. (a) J.C. Rodriguez-Ubis, B. Alpha, D. Plancherel and J.M. Lehn, *Helv. Chim. Acta* 67 (1984) 2264; (b) J.P. Konopelski, F. Kotzyba-Hibert, J.M. Lehn, J.P. Desvergne, F. Fages, A. Castellar and H. Bouas-Laurent, *J. Chem. Soc. Chem. Commun.* (1985) 433.
61. A. Caron, J. Guilhelm, C. Riche, C. Pascard, B. Alpha, J.M. Lehn and J.C. Rodriguez-Ubis, *Helv. Chim. Acta* 68 (1985) 1577.
62. O. Juanes, J. de Mendoza and J.C. Rodriguez-Ubis, *J. Chem. Soc. Chem. Commun.* (1985) 1765.
63. M.W. Hosseini and K.N. Raymond, Poster Abstract from *Proc. XII Int. Symp. on Macrocyclic Chemistry*, Florence 1986, 150.
64. D.G. Parsons, *J. Chem. Soc. Perkin Trans.* 1 (1978) 451.
65. A.C. Coxon and J.F. Stoddart, *J. Chem. Soc. Perkin Trans.* 1 (1977) 767.
66. G. Oepen and F. Vogtle, *Justus Liebigs Ann. Chem.* (1979) 1094.
67. P.D. Beer, P.J. Hammond, C. Dudman, J. Elliot and C.D. Hall, *J. Organometall. Chem.* 263 (1984) C37.
68. P.D. Beer, C.G. Crane, A.D. Keefe and A.R. Whyman, *J. Organometall. Chem.* 314 (1986) C9.
69. I.I. Creaser, J. Harrowfield, A.M. Herlt, A.M. Sargeson, J. Springborg, R.J. Geue and M.R. Snow, *J. Amer. Chem. Soc.* 99 (1977) 3181.
70. I.I. Creaser, J. Harrowfield, A.J. Herlt, A.M. Sargeson, J. Springborg, R.J. Geue and M.R. Snow, *J. Amer. Chem. Soc.* 104 (1982) 6016.
71. B. Dietrich, J.M. Lehn and J. Simon, *Angew. Chem. Int. Edit. Engl.* 13 (1974) 406.

72. J.S. Bradshaw and P.E. Stott, *Tetrahedron* **36** (1980) 461.
73. (*a*) R.A. Schultz, D.M. Dishong and G.W. Gokel, *J. Amer. Chem. Soc.* **104** (1982) 625; (*b*) D.M. Dishong, C.J. Diamond, M.I. Cinoman and G.W. Gokel, *J. Amer. Chem. Soc.* **105** (1983) 586; (*c*) C.J. Diamond, D.M. Dishong and G.W. Gokel, *J.C.S. Chem. Commun.* (1980) 1053.
74. A. Masuyama, Y. Nakatsiyi, I. Ikeda and M. Okahara, *Tetrahedron Lett.* **22** (1981) 4665.
75. Y. Nakatsiyi, M. Yonetani and M. Okahara, *Chem. Letts.* (1984) 2143.
76. A. Kaifer, H.D. Durst, L. Echegoyen, D.M. Dishong, R.A. Schultz and G.W. Gokel, *J. Org. Chem.* **47** (1982) 3195.
77. F.R. Fronczek, V.G. Gatto, R.A. Schultz, S.J. Jurryk, W.J. Collucci, R.D. Gandour and G.W. Gokel, *J. Amer. Chem. Soc.* **105** (1983) 6717.
78. R.D. Gandour, F.R. Fronczek, V.J. Gatto, C. Minganti, R.A. Schultz, B.D. White, K.A. Arnold, D. Mazzocchi, S.R. Miller and G.W. Gokel, *J. Amer. Chem. Soc.* **108** (1986) 4078.
79. V.J. Gatto and G.W. Gokel, *J. Amer. Chem. Soc.* **106** (1984) 8240.
80. (*a*) G.M. Lein and D.J. Cram, *J.C.S. Chem. Commun.* (1982) 301; (*b*) D.J. Cram, T. Kaneda, R.C. Helgeson, G. M. Lein, *J. Amer. Chem. Soc.* **101** (1979) 6752; (*c*) D.J. Cram, G.M. Lein, T. Kaneda, R.C. Helgeson, C.B. Knobler, E. Maverick and K.N. Trueblood, *J. Amer. Chem. Soc.* **103** (1981) 6228.
81. D.J. Cram and K.N. Trueblood, *Top. Curr. Chem.* **98** (1981) 43.
82. K.E. Koenig, G.M. Lein, P. Stuckler, T. Kaneda and D.J. Cram, *J. Amer Chem. Soc.* **101** (1979) 3553.
83. (*a*) D.J. Cram, S.P. Ho, C.B. Knobler, E. Maverick and K.N. Trueblood, *J. Amer. Chem. Soc.* **108** (1986) 2989; (*b*) D.J. Cram and S.P. Ho, *J. Amer. Chem. Soc.* **108** (1986) 2998.
84. (*a*) E. Graf and J.M. Lehn, *J. Amer. Chem. Soc.* **97** (1975) 5022; (*b*) E. Graf and J.M. Lehn, *Helv. Chim. Acta* **64** (1981) 1040.
85. E. Graf, J.P. Kintzinger, J.M. Lehn and J. LeMoigne, *J. Amer. Chem. Soc.* **104** (1982) 1672.
86. R.C. Hayward, *Chem. Soc. Rev.* **12** (1983) 285.
87. D.J. Cram and J.M. Cram, *Acc. Chem. Res.* **11** (1978) 8.
88. J.M. Lehn and P. Vierling, *Tetrahedron Lett.* **21** (1980) 1323.
89. J.P. Behr, J.M. Lehn and P. Vierling, *Helv. Chim. Acta* **65** (1982) 1853.
90. J.M. Lehn, *Science* **227** (1985) 849.
91. J.P. Behr, J.M. Lehn and P. Vierling, *J.C.S. Chem. Commun.* (1976) 621.
92. (*a*) F. Kotzyba-Hibert, J.M. Lehn and P. Bierling. *Tetrahedron Lett.* (1980) 941; (*b*) J.P. Kintzinger, F. Kotzyba-Hibert, J.M. Lehn, A. Pagelot and K. Saigo, *J.C.S. Chem. Commun.* (1981) 833; (*c*) C. Pascard, C. Riche, M. Cesario, F. Kotzyba-Hibert and J.M. Lehn, *J.C.S. Chem. Commun.* (1982) 557.
93. (*a*) N.F. Jones, A. Kumar and I.O. Sutherland, *J.C.S. Chem. Commun.* (1981) 990; (*b*) I.O. Sutherland (Tilden Lecture), *Chem. Soc. Rev.* **15** (1986) 83.
94. H.M. Colquhoun, E.P. Goodings, J.M. Maud, J.F. Stoddart, J.B. Wolstenholme and D.J. Williams, *J. Chem. Soc. Perkin Trans.* **2** (1985) 607.
95. B.L. Allwood, F.H. Kohnke, J.F. Stoddart and D.J. Williams, *Angew. Chem. Int. Ed. Engl.* **24** (1985) 581.
96. J.M. Lehn, *Pure Appl. Chem.* **50** (1978) 871.
97. C.S.G. Philips and R.J.P. Williams, *Inorganic Chemistry*, Oxford University Press (1985) 159.
98. L. Radom, *Austr. J. Chem.* **29** (1976) 1635.
99. G. Taborsky and K. McCollum, *J. Biol. Chem.* **254** (1979) 7069.
100. F.A. Cotton, E.E. Hazen Jr. and M.J. Legg, *Proc. Natl. Acad. Sci. USA* **76** (1979) 2551.
101. C.H. Park and H.E. Simmons, *J. Amer. Chem. Soc.* **90** (1968) 2431.
102. R.A. Bell *et al.*, *Science* **190** (1975) 151.
103. E. Graf and J.M. Lehn, *J. Amer. Chem. Soc.* **98** (1976) 6403.
104. B. Metz, J.M. Rosalky and R. Weiss, *J.C.S. Chem. Commun.* (1976) 533.
105. (*a*) F.P. Schmidtchen, *Angew. Chem. Int. Edit. Engl.* **16** (1977) 720; (*b*) F.P. Schmidtchen, *Chem. Ber.* **113** (1980) 864; (*c*) F.P. Schmidtchen, *Chem Ber.* **114** (1981) 597.
106. F.P. Schmidtchen and G. Muller, *J.C.S. Chem. Commun.* (1984) 1115.
107. J.M. Lehn, E. Sonveaux and A.K. Willard, *J. Amer. Chem. Soc.* **100** (1978) 4914.
108. B. Dietrich, J. Guilhem, J.M. Lehn, C. Pascard and E. Sonveaux, *Helv. Chim. Acta* **67** (1984) 91.
109. B. Dietrich, M.W. Hosseini, J.M. Lehn and R.B. Sessions, *Helv. Chim. Acta* **66** (1983) 1262.
110. B. Dietrich, M.W. Hosseini, J.M. Lehn and R.B. Sessions, *J. Amer. Chem. Soc.* **103** (1981) 1282.

111. H.M Colquhoun, J.F. Stoddart and D.J. Williams, *Angew. Chem. Int. Ed. Engl.* **25** (1986) 487.
112. F. Peter, M. Gross, M.W. Hosseini, J.M. Lehn and R.B. Sessions, *J. Chem. Soc. Chem. Commun.* (1981) 1067.
113. (a) M.F. Manfrin, N. Sabbatini, L. Muggi, V. Balzani, M.W. Hosseini and J.M. Lehn, *J.C.S. Chem. Commun.* (1984) 555; (b) M.F. Manfrin, N. Sabbatini, L. Muggi, V. Balzani, M.W. Hosseini and J.M. Lehn, *J. Amer. Chem. Soc.* **107** (1985) 6888.
114. (a) M.W. Hosseini and J.M. Lehn, *J. Amer. Chem. Soc.* **104** (1982) 3525; (b) M.W. Hosseini and J.M. Lehn, *Helv. Chim. Acta* **69** (1986) 587.
115. (a) E. Kimura, A. Sakonaka, T. Yatsunami and M. Kodama, *J. Amer. Chem. Soc.* **103** (1981) 3041; (b) E. Kimura, *Top. Curr. Chem.* **128** (1985) 113.
116. B. Dietrich, T.M. Fyles, J.M. Lehn, L.G. Pease and D.L. Fyles, *J.C.S. Chem. Commun.* (1978) 934.
117. B. Dietrich, D.L. Fyles, T.M. Fyles and J.M. Lehn, *Helv. Chim. Acta* **62** (1979) 2763.
118. K. Takemoto and N. Sonoda, *Inclusion Compounds*, Vol. 2, Academic Press, New York (1984) 47.
119. D.D. MacNicol, *Inclusion Compounds*, Vol. 2, Academic Press, New York (1984) 1.
120. E. Giglio, *Inclusion Compounds*, Vol. 2, Academic Press, New York (1984) 207.
121. (a) M.L. Bender and M. Komiyama, *Cyclodextrin Chemistry*, Springer Verlag, New York (1978); (b) W. Saenger, *Inclusion Compounds*, Vol. 2, Academic Press, New York (1984) 231.
122. (a) F. Vogtle, H. Sieger and W.M. Muller, *Top. Curr. Chem.* **98** (1981) 107; (b) I. Goldberg, *Inclusion Compounds*, Vol. 2, Academic Press, New York (1984) 261.
123. H. Stetter and E.E. Roos, *Chem. Ber.* **88** (1955) 1390.
124. R. Hilgerfield and W. Saenger, *Angew. Chem. Int. Ed. Engl.* **21** (1982) 787.
125. G.D. Gutsche, *Top. Curr. Chem.* **123** (1984) 1.
126. G.D. Andreetti, R. Ungaro and A. Pochini, *J. Chem. Soc. Chem. Commun.* (1979) 1005.
127. G.D. Gutsche, M. Iqbal and D. Stewart, *J. Org. Chem.* **51** (1986) 742.
128. G.D. Andreetti, A. Pochini and R. Ungaro, *J. Chem. Soc. Perkin II* (1983) 1773.
129. R. Ungaro, A. Pochini, G.D. Andreetti and P. Domiano, *J. Chem. Soc. Perkin II* (1985) 197.
130. G.D. Andreetti, A. Mangia, A. Pochini and R. Ungaro, *Abstr. 2nd Int. Sym. on Clathrate Compounds and Molecular Inclusion Phenomena*, Parma, Italy (1982) 42.
131. J.W. Cornforth, P. D'Arcy Hart, G.A. Nicholls, R.J.W. Rees and J.A. Stock, *Brit. J. Pharmacol.* **10** (1955) 73.
132. (a) H. Kammerer, G. Happel and F. Caesar, *Makromol. Chem.* **162** (1972) 179; (b) G. Happel, B. Mathiasch and H. Kammerer, *Makromol. Chem.* **176** (1975) 3317.
133. C.D. Gutsche, B. Dhawan, J.A. Levine, K.H. No and L.J. Bauer, *Tetrahedron* **39** (1983) 409.
134. C. Rizzoli, G.D. Andreetti, R. Ungaro and A. Pochini, *J. Mol. Struct.* **82** (1982) 133.
135. V. Bohmer, H. Goldmann and W. Vogt, *J. Chem. Soc. Chem. Commun.* (1985) 667.
136. D.J. Cram, *Science* **219** (1983) 1177.
137. D.J. Cram, K.D. Stewart, I. Goldberg and K.N. Trueblood, *J. Amer. Chem. Soc.* **107** (1985) 2574.
138. (a) J. Canceill, L. Lacombe and A. Collet, *J. Amer. Chem. Soc.* **107** (1985) 6993; (b) J. Canceill, L. Lacombe and A. Collet, *J. Amer. Chem. Soc.* **108** (1986) 4230.
139. J. Gabard and A. Collet, *J. Chem. Soc. Chem. Commun.* (1981) 1137.
140. (a) J. Canceill, M. Cesario, A. Collet, J. Guilhem, C. Riche and C. Pascard, *J. Chem. Soc. Chem. Commun.* (1986) 339; (b) J. Canceill, M. Cesario, A. Collet, J. Guilhem, C. Riche and C. Pascard, *J. Chem. Soc. Chem. Commun.* (1985) 361.
141. I. Tabushi and K. Yamamura, *Top. Curr. Chem.* **113** (1983) 145.
142. K. Odashima, A. Itai, Y. Iitaka and K. Koga, *J. Org. Chem.* **50** (1985) 4478.
143. (a) I. Tabushi, Y. Kimura and K. Yamamura, *J. Amer. Chem. Soc.* **100** (1978). 1304; (b) I. Tabushi, Y. Kimura and K. Yamamura, *J. Amer, Chem. Soc.* **103** (1981) 6486.
144. I. Tabushi, H. Sasaki and Y. Kuroda, *J. Amer. Chem. Soc.* **98** (1976) 5727.
145. S. Shinkai, S. Mori, T. Tsubaki, T. Sone and O. Manabe, *Tetrahedron Lett.* **25** (1984) 5313.
146. S. Shinkai, S. Mori, H. Koreishi, T. Tsubaki and O. Manabe, *J. Amer. Chem. Soc.* **108** (1986) 2409.

3

IMPLEMENTING MOLECULAR AND IONIC RECOGNITION

3 Organic sensor materials in entangled and polymer-bound matrices for ion-selective electrodes

G.J. MOODY and J.D.R. THOMAS

3.1 Introduction

Since ion-selective electrodes (ISEs) for fluoride (1), calcium (2), potassium (3, 4) and other halides (5) became available in the mid-1960s, a diverse range of materials suitable for other sensor electrodes has been reported (6–8). Their sensor membranes are conveniently classified on the basis of fixed or mobile exchanger sites respectively. This review concerns some fundamental aspects (particularly regarding the synthetic design of sensor material) of a group of ISEs which comprise neutral carrier complex and liquid ion-exchangers, each in conjunction with the appropriate solvent mediator which are normally entangled in poly(vinyl chloride) (PVC) sensor membrane supports. The sensitivity of other polymer support materials is also briefly considered.

The sensor materials in question are organic in nature, and with notable exceptions, such as the naturally occurring actins and valinomycin, and the ubiquitous enzymes, the neutral carrier and ion-exchanger materials, like their associated solvent mediators, must be synthesized.

The original calcium ISE, No 92-20, marketed by Orion Incorporated in 1966, was a complex plastics engineering unit requiring a considerable load of expensive sensor cocktail and whose supporting porous membrane was difficult to position prior to its final assembly with a screw-on head (Figure 3.1a). This system was greatly simplified (9) by casting thin, low-cost, flexible sensor membranes from PVC and the Orion 92-20-02 calcium liquid ion-exchanger cocktail dissolved in tetrahydrofuran by controlled evaporation over two days (Figure 3.2a, b). A small disc (6 mm diameter) of the resulting sensor master membrane (Figure 3.2c) was then sealed to a hollow, rigid PVC tube and the ISE fabricated with an internal silver/silver chloride reference electrode immersed in the internal filling solution of 0.1 M calcium chloride (Figure 3.1b). ISEs for many other materials, ranging from simple cations and anions to large organic molecules, have since been constructed from an appropriate cocktail based on either liquid ion-exchanger or neutral carrier (10–21). The thin PVC membrane comprising such sensor cocktails which invariably contain the sensor compound* in conjunction with one, or

---

* The term 'sensor' is really a misnomer, since electrochemical response of this class of electrodes demands the presence of the so-called sensor, e.g. valinomycin, as well as the appropriate solvent mediator.

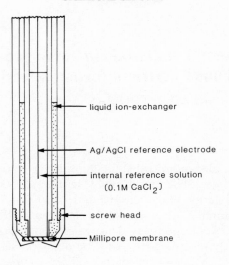

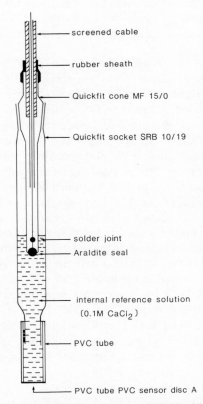

Figure 3.1   Longitudinal sections of (a) Orion 92-2- and (b) PVC calcium ion-selective electrodes.

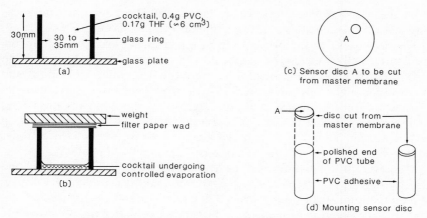

Figure 3.2 Casting and cutting of a PVC master membrane.

occasionally two plasticizing solvent mediators is formally analogous to the classical thin, glass pH-selective membrane.

3.2 Calcium-ion selective electrodes

The huge number of PVC ISEs reported since 1970 means that the selection of sensor topics for this chapter must be restricted. Much of the post-1970 development work with the second class of ISEs has centred on many aspects of PVC calcium electrodes. It is therefore appropriate, in the first instance, to discuss sensor developments in terms of the calcium studies. Indeed, the original electrode model (9) constitutes the basis for the subsequent diverse developments in sensor design wherein the appropriate materials are generally entangled in a polymer matrix (10).

3.2.1 *Liquid ion-exchanger calcium electrodes*

The fact that phosphate and polyphosphate ions formed relatively stable complexes with the Group II, but not Group I, cations led to the first commercial calcium ISE (2). This was based on a liquid ion-exchanger comprising at least calcium bis(didecylphosphate) (**1**) and di-octylphenyl-phosphonate (**2**).

$$[(C_{10}H_{21}O)_2\overset{\overset{\displaystyle O}{\|}}{P}O]_2Ca \qquad\qquad (C_8H_{17}O)_2\overset{\overset{\displaystyle O}{\|}}{P}\text{-}C_6H_5$$

$$\textbf{(1)} \qquad\qquad\qquad\qquad \textbf{(2)}$$

In this Orion model the water-immiscible exchanger solution was essentially supported by a thin Millipore membrane and constantly refurbished from the reservoir of the exchanger until, of course, the supply was exhausted. Apart from the practical inconvenience in design, the Millipore membrane was physically weak and subject to stirring and pressure effects.

The drawbacks of the original Orion liquid membrane electrode model were successfully overcome by simply incorporating the neat exchanger within a thin matrix of PVC (9). Besides having more favourable mechanical and physical characteristics, these novel PVC electrodes had prolonged functional lifetimes and a performance, in terms of Nernstian calibration, response times and selectivity coefficients, on a par with the original Orion 92-20 model. In fact, the mean content of calcium determined with the respective electrodes in tap water on 10 consecutive days was 31.55 and 31.57 ppm (9).

Moreover, its design offered considerable economy of sensor cocktail and the electrode was simple to fabricate. The same quantity of exchanger (~ 0.4 g) is required for a single load of the 92-20 model electrode as the casting of a PVC master membrane (diameter ~ 30 mm) but the latter can provide more than a dozen individual membranes (diameter ~ 6 mm) for the novel PVC electrode (Figure 3.2c). The 'holed' master membrane represents about 50% of the original cocktail and can be recast if necessary.

The utility of the PVC model electrode for calcium has been long since reflected in the reports of many other, diverse sensing systems similarly incorporating appropriate cocktails in PVC matrices (10). Thin-layer chromatography (TLC) of the Orion 92-20-02 exchanger (11) revealed at least three compounds in addition to the expected calcium salt and mediator (Table 3.1).

Differences between five batches of the cocktail have also been inferred from diffusion studies with ^{45}Ca tracer through PVC membranes (12).

Generally the performances of *all* PVC ISEs, particularly in terms of selectivity, depend on the sensor, the solvent (mediator) and their relative masses. Therefore meaningful fundamental studies on the development of organic sensors can only be made with high-quality sensor and mediator and a compatible polymer matrix. Such high-quality sensors/mediators are

Table 3.1 Chromatographic patterns of some liquid ion-exchanger materials (11). Silica gel supports. Developer, benzene–acetic acid (9:1 v/v); visualizer, ammonium molybdate/tin (II) chloride

Material	R_f values
Corning divalent cation cocktail	0.67[a] (intense); 0.47
Orion divalent cation cocktail	0.75 (intense); 0.68[a]; 0.41; 0.0
Orion calcium cocktail	0.76 (intense); 0.48; 0.41; 0.19[b]; 0.0
Corning calcium cocktail	0.79 (intense); 0.67[a]; 0.41; 0.22[b]
Ca-di-2-ethylhexylphosphate	0.47
Ca-didecylphosphate	0.48
Ca-bis{di[4-(1,1,3,3-tetramethylbutyl) phenyl] phosphate}	0.21
Di-octylphenyl phosphonate (DOPP)	0.77 (intense); 0.22[b] (faint)

[a] R_f Decan-1-ol $= 0.67$ (visualizer 30% H_2SO_4)
[b] Probably the monoester

Table 3.2 Characteristics of PVC calcium ISEs based on calcium bis[di(4-alkylphenyl) phosphate] sensors and various mediators (16)

Membrane number	Sensor (calcium salt)	Solvent mediator	Slope (mV decade^{-1})	Detection limit (M)	$k_{Ca,B}^{pot}$ for [B] = 0.1 M			pH range for 10^{-3} M Ca^{2+} in 10^{-1} M Na$^+$
					B = Na	B = K	B = Mg	
1		Dioctylphenyl phosphonate	29.5	2.7×10^{-6}	1.1×10^{-3}	1.0×10^{-3}	4.9×10^{-4}	4.7–9.0
2		Dipentylphenyl phosphonate	29.3	5.5×10^{-6}	2.4×10^{-3}	2.2×10^{-3}	2.7×10^{-3}	4.8–8.9
3		Dihexylphenyl phosphonate	29.5	4.0×10^{-6}	3.2×10^{-3}	2.2×10^{-3}	1.2×10^{-3}	4.5–8.8
4	Di(4-octylphenyl)- phosphate	Diheptylphenyl phosphate	29.5	6.2×10^{-6}	4.8×10^{-3}	2.1×10^{-3}	8.6×10^{-4}	4.9–9.1
5		Dinonylphenyl phosphonate	28.5	4.1×10^{-6}	7.9×10^{-3}	4.8×10^{-3}	4.2×10^{-4}	5.0–9.1
6		Didecylphenyl phosphonate	28.5	4.7×10^{-6}	3.7×10^{-3}	1.6×10^{-3}	1.6×10^{-3}	5.1–9.3
7		Diundecylphenyl phosphonate	30.0	1.1×10^{-5}	8.5×10^{-3}	7.9×10^{-3}	5.5×10^{-3}	5.5–8.9
8		Decan-1-ol	27.5	3.0×10^{-6}	7.0×10^{-2}	7.0×10^{-2}	1.6	5.6–8.9[a]

Table 3.2 (continued)

Membrane number	Sensor (calcium salt)	Solvent mediator	Slope (mV decade^{-1})	Detection limit (M)	$k_{Ca,B}^{pot}$ for [B] = 0.1 M			pH range for 10^{-3} M Ca^{2+} in 10^{-1} M Na$^+$
					B = Na	B = K	B = Mg	
9	Di[4-(1,1,3,3-tetramethylbutyl)phenyl] phosphate		29.6	1.9×10^{-6}	1.0×10^{-3}	4.6×10^{-4}	3.3×10^{-4}	4.9–8.9
10	Di(4-hexylphenyl) phosphate	Dioctylphenyl phosphonate	28.5	2.3×10^{-6}	3.0×10^{-3}	6.1×10^{-4}	5.1×10^{-4}	4.8–9.0
11	Di[4-(1,1,3,3-tetramethylbutyl)-2,6-dinitrophenoxide]		30.5	1.1×10^{-5}	1.1×10^{-2}	9.4×10^{-3}	6.2×10^{-2}[b]	4.7–10[c]
12	Orion 92-20-02 liquid ion exchanger	—	30.5	3.1×10^{-5}	3.1×10^{-2}	9.0×10^{-3}	1.2×10^{-2}	5.5–9.0[a]

[a] In the absence of 10^{-1} M NaCl
[b] In 5×10^{-4} M MgCl$_2$
[c] In 10^{-2} M CaCl$_2$

Table 3.3 Characteristics of PVC ISEs with calcium bis{di-[4-(1,1,3,3-tetramethylbutyl) phenyl] phosphate} sensor and various mediators (17). Viscosity and relative permittivity at 25°C are listed below each solvent mediator

Membrane number	Liquid ion-exchanger system Solvent mediator Viscosity (mN s m^{-2})	Relative permittivity (ε)	Sensor (g)	Lower limit of detection Ca^{2+}/M	Slope (mV decade^{-1})	Operational lifetime	pH range for 10^{-3} M CaCl$_2$
13	Orion 92-20-02 calcium liquid ion-exchanger		0.40	7.5×10^{-6}	33.5	3 months	6.1–8.3
14	Dioctylphenyl phosphonate (16.7)	(6.2)	0.0075	3.2×10^{-6}	30.5 ⎫	⎫	4.5–8.8
15			0.015	5.8×10^{-6}	30.0 ⎬	⎬ 3 months	4.5–8.7
16			0.025	4.1×10^{-6}	30.5 ⎪	⎪	4.5–8.5
17			0.036	3.0×10^{-6}	30.5 ⎭	⎭	4.8–8.8
18	Tributyl phosphate (3.25)	(8.3)	0.0075	2.7×10^{-5}	30.5 ⎫	⎫	5.5–8.7
19			0.015	2.6×10^{-5}	30.0 ⎬	⎬ 1 week	4.5–9.0
20			0.025	2.0×10^{-5}	31.0 ⎪	⎪	5.0–8.9
21			0.036	1.2×10^{-5}	30.5 ⎭	⎭	4.5–8.9
22	Tripentyl phosphate (4.64)	(7.0)	0.0075	4.4×10^{-6}	30.5 ⎫	⎫ 7 weeks	5.0–8.4
23			0.015	6.0×10^{-6}	31.5 ⎬	⎬	4.9–8.1
24			0.025	2.0×10^{-6}	31.0 ⎪	⎪ 3 months	4.6–8.0
25			0.036	6.0×10^{-6}	31.0 ⎭	⎭	4.8–8.2
26	Trioctyl phosphate (11.8)	(5.1)	0.0075	9.0×10^{-6}	29.0 ⎫	⎫	5.5–8.6
27			0.015	9.0×10^{-6}	30.5 ⎬	⎬ 4 weeks	5.2–8.5
28			0.025	8.5×10^{-6}	31.0 ⎪	⎪	4.7–8.5
29			0.036	1.1×10^{-5}	30.0 ⎭	⎭	4.9–8.3

Table 3.3 (*continued*)

Membrane number	Liquid ion-exchanger system			Relative permittivity (ε)	Sensor (g)	Lower limit of detection Ca^{2+}/M	Slope (mV decade^{-1})	Operational lifetime	pH range for 10^{-3} M $CaCl_2$
	Solvent mediator								
	Viscosity (mN s m^{-2})								
30	Tri-(1,1,3,3-tetramethylbutyl)				0.0075	4.5×10^{-6}	30.5		5.1–7.7
31	phosphate				0.015	6.5×10^{-6}	31.5	3 months	5.0–8.9
32	(14.3)			(5.0)	0.0025	5.6×10^{-6}	31.0		5.3–8.1
33					0.036	6.0×10^{-6}	31.5		5.2–8.2
34	Dioctylphenyl phosphonate (16.7) with calcium bis[di-(4-octyl-phenyl) phosphate]			(6.2)	0.036	2.7×10^{-6}	30.5		4.6–8.2

becoming increasingly available, but the nature of the viable supporting polymer (PVC) matrix is unfortunately often obscure (section 3.4.1).

In an attempt to establish the optimal composition of PVC membranes for calcium electrodes, master membranes were cast using different quantities of PVC (Breon S110/10), di-octylphenyl phosphonate and monocalcium dihydrogentetra(didecylphosphate) (11). Of the twelve membrane types examined, the most viable related to one comprising 28.8% PVC and 71.2% solvent/calcium salt in the mass ratio 10:1. Cocktails of comparable mass composition made up in tetrahydrofuran, or cyclohexanone, have since been used for casting master membranes for other ISEs. However, this is not a universal prescription for sensor cocktails, for smaller quantities of the sensor component are frequently sufficient for viable membranes and further conserve the relatively expensive compounds.

Nature of cocktail materials in calcium ISEs. The electrochemical performances of calcium ISEs (and all other ISEs based on organic-type sensors) depend on the structure of the sensor as well as mediator. Thus, calcium ISEs fabricated from calcium salts of dialkylphosphoric acids depend on the solvent mediator, e.g. di-octylphenyl phosphonate (DOPP) yields a calcium ISE whereas decan-1-ol produces the so-called 'water hardness' (or calcium/magnesium) ISE. Several fundamental studies have been made on the molecular variation of these two critical constituents (13–21).

The dip in the pH/potential plots below pH ~ 5 is a characteristic feature of calcium ISEs based on calcium bis-dialkylphosphate type sensors, although it was absent from the original Ross paper (2). Ruzicka *et al.* proposed that the introduction of electrophilic groups into the alkyl substituents of the phosphate sensors (e.g., pK of didecylphosphoric acid = 3.33) would increase the lability of the acid proton (18). Thus, in terms of calcium ISEs the region of existence of the free acid in the membrane phase should shift to lower pH values and hence improve the hydrogen ion selectivity parameter and extend the useful operational pH range. The minima in the pH–emf curves were indeed displaced towards the acid region using calcium salt of di(octylphenyl) phosphoric acid (3) or its isomer (4) with DOPP mediator (Table 3.2) (16, 18).

$$(C_8H_{17}C_6H_5O)_2P(O)OH \qquad (CMe_3CH_2CMe_2C_6H_5O)_2P(O)OH$$
$$(3)\,(pK = 1.66) \qquad\qquad (4)\,(pK = 2.06)$$

The overall improvement in performance of PVC calcium ISEs based on either (3) or (4) compared with the original didecylphosphoric acid sensor (1), is evident from Tables 3.2–3.4, although the dramatic (unexpected) improvement in sodium selectivity could not have been predicted. There is little difference in the performances of these aryl type phosphate sensors (membranes 9 *v*. 13, Table 3.2) but the isomer (4) was the easier to synthesize since the branched chain 4-(1, 1, 3, 3-tetramethylbutyl)phenol starting compound needed for its synthesis (22) was much more readily available

D

Table 3.4 Selectivity coefficients of PVC calcium ISEs based on calcium bis {di-[4-(1, 1, 3, 3-tetramethylbutyl) phenyl] phosphate} and various mediators (17)

Membrane number	Selectivity coefficient, $k_{Ca,B}^{pot}$ (mixed solution method) $[MCl] = 0.05\,M$; $[MCl_2] = 5 \times 10^{-4}\,M$							
	B = Na	B = K	B = Mg	B = Sr	B = Ba	B = Cu	B = Ni	B = Zn
13	4.5×10^{-2}	6.2×10^{-2}	1.3×10^{-1}	1.4×10^{-1}	5.8×10^{-2}	1.6×10^{-1}	9.6×10^{-2}	a
14	2.6×10^{-2}	2.1×10^{-2}	3.3×10^{-2}	2.6×10^{-2}	2.6×10^{-2}	2.6×10^{-2}	2.0×10^{-2}	4.8×10^{-1}
15	1.2×10^{-2}	9.3×10^{-3}	4.5×10^{-2}	5.5×10^{-2}	2.4×10^{-2}	4.1×10^{-2}	2.6×10^{-2}	4.1×10^{-1}
16	2.2×10^{-2}	1.8×10^{-2}	4.5×10^{-2}	8.1×10^{-2}	4.1×10^{-2}	4.8×10^{-2}	2.9×10^{-2}	5.6×10^{-1}
17	1.7×10^{-2}	1.8×10^{-2}	2.1×10^{-2}	4.1×10^{-2}	9.0×10^{-3}	1.4×10^{-2}	1.3×10^{-2}	3.0×10^{-1}
18	3.3×10^{-2}	2.4×10^{-2}	4.8×10^{-2}	6.4×10^{-2}	2.9×10^{-2}	3.1×10^{-2}	3.8×10^{-2}	4.3×10^{-1}
19	2.4×10^{-2}	1.6×10^{-2}	4.8×10^{-2}	7.2×10^{-2}	3.2×10^{-2}	6.9×10^{-2}	3.3×10^{-2}	3.4×10^{-1}
20	1.6×10^{-2}	1.3×10^{-2}	3.3×10^{-2}	6.0×10^{-2}	3.6×10^{-2}	2.7×10^{-2}	1.9×10^{-2}	2.3×10^{-1}
21	2.4×10^{-2}	1.9×10^{-2}	4.3×10^{-2}	3.6×10^{-2}	3.0×10^{-2}	5.5×10^{-2}	3.1×10^{-2}	4.3×10^{-1}
22	2.5×10^{-2}	1.6×10^{-2}	5.3×10^{-2}	8.1×10^{-2}	3.8×10^{-2}	6.0×10^{-2}	3.8×10^{-2}	3.6×10^{-1}
23	2.1×10^{-2}	1.7×10^{-2}	4.3×10^{-2}	6.7×10^{-2}	5.5×10^{-2}	2.9×10^{-2}	3.6×10^{-2}	4.3×10^{-1}
24	9.9×10^{-2}	8.9×10^{-2}	3.1×10^{-2}	5.5×10^{-2}	2.0×10^{-2}	2.1×10^{-2}	1.8×10^{-2}	2.2×10^{-1}
25	2.1×10^{-2}	2.2×10^{-2}	6.2×10^{-2}	9.1×10^{-2}	4.3×10^{-2}	8.6×10^{-2}	8.2×10^{-2}	4.8×10^{-1}
26	1.2×10^{-1}	5.6×10^{-2}	1.5×10^{-1}	1.1×10^{-1}	9.6×10^{-2}	9.6×10^{-2}	6.9×10^{-2}	a
27	1.2×10^{-1}	7.5×10^{-2}	1.3×10^{-1}	1.2×10^{-1}	6.7×10^{-2}	1.6×10^{-1}	6.9×10^{-2}	
28	9.7×10^{-2}	5.6×10^{-2}	1.8×10^{-1}	1.2×10^{-1}	8.1×10^{-2}	2.0×10^{-1}	6.9×10^{-2}	
29	4.3×10^{-2}	3.5×10^{-2}	1.9×10^{-1}	1.4×10^{-1}	7.7×10^{-2}	1.8×10^{-1}	1.0×10^{-1}	
30	2.6×10^{-2}	2.0×10^{-2}	8.1×10^{-2}	1.1×10^{-1}	3.6×10^{-2}	4.3×10^{-2}	6.2×10^{-3}	4.8×10^{-1}
31	3.0×10^{-2}	2.6×10^{-2}	4.3×10^{-2}	6.0×10^{-2}	3.3×10^{-2}	2.9×10^{-2}	3.2×10^{-2}	3.6×10^{-1}
32	2.6×10^{-2}	2.2×10^{-2}	4.5×10^{-2}	6.9×10^{-2}	2.9×10^{-2}	6.0×10^{-2}	4.3×10^{-2}	5.1×10^{-1}
33	2.0×10^{-2}	1.9×10^{-2}	5.3×10^{-2}	6.4×10^{-2}	3.3×10^{-2}	4.1×10^{-2}	3.6×10^{-2}	5.3×10^{-1}
34	1.2×10^{-2}	9.3×10^{-2}	6.2×10^{-2}	9.0×10^{-2}	4.1×10^{-2}	3.8×10^{-2}	2.0×10^{-2}	1.2×10^{-1}

^{a} $k_{Ca,Zn}^{pot}$ not determined because the calibration plot for calcium/zinc standards never coincided with the normal calcium calibration than the linear alkyl chain isomer. Shortening the alkyl chain from eight carbons in (3) to six carbon atoms (sensor of membrane 10) does not adversely affect its selectivity or linear detection limit (Table 3.2).

Varying the proportion of sensor (3) to mediator shows that master membranes with as little as 0.0075 g sensor and 0.36 g of solvent mediator function well and there is no regular feature to suggest an optimum ratio of sensor to mediator (17). However, gel-like inclusions tend to form at higher sensor to mediator ratios.

As well as having improved selectivity performances, the new aryl phosphate-based PVC calcium ISEs behave well during repeated calibrations, and the use of ion-buffers (16, 18) enables the detection limit to be extended from typically $\sim 2 \times 10^{-6}$ M to $\sim 10^{-8}$ M (16), (Figures 3.3, 3.4) and the calibration profile of a PVC electrode based on a neutral carrier sensor (ETH 1001) is included for comparison (23). This constitutes some improvement on the corresponding values of $\sim 7 \times 10^{-7}$ M quoted (19) for a calcium ISE containing the same dialkylphenylphosphate sensor/mediator but which had not been subjected to a sufficiently rigorous purification (22). The importance of using high-grade material, wherever possible, is therefore well exemplified.

The selectivity coefficients of ISEs based on immiscible organic liquid ion-exchangers can be correlated with data for liquid extraction systems (24–27) by computing the degree of saturation of the organic phase by the ion to be measured from the extraction constants or distribution coefficients. Since the ion transference occurs through the membrane and associated diffusion layers, it is probable that both equilibrium ion-exchange and relative

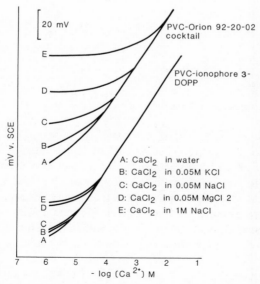

Figure 3.3 Calibrations of PVC calcium ion-selective electrodes in various electrolytes.

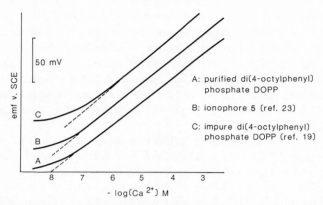

Figure 3.4 Calibrations of PVC calcium ion-selective electrodes in calcium buffers (ionic strength = 0.1 M).

Table 3.5 Extraction coefficients of divalent cations from 4M sodium nitrate into di(2-ethylhexyl) phosphoric acid in benzene, and selectivity coefficients of PVC ISEs with dialkylphosphate sensors (24)

			Selectivity coefficient, $k_{Ca,B}^{pot}[BCl_2] = 4 \times 10^{-4}$ M		
			Ca di(2-ethylhexyl)-phosphate with	Orion 92-20-02	
	Log extraction coefficients		decan-1-ol	liquid ion-exchanger	
B	pH = 7.0	pH = 5.2	(25°C)	(25°C)	(35°C)
Mg	1.2	1.4	1.0	0.01	0.055^a
					0.052
Ca	0.9	2.2	1.0	1.0	1.0
Sr	− 0.2	0.1	0.5	0.05	0.09
Ba	− 0.6	− 0.5	0.5	0.002	0.01
Zn					0.13^a
					0.12

^{a} These $k_{Ca,B}^{pot}$ values were determined with metal sulphates; all others refer to metal chlorides

ionic mobilities are interrelated with respect to selectivity. However, the mobilities of the associated ion/site pairs for long-chain ion-exchangers are about equal for all the counter-ions, so that electrode selectivities may be simply related to just equilibrium-ion exchange (25). This is illustrated by the large extraction coefficients for calcium and magnesium respectively into di(2-ethylhexyl)phosphate in benzene on the one hand and the smaller selectivity coefficients for strontium and barium of the same sensor (Table 3.5). Ideally, of course, the extraction data should be measured using decan-1-ol, not benzene.

Co-ordinating solvents such as alcohols, trialkylphosphates and phosphonates, unlike weakly coordinating solvents, produce functional liquid

Table 3.6 Characteristics of some PVC ISEs (14)

Membrane number	Liquid ion-exchanger system		Lower limit of detection for Ca^{2+} (M)[a]	Remarks
	Solvent mediator	Sensor		
35	Orion (92-20-02) calcium ion-exchanger		4×10^{-6}	Pale yellow transparent, soft and rubbery membrane with no inclusions. Good operational lifetime of > 2 weeks.
36	Corning (476235) divalent ion-exchanger		4.5×10^{-5} for Ca^{2+} and Mg^{2+}	Clear transparent, semi-rigid membrane with surface exudate. Good operational lifetime of > 2 weeks.
37	Orion (92-32-02) divalent ion-exchanger		10^{-4} for $\begin{cases} Ca^{2+} \\ Mg^{2+} \\ Mn^{2+} \end{cases}$	Pale pink transparent, semi-rigid membrane with surface exudate. Good operational lifetime of > 2 weeks.
38	Dioctylphenyl phosphonate	Calcium bis (didecylphosphate)/ didecylphosphoric acid (1:1 m/m)	6×10^{-7}	Transparent soft and rubbery but with tendency for gel-like inclusions in parts.
39	Dinonyl phthalate			More rigid than membrane 38. Unsteady readings in presence of interferents.
40	Decan-1-ol			Rigid membrane with extensive crystalline inclusions and considerable surface exudate. Unsuitable for electrodes.
41	Decan-1-ol (0.18g) plus di-octylphenyl phosphonate (0.18 g)		8.5×10^{-6}	Transparent semi-rigid membrane with gel-like inclusions. Some surface exudate.

Table 3.6 (continued)

Membrane number	Liquid ion-exchanger system		Lower limit of detection for Ca²⁺ (M)ᵃ	Remarks
	Solvent mediator	Sensor		
42	Dioctylphenyl phosphonate		2.2×10^{-5}	Clear transparent, soft and rubbery membrane with no inclusions. Good operational lifetime of > 2 weeks.
43	Dinonyl phthalate		1.2×10^{-6}	Soft, rubbery, but opaque membrane. Susceptible to electrical interference, i.e., noisy output and with low slope; is generally unsuitable for ISEs.
44	Octan-1-ol	Calcium bis (di-2-ethylhexylphosphate)	6×10^{-6}	The alcohol readily exudes from the membrane which has extensive crystalline deposits. Lifetime ∼ 4 days.
45	Decan-1-ol		$3 \times 10^{-5} \begin{cases} Ca^{2+} \\ Mg^{2+} \end{cases}$	Cloudy white shrunken membrane with no exudate. Good operation lifetime of ∼ 2 weeks.
46	Dodecan-1-ol		Erratic response	State of membrane dependent on ambient conditions since the alcohol component solidifies with a drop in temperature.
47	Decan-1-ol(0.18 g) plus di-octylphenyl phosphonate (0.18 g)		8×10^{-5}	Semi-rigid but completely transparent membrane. Some exudate on surface.

ᵃ Lower limit of detection is that activity of calcium ions (or ions of interest) at which the emf deviates by 9 mV (that is, $18/z_j$, where z_j is the valence of the ion of interest) from the extrapolated linear section of the calibration plot membrane calcium ISEs (13). The mediator influences the formation constant and partition coefficient of this system which define the ion-exchange between the calcium species in the liquid ion-exchanger and the other cations in the sample solutions.

That different solvents, or mixtures of solvents, are critically associated with differences in selectivity coefficients of calcium (or indeed all other types) of PVC ISEs based on organo-sensors is evident from Tables 3.2–3.10. It is highly desirable that the mediator should additionally function as a plasticizer for the PVC membrane. In view of the extensive use of PVC for fabricating the organic sensor type of ISEs, it is indeed fortunate that many of the necessary solvents fulfil this additional role. Thus di-octylphenyl phosphonate with, or without, a calcium salt is perfectly compatible with PVC. This requirement is rarely observed with alkan-1-ols which fail to plasticize the master membranes, and surface exudations are common (13). Despite this drawback, functional PVC ISEs can sometimes be fabricated (Tables 3.8, 3.10) and extensive studies have been made on various solvents (Tables 3.2–3.10).

Apart from the fact that very-low-viscosity mediators give poor PVC–matrix ISEs, investigations of the viscosity and permittivity of solvent mediators do not lead to any regular conclusions on quality. Di-octylphenyl phosphonate ($\varepsilon = 6.2$) is much used for promoting calcium ion selectivity of calcium ISEs based on sensors (3) and (4). Equally good calcium ISEs can also be fabricated with other members of the homologous series $(RO_2)_2P(O)C_6H_5$ for R = 5–11, the permittivities for the pentyl, hexyl and heptyl compounds being 8.5, 8.1 and 7.1, respectively (Table 3.2).

The influence of decan-1-ol ($\varepsilon = 8.1$) and DOPP ($\varepsilon = 6.2$) on the $k^{pot}_{Ca,Mg}$ parameters of calcium sensors provides a remarkable example of solvent effects despite their similar permittivities (Tables 3.6, 3.7 and 3.10). Thus, membrane 45 based on decan-1-ol exhibits little selectivity among the divalent cations while the DOPP membrane 42 is clearly selective for calcium. Membrane 47 with equal masses of the two mediators gives an intermediate selectivity pattern. These trends also accord with the selectivity patterns of the commercial cocktails (membranes 35–37) and TLC (Table 3.1).

A more detailed experiment with various amounts of these classical solvents, but the alternative aryl-substituted sensor, showed a gradation in selectivity coefficients on going from a high fraction of decan-1-ol to a high fraction of DOPP (membranes 68–70, Table 3.10). Thus $k^{pot}_{Ca,Mg}$ fell from 1.6 for the PVC electrode based exclusively on decan-1-ol to 4.9×10^{-4} on changing completely to DOPP. A similar but less dramatic trend was shown for the sodium selectivities (15).

Ion diffusion through membranes. The diffusion of Mg^{2+} and Ca^{2+} ions through the PVC sensor membrane of the ideal divalent ISE should be closely similar, since $k^{pot}_{Ca,Mg}$ is expected to be unity. Thus, it is interesting that the migration profile of $^{45}Ca^{2+}$ between identical pairs of calcium chloride

Table 3.7 Selectivity coefficients of the PVC ISEs in Table 3.6 (14)

Membrane number	Selectivity coefficient, $k_{Ca,B}^{pot}$ (mixed solution method)						
	B = Na	K	Mg	Sr	Ba	Cu	Ni
35	1.05 $(10^{-2})^a$	1.2 (10^{-2})	8.6×10^{-2} (10^{-2})	5.5×10^{-2} (10^{-3})	1.4×10^{-3} (10^{-3})	0.16 (10^{-3})	0.12 (10^{-3})
36	3.3 (10^{-2})	3.6 (10^{-2})	1 (all levels)	0.80 (10^{-3})	0.92 (10^{-3})	1.4 (10^{-3})	1.3 (10^{-3})
37	1.3 (10^{-2})	1.5 (10^{-2})	1 (all levels)	0.63 (10^{-3})	0.65 (10^{-3})	2.6 (10^{-3})	2.3 (10^{-3})
38	0.19 (10^{-2})	0.56 (10^{-2})	5.2×10^{-2} (10^{-3})	3.1×10^{-2} (10^{-3})	4.0×10^{-2} (10^{-3})	0.21 (10^{-3})	5.7×10^{-2} (10^{-3})
39	0.56 (10^{-2})	2.4 (10^{-2})	0.14 (10^{-3})	0.14 (10^{-3})	0.76 (10^{-3})	13 (10^{-3})	0.76 (10^{-3})
40	—	—	—	—	—	—	—
41	1.6 $(10.^{-2})$	0.52 (10^{-2})	0.55 (10^{-3})	0.21 (10^{-3})	0.29 (10^{-3})	0.86 (10^{-3})	0.71 (10^{-3})
42	46 (10^{-2})	14 (10^{-2})	5.7×10^{-2} (10^{-3})	4.0×10^{-2} (10^{-3})	4.8×10^{-2} (10^{-3})	1.7 (10^{-3})	0.18 (10^{-3})
43	—	—	—	—	—	—	—
44	—	—	—	—	—	—	—
45	2.4 (10^{-2})	2.4 (10^{-2})	1 (all levels)	0.50 (10^{-3})	0.50 (10^{-3})	2 (10^{-3})	2 (10^{-3})
46	—	—	—	—	—	—	—
47	1.4 (10^{-2})	1.4 (10^{-2})	0.52 (10^{-3})	0.17 (10^{-3})	0.17 (10^{-3})	0.66 (10^{-3})	0.26 (10^{-3})

^{a} Parenthesis values are levels (M) of B at which the coefficients were determined

solutions (10^{-3} M) positioned on either side of a PVC membrane incorporating the Orion 92-32-02 divalent cation exchanger matches that when one of the calcium chloride solutions is replaced with magnesium chloride (10^{-3} M). When C'' is the concentration of $^{45}Ca^{2+}$ on the initially inactive side and C' is the concentration on the initially active side, then $d(C''/C')/dt \sim 32 \times 10^{-7} s^{-1}$. The corresponding time-independent diffusion values for identical experiments undertaken with the calcium bis{di[4-(1, 1, 3, 3-tetramethyl-butyl)phenyl]phosphate} and DOPP were $29 \times 10^{-7} s^{-1}$ and $4.3 \times 10^{-7} s^{-1}$ respectively. This difference accords fully with a PVC sensing system for which $k_{Ca,Mg}^{pot} < 1$ (15).

The extent of ion permeation in such tracer studies obviously relates to selectivity, and membranes with calcium ion-sensors and solvent mediators which promote good quality calcium ISEs in turn show high $d(C''/C')/dt$ values. The largest value of $59 \times 10^{-7} s^{-1}$ has been reported for a tripentyl phosphate/calcium bis{di[4-(1, 1, 3, 3-tetramethylbutyl)phenyl] phosphate} based sensor [12].

Unusual interferences. PVC calcium ISEs based on the same di[4-(1, 1, 3, 3-tetramethylbutyl)phenyl]phosphate sensor but with trioctyl phosphate solvent mediator are superior to any others examined, including DOPP, in

Table 3.8a Characteristics of some PVC ISEs based on calcium bis{di-[4-(1,1,3,3-tetramethylbutyl) phenyl] phosphate} (15)

Membrane number[a]	Solvent mediator Type	Mass (g)	Slope (mV decade^{-1})	Lower detection limit (M^b)	Linear range (M)	Remarks
48	Dibutyl sebacate	0.36	33	1.6×10^{-5}	$10^{-1} - 5 \times 10^{-4}$	Colourless, transparent, soft and rubbery Gel-like inclusions.
49	Dibutyl sebacate/Decan-1-ol	0.36 0.36	26.5	2×10^{-6}	$10^{-1} - 7 \times 10^{-5}$	Semi-rigid, shrunken membrane with surface exudate. Extensive crystalline inclusions. Lifetime 6 weeks.
50	Dibutyl sebacate/Dioctylphenyl phosphonate	0.36 0.36	32	1.9×10^{-5}	$10^{-1} - 3 \times 10^{-4}$	Clear, transparent, soft and rubbery with no inclusions. Lifetime > 2 months.
51	Di-n-butyl sebacate/Decan-1-ol	0.18 0.18	30	4×10^{-6}	$10^{-1} - 3 \times 10^{-5}$	Semi-rigid, shrunken membrane. Surface exudate with extensive crystalline inclusions. Lifetime 1 week.
52	Dibutyl sebacate/Dioctylphenyl phosphonate	0.18 0.18	31	6×10^{-6}	$10^{-1} - 4 \times 10^{-5}$	Clear, transparent, soft and rubbery with no inclusions. Lifetime > 6 weeks.
53	Diethyl adipate	0.36	30	1.3×10^{-6}	$10^{-1} - 5 \times 10^{-5}$	Opaque, soft membrane, shrunken with gel-like inclusions.
54	Diethyl adipate/Decan-1-ol	0.36 0.36	32	1.6×10^{-5}	$10^{-1} - 9 \times 10^{-5}$	Opaque, semi-rigid, large shrunken membrane. Surface exudate with extensive crystalline deposit. Lifetime 1 day.
55	Diethyl adipate/Dioctylphenyl phosphonate	0.36 0.36	24	1.1×10^{-6}	$10^{-1} - 6 \times 10^{-6}$	Clear, transparent, soft and rubbery with no inclusions. Lifetime > 2 months.
56	Diethyl adipate/Decan-1-ol	0.18 0.18	26	1.6×10^{-6}	$10^{-2} - 3 \times 10^{-5}$	Opaque, semi-rigid, large shrunken membrane, surface exudate with extensive crystalline deposit. Lifetime 2 days.
57	Diethyl adipate/Dioctylphenyl phosphonate	0.18 0.18	31.5	1×10^{-5}	$10^{-1} - 3 \times 10^{-4}$	Clear, transparent, soft and rubbery with no inclusions. Lifetime > 5 weeks.

[a] Each membrane contains 0.04 g of calcium bis{di-[-(1,1,3,3-tetramethylbutyl) phenyl] phosphate}
[b] As defined in Table 3.6

Table 3.8b Solvent mediator effects on properties of some PVC ISEs based on calcium bis-decylphosphate (15)

Membrane number[a]	Solvent mediator Type	Mass (g)	Slope (mV decade^{-1})	Lower detection limit (M[b])	Linear range (M)	Remarks
58	Dibutyl sebacate	0.36	35	3.2×10^{-5}	10^{-1}–5×10^{-4}	White transparent, soft and rubbery membrane.
59	Dibutyl sebacate/ Decan-1-ol	0.36	32	3.2×10^{-5}	10^{-1}–7×10^{-4}	White, opaque, semi-rigid, shrunken membrane, some exudate on the surface with extensive crystalline inclusions. Lifetime 1 week.
60	Dibutyl sebacate/ Dioctylphenyl phosphonate	0.36 0.36	36	$2 \times 5 \times 10^{-5}$	10^{-1}–3×10^{-4}	Colourless, transparent, soft and rubbery membrane. Lifetime > 2 months.
61	Dibutyl sebacate/ Decan-1-ol	0.18 0.18	32	4×10^{-5}	10^{-1}–6×10^{-4}	White, opaque, semi-rigid membrane with crystalline inclusions. Lifetime 1 week.
62	Dibutyl sebacate/ Dioctylphenyl phosphonate	0.18 0.18	40	2.3×10^{-5}	10^{-1}–10^{-4}	Colourless, transparent, soft and rubbery membrane, with crystalline inclusions. Lifetime > 7 weeks.
63	Diethyl adipate	0.36	28.6	1.1×10^{-4}	10^{-1}–10^{-3}	Colourless soft membrane.
64	Diethyl adipate/ Decan-1-ol	0.36 0.36	32	1.7×10^{-5}	10^{-1}–7×10^{-4}	White, opaque, rigid, shrunken membrane with surface exudate. Lifetime 2 weeks.
65	Diethyl adipate/ Dioctylphenyl phosphonate	0.36 0.36	38	1.1×10^{-5}	10^{-1}–7×10^{-5}	Colourless, transparent, soft membrane with gel-like inclusions. Lifetime 4 weeks.
66	Diethyl adipate/ Decan-1-ol	0.18 0.18	30	3×10^{-6}	10^{-2}–5×10^{-5}	White, opaque, rigid, shrunken membrane surface exudate with crystalline inclusions. Lifetime 2 days.
67	Diethyl adipate/ Dioctylphenyl phosphonate	0.18 0.18	39	2.8×10^{-6}	10^{-1}–3×10^{-5}	Colourless, transparent, soft membrane with gel-like inclusions. Lifetime > 6 weeks.

[a] Each membrane contains 0.04 g of calcium bis-didecylphosphate
[b] As defined in Table 6

Table 3.9 Selectivity coefficients of some PVC calcium ISEs (15)

Selectivity coefficient $k_{Ca,B}^{pot}$ (mixed solution method) $[BCl] = 5 \times 10^{-3} M$
$[BCl_2] = 5 \times 10^{-4} M$

Membrane number	B = Na	K	Mg	Sr	Ba	Cu	Ni	Zn
48	0.5	0.5	3×10^{-2}	3×10^{-2}	4×10^{-2}	5×10^{-2}	0.3	3×10^{-2}
49	0.5	0.6	4×10^{-2}	3×10^{-2}	4×10^{-2}	0.1	9×10^{-2}	4×10^{-2}
50	0.7	0.7	0.3	9×10^{-3}	4×10^{-3}	3×10^{-2}	3×10^{-2}	3×10^{-2}
51	0.6	0.5	0.1	7×10^{-2}	7×10^{-2}	0.7	0.2	0.6
52	0.4	0.4	3×10^{-2}	6×10^{-2}	1×10^{-2}	2×10^{-2}	3×10^{-2}	2×10^{-2}
53	0.3	0.2	4×10^{-2}	3×10^{-2}	2×10^{-2}	2×10^{-2}	0.5	[b]
54[a]	—	—	—	—	—	—	—	—
55	0.5	0.6	3×10^{-2}	6×10^{-2}	1×10^{-2}	2×10^{-2}	0.2	0.1
56[a]	—	—	—	—	—	—	—	—
57	0.3	0.1	2×10^{-2}	4×10^{-2}	9×10^{-2}	2×10^{-2}	3×10^{-2}	0.2
58	6.7	8.7	0.3	0.2	0.3	1.8	0.4	[b]
59	2.4	2.1	[b]	0.2	1.8	[b]	[b]	[b]
60	4.4	0.8	2×10^{-2}	4×10^{-2}	2×10^{-2}	3×10^{-2}	0.3	[b]
61	0.6	4.9	2.4	0.4	0.8	17.7	12.0	[b]
62	2.5	0.4	2×10^{-2}	1×10^{-2}	8×10^{-3}	9×10^{-2}	3×10^{-3}	[b]
63	2.0	17.4	0.3	1.4	0.7	[b]	3×10^{-2}	0.6
64	0.4	1.2	0.6	0.3	0.4	3.5	0.7	2.4
65	0.4	0.4	3×10^{-2}	2×10^{-2}	1×10^{-2}	4×10^{-2}	0.53	[b]
66	0.4	1.3	0.8	0.2	1.5	1.3	0.7	[b]
67	0.4	0.2	1×10^{-2}	1×10^{-2}	8×10^{-3}	7×10^{-2}	2×10^{-2}	[b]

[a] Electrode quality too poor for selectivity evaluation
[b] $k_{Ca,B}^{pot}$ could not be determined because the calibration plots for calcium-interference standards never coincided with the normal calcium ion calibration

Table 3.10 Some characteristics of PVC ISEs with mixed-mediator membranes (15)

Membrane[a] number	Decan-1-ol (g)	Dioctylphenyl phosphonate (g)	Electrode slope (mV decade^{-1})	Detection limit (M)	$k_{Ca,Mg}^{pot}$ [b]	$k_{Ca,Na}^{pot}$ [b]
68[c]	0.36	—	27.5	3×10^{-6}	1.6	7.0×10^{-2}
69[c]	0.30	0.06	28.5	1.6×10^{-5}	7.9×10^{-1}	3.0×10^{-2}
70[c]	0.24	0.12	28.5	1.4×10^{-5}	1.0×10^{-1}	1.3×10^{-2}
71[c]	0.18	0.18	28.7	8.0×10^{-6}	8.6×10^{-3}	6.1×10^{-3}
72	0.12	0.24	28.5	3.8×10^{-6}	1.6×10^{-3}	3.7×10^{-3}
73	0.06	0.30	28.0	2.9×10^{-6}	7.1×10^{-4}	2.1×10^{-3}
74	—	0.36	29.5	2.7×10^{-6}	4.9×10^{-4}	1.1×10^{-3}

[a] Each membrane contained PVC (0.17 g) and calcium bis{di[4-(octyl) phenyl] phosphate} (0.04 g)
[b] $[MgCl_2] = [NaCl] = 5 \times 10^{-2} M$, using mixed solution method
[c] Surface exudates observed on membrane surfaces

Table 3.11 Potential changes caused by biochemical components, in calcium chloride in 0.15 M sodium chloride solutions, on calcium ISE response (21)

| Biochemical component (0.05 M) | ΔE caused by added components to solutions/mV | | | |
	DOPP[a] (10^{-2} M)	TPP[b] (10^{-2} M)	TOP[c] (10^{-2} M)	Philips IS561/SP electrode (10^{-2} M)
SDS[d] } for comparison	− 85	− 40.0	− 2.8	− 26.5
DBSS[e] }	− 12.5	− 41.5	− 3.8	− 39.8
Cholic acid	− 13.1	− 6.8	0.5	− 1.6
Cholesterol	4.1	− 0.6	1.7	0.0
Lecithin	− 5.6	− 10.5	0.2	0.4
Vitamin D$_2$	− 3.0	− 6.5	− 1.0	5.7
Urea	− 0.9	0.8	− 0.2	− 0.1
Glucose	− 0.2	− 1.1	− 0.3	− 0.5

All electrodes based on calcium bis{di[4-(1,1,3,3-tetramethylbutyl) phenyl] phosphonate} except for Philips model in PVC with solvent mediator indicated
[a] Di-octylphenyl phosphonate; [b] Tripentyl phosphate; [c] Trioctyl phosphate; [d] Sodium dodecyl-sulphate; [e] Dodecylbenzene sodium sulphonate

resisting interference by anionic surfactants, and also facilitate measurements of calcium in wash liquors (19, 20). Decan-1-ol, dodecan-1-ol and tetradecan-1-ol also offer some improvement over DOPP but their use impairs calcium selectivity and plasticizer qualities of the PVC membranes (19).

To complement the observations on anionic surfactants, the effects of various biochemical components on PVC calcium ISEs have been investigated (21). The electrodes with trioctyl phosphate mediator are again the most resistant (Table 3.11).

3.2.2 Neutral carrier calcium electrodes

High-quality calcium ISEs can also be fabricated with either acylic (6, 7, 23, 28–31) or cyclic (32–34) neutral carrier compounds (Table 3.12). Thus N, N'-di-[-(11-ethoxycarbonyl)undecyl]N, N'-4, 5-tetramethyl 3, 6-dioxa-octane diamide (ETH 1001) (5) now constitutes the basis of many commercial cocktails (Table 3.12). As demonstrated for a new pH-buffered calcium ion calibrant based on a mixed ligand calcium ion-buffered system, it may be

(5)

(6)

Table 3.12 Some characteristics of PVC neutral carrier calcium ISEs

Property		Membrane composition/mass %			
		Acyclic ionophore (31)		Cyclic ionophore (33)	
Selectivity coefficient: $k_{Ca,B}^{pot}$	Ionophore:	**(5)**1	**(6)** 1	**(7)** 3	**(8)**3
	o-NPOE:	64	65.6	66	66
	PVC:	34.5	32.8	30	30
	KTpClPB:	0.5	0.6	1	1
B = Li		1.58×10^{-3a}	5.0×10^{-4a}	4×10^{-4d}	3×10^{-4d}
B = Na		3.98×10^{-4a}	1.99×10^{-4a}	9×10^{-5}	2×10^{-4}
		7.9×10^{-7b}	3.9×10^{-8b}		
B = K		1.58×10^{-4a}	10^{-4a}	3×10^{-5}	2×10^{-4}
		2.51×10^{-7a}	10^{-8c}		
B = Mg		3.98×10^{-5a}	1.25×10^{-5b}	$< 10^{-5}$	2×10^{-4}
Detection limits in calcium buffers of:					
(i) 0.125 M KCl		1.99×10^{-9} M	7.9×10^{-11} M		
(ii) 0.094 M NaCl		3.98×10^{-9} M	1.99×10^{-10} M		

[a] Separate solution method in 0.1 M unbuffered metal chlorides
[b] Mixed solution method in calcium buffer [NaCl = 0.094 M]
[c] Mixed solution method in calcium buffer [KCl = 0.125 M]
[d] Mixed solution method in all four cases for 0.5 M unbuffered chlorides

used down to 10^{-8} M calcium in the presence of millimolar amounts of magnesium (35) and interestingly, like ionophore **(6)**, dips are absent in the pH/emf calibrations.

Such neutral carriers (also called ionophores), which carry no net charge until complexed with a cation, have led to ISEs with a wide range of selectivities (7). Three fundamental requirements are mandatory in order to realize such potentiometric sensors (6):

(i) The ionophore has to be highly lipophilic
(ii) The free energy of ligand exchange must be relatively low
(iii) The ionophore needs to induce ion-permeability selectivities in sensor membranes. This depends critically on the structure and free concentration of the ionophore in the membrane phase and the partitioning capacity of the solvent mediator.

Calcium phosphate sensors are commonly based on solvents with low permittivities, e.g. DOPP ($\varepsilon = 6.2$), whereas neutral carrier types may require highly polar solvents, typically 2-nitrophenyl octylether (2-NPOE), $\varepsilon = 23.6$ (28). Another characteristic feature of neutral carrier systems (Table 3.12) is the presence of lipophilic anions, e.g. tetraphenylborate (KTPB) which was first incorporated to reduce the interference by lipophilic sample anions (36) and so dislodge interferents from the membrane phase. Today tetrakis(4-chlorophenyl)borate, KTpClPB, is generally preferred because its water solubility is smaller than that of KTPB by a factor of about 1000, and its presence may produce significantly, but not necessarily beneficial, changes in selectivity (7, 31). This is amply demonstrated for the latest calcium ionophore, N, N, N', N-tetracyclohexyl-3-oxapentane diamide (6), where the optimal sodium and potassium selectivities are observed for a concentration of about 0.5 mass % KTpClPB, that is, about 50 mol % with respect to ionophore. The performance of both acyclic sensors is considered superior to the phosphate counterparts, particularly with respect to magnesium selectivity. It must be emphasized, however, that the excellent selectivity and low detection limits relate to evaluation using calcium-ion-buffered media with a high background of interferents (31).

Another benefit is the associated reduction in membrane resistance which is crucial for the fabrication of functional microelectrodes with tip diameters of $\approx 1 \, \mu m$ (7). This feature, together with the good selectivities and low detection limits of sensor (6) (Table 3.12) could provide microelectrodes for reliable intracellular studies (31).

Macrocyclic polyethers (or crown ethers) also possess attractive cation complexing abilities resulting in the formation of lipophilic species. Not surprisingly, such ionophores have been studied with respect to calcium sensors (32–34), and two examples are shown in Table 3.12. The macrobicyclic polyether-amide (7) provides a better calcium ISE than its monocyclic analogue (8).

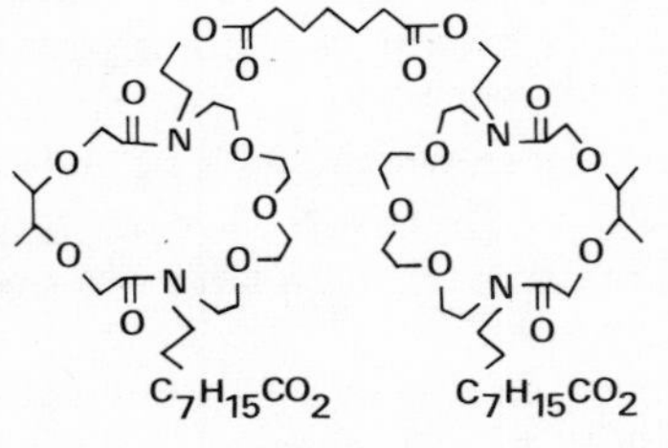

(7)

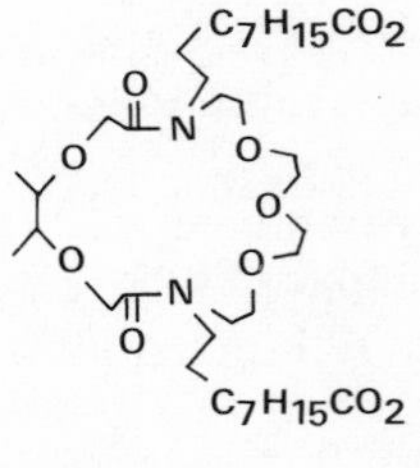

(8)

3.2.3 Magnesium electrodes

The ionophore N,N'-diheptyl-N,N'-dimethylsuccinic acid diamide (9) is considered the best available lipophilic ligand regarding selectivity for magnesium over the physiologically important ions of sodium, potassium and calcium (37). Glass microelectrodes filled with (9) (20 mass %), propylene

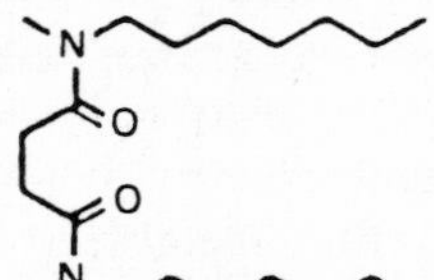

(9)

carbonate (79 mass %) and sodium tetraphenylborate (1 mass %) have facilitated intracellular magnesium studies in sheep Purkinje fibres despite a value of $k_{Mg,Ca}^{pot}$ of 12.6. This is not an entangled sensor system, but it should be feasible to incorporate this cocktail (in the usual PVC mode) into the back-filled epoxy type of microelectrode recently described (38).

Despite much effort there is still no meaningful magnesium ISE. The current approach utilizes the divalent ISE (10, 35–39). In principle, the magnesium may be found from the difference in samples assayed respectively with the divalent ISE and the calcium (only) ISE (10). However, serious errors can arise with ready-to-use cocktails, e.g. Orion 92-32-02 contains appreciable amounts of di-octylphenyl phosphonate (Table 3.1). Selectivity values for $k_{Mg,Ca}^{pot}$ deviate considerably from unity, not only for PVC-Orion 92-32 electrodes, but also for PVC ISEs based on pure di-decylphosphate/decan-1-ol (39).

More recently, a PVC ISE based on the [4-(1,1,3,3-tetramethylbutyl) phenyl] phosphate/decan-1-ol system in conjunction with a chemometric technique has facilitated magnesium measurements (35). Attempts to make magnesium sensors with magnesium adducts of poly(propylene glycol) in conjunction with tributyl phosphate, di-octylphenyl phosphonate, its 3-nitrophenyl derivative, 2-nitroethyl benzene and 2-nitrophenyl phenyl ether solvents produced instead calcium sensors with $k_{Ca,Mg}^{pot} = 0.06$–0.08. Again, the classical pH–emf dips reported for phosphate-type calcium ISEs were absent (40).

3.2.4 Potassium electrodes

Valinomycin (10) is a cyclic dodecadepsipeptide in which the sequence L-lactic acid, L-valine, D-valine and D-α-hydroxyisovaleric acid is repeated to give a molecule with threefold symmetry. This naturally occurring antibiotic selectively transports K^+ across lipophilic membranes of mitochondria. Complexation with potassium cations is established by just six donor sites based on oxygen atoms in the 36-membered ring, but ligand conformation is reinforced by intramolecular hydrogen bonds. Some characteristics of this widely acclaimed PVC sensor are given in Table 3.13 (41, 42).

Typical valinomycin cocktails contain 65–70 mass % of mediator which provides an ideally flexible PVC membrane for macro ISEs but shows poor adhesion characteristics for loading gates of ion-sensitive field effect transistor (ISFET) devices. Attempts have thus been made to improve adhesion by lowering the solvent content of the sensor cocktails without unduly compromising the usual electrode performances. Cocktails with 50 mass % solvent gave usable ISFETs but no improvement in adhesion. An ISFET with a gate membrane of 35.2 mass % solvent was noisy and showed no real K^+ response (43).

Each of the valinomycin cocktails in Table 3.13 used in ISEs without

Table 3.13 Some characteristics of PVC potassium ISEs.

	Membrane composition (mass %)					
Ionophore:	(10)(1)	(10)(1.3)	(10)(2.8)	(13)(2.8)	(14)(2.8)	(15)(3.2)
PVC:	(33)	(30.4)	(27.8)	(27.8)	(27.8)	(32.2)
Property	Solvent: BPA(66)	EHS(68.3)	2-NPOE(69.4)	2-NPOE(69.4)	2-NPOE(69.4)	2-NPOE(64.6)
Selectivity coefficient: $k_{K,B}^{pot}$						
B = Li	5.0×10^{-5}	1.99×10^{-5}				
B = Na	10^{-4}	1.99×10^{-4}	2×10^{-4}	3×10^{-4}	10^{-3}	2×10^{-4}
B = Rb	1	0.4	1.4	0.2	0.7	0.2
B = NH$_4$	10^{-2}	1.25×10^{-2}	0.1	10^{-2}	5×10^{-2}	4×10^{-2}
B = Mg	1.58×10^{-5}	2.51×10^{-5}				
B = Ca	2.51×10^{-5}	1.58×10^{-5}				
Slope (mV decade^{-1})	59.8 ± 0.1	59.2 ± 0.1	59	58	58	55
Detection limit (M)	5×10^{-6}	5×10^{-6}	3.16×10^{-5}	10^{-5}	10^{-5}	10^{-4}
Resistance (MΩ)	98	320				
Reference	(41)	(41)	(42)	(42)	(42)	(47)

BPA: bis (1-butylpentyl) adipate; EHS: bis (2-ethylhexyl) sebacate; 2-NPOE: 2-nitrophenyloctyl ether

(10)

(11)

(12)

(13)

(14)

(15)

(16)

(17)

D = digoxin

KTpClPB exhibit high resistances, but these fall from 4.9 MΩ to 35 kΩ with added KTpClPB (44).

Attempts (42, 45–47) to improve on the performance of the valinomycin sensor relate to synthetic cyclic ionophores (11–16) wherein high lipophilicity may be endowed through the dodecyl group as in (15). Petránek and Ryba (45) examined PVC potassium ISEs based on 20 cyclic ethers in dipentyl phthalate. The size of the macrocyclic ring and the nature of attached groups and the number of oxygen atoms exert pronounced effects on sodium selectivity. Thus $k_{K,Na}^{pot}$ was 7.7×10^{-2} for dicyclohexyl-18-crown-6 (11) and 1.1×10^{-2} for dibenzo-18-crown-6 (12), while overall the best selectivities were recorded for ionophores with either six oxygens in 18-membered rings or ten oxygens in 30-membered rings (45). The preference of potassium over sodium for the bis crown ethers (13) and (15) exceeds that for the corresponding monocyclic ionophores (14) and (16) (42, 46). These alternative potassium sensors often compare favourably with the valinomycin model regarding selectivity and response times and are more easily synthesized (42). However, low-cost valinomycin is readily available.

While the PVC/dibutyl sebacate/digoxin-benzo-15-crown-5 conjugate system (17) responds to potassium, its adoption for measuring antibodies or digoxin is more interesting (48). Potential modulation relates to selective and reversible binding of the digoxin antibodies to the digoxin segment of the conjugate sensor molecule at the membrane interface.

3.2.5 Ammonium electrodes

All ammonium ISEs are based on the macrotetrolide antibiotics and their major applications relate to enzyme studies. PVC membranes with nonactin (18) (containing up to 28% monactin 19) have been fabricated with a wide range of solvents, namely diethyl-, dibutyl- and dioctyl phthalates; dioctyl adipate; dibutyl-, dioctyl- and diethylhexyl sebacates; diethylhexyl- and tripentyl phosphates; di-octylphenyl phosphonate and 2-nitrophenyl phenyl ether (49, 50). In terms of selectivity, dioctyl adipate (DOA) was the best solvent, and the nonactin sensor could be conserved by fixing the mass ratios of nonactin:DOA:PVC at 1:300:160. The Nerntian slope was 57.5 mV decade^{-1} and $k_{NH_4,K}^{pot} = 0.11$ (50).

Nonactin (18) $R_1 = R_2 = R_3 = R_4 = CH_3$
Monactin (19) $R_1 = R_2 = R_3 = CH_3$, $R_4 = C_2H_5$

Electrodes with membranes containing KTpClPB-nonactin ranging from 67 to 235 mole % ratios showed no selectivity enhancement; in fact, $k^{pot}_{NH_4,K}$ ranged from 2.01 to 2.18 (50). PVC sensor membranes with up to 25 mass % nonactin were opaque, indicating minimal solubility of the sensor in the matrix, and their resistances (≈ 200 MΩ) and time responses were all similar (49).

Trials with salinomycin, an antibiotic extracted from *Streptomyces albus*, indicate some interference from potassium—$k^{pot}_{NH_4,K} = 0.18$, when incorporated in PVC-DOA membranes (50).

3.2.6 *Sodium electrodes*

Despite the high selectivity of sodium glass ISEs, problems arise in clinical use, and the contaminated membrane surfaces need special treatment at regular intervals (7, 51). Polymeric membrane ISEs with ionophores (**20**) and (**21**) are less affected and permit sodium measurements in undiluted serum and urine (51).

Some characteristic features of PVC sodium electrodes based on the acyclic ionophore (**20**) and the cyclic bis-12-crown-4 with aliphatic bridging (**21**) are compared with those of a glass electrode in Table 3.14.

(**20**)　　　　　　(**21**)

3.2.7 *Hydrogen electrodes*

Glass pH electrodes possess outstanding selectivities and the need for alternative H$^+$ ISEs might seem superfluous. Nonetheless, the contamination in biological media and high membrane resistances—particularly for micro glass sensors—pose problems (7). The pH behaviour of PVC sensor membranes comprising eleven different neutral ionophores in conjunction with KTpClPB and solvent mediator in the usual proportions is of some interest (53, 54).

The selectivity coefficients for an electrode based on (**22**) with 2-NPOE for K$^+$, and Na$^+$ and Li$^+$ were 3.98×10^{-10}, 3.16×10^{-11} and $< 6.3 \times 10^{-12}$ respectively (53).

Corresponding selectivities for a PVC tridodecylamine-bis(2-ethylhexyl)

Table 3.14 Some characteristics of sodium ISEs (51, 52)

Property	Composition of PVC membrane/mass %			Glass ISE NAS 11-18
	Ionophore: PVC: Solvent	(20)(1) (33) BPA^a(66)	(21)(3.2) (32.5) 2-NPOE (64.3)	
$k_{Na,B}^{pot}$				
B = K		10^{-2}	10^{-2}	10^{-3}
B = NH$_4$		1.99×10^{-2}	6×10^{-3}	
B = Mg		6.3×10^{-3}	10^{-4}	3.16×10^{-5}
B = Ca		3.1×10^{-2}	10^{-4}	
Detection limit (M)		7.94×10^{-6}		
Slope (mV decade^{-1})		60.1 ± 0.7	53	
Resistance (MΩ)		21		

^{a} Bis(1-pentyl adipate)

(22)　　　　　　　　　　　　　　　　　　(23)

sebacate membrane were 1.58×10^{-10}, 3.98×10^{-11} and 7.94×10^{-12} (54). Such impressive values are still inferior compared with certain classical glass sensors. Dramatic effects are exerted by the mediator, as exemplified by (23). Thus, tri-ethylhexyl phosphate leads to a sensor with a poor detection limit ($\sim$ pH 3) in a background of 0.06 M lithium, but extends to $\sim$ pH 8 for a 2-NPOE-based sensor, again in 0.06 M lithium.

The set of eleven pH sensors had Nernstian slopes of 56.6–58.7 mV decade^{-1}, while the resistances lying in the range 30 kΩ–1 MΩ must relate to the incorporated anionic sites. Such favourable features, combined with the lipophilicity of the sensor molecules, have facilitated measurements with a miniprobe in gastric fluids (53).

To realize broad detection limits, the sensor molecule should contain no structural features with coordination sites such as ether, amide or ester groups, so as to offset cation interactions. The pK of the mandatory basic nitrogen atom fixes the position of the dynamic range of the electrodes. Thus, the pKs for tridodecylamine and its C$_{14}$ and C$_{16}$ homologues are identical, 10.6, and the dynamic range of their PVC sensor are virtually identical as predicted (54). Finally, high lipophilicity is also required to cope with the similarly highly lipophilic environment of clinical samples (53).

3.2.8 *Lithium electrodes*

Lithium is an effective therapeutic agent for controlling manic depression, and its simple and rapid assay in patient's blood at levels around 0.5–1 mM is clinically desirable. Thus, a viable lithium ISE should have adequate selectivity to handle the high sodium background—typically 140 mM—as well as other interference sources.

Lithium glass electrodes such as LAS 15–25B are quite inadequate since $k_{Li,Na}^{pot}$ is ~ 0.3. The research effort to realize a high performance lithium ISE has centred on the synthesis and evaluation of dozens of cyclic and acyclic ionophores (55–59).

Complexation between lithium and ionophore (L) may be represented by

$$[Li(H_2O)_n]^+ + L \rightleftharpoons [LiL]^+ + nH_2O; \Delta H = 510\,kJ\,mol^{-1}$$

and the reversibility and extend of this exchange, as well as the one with any interferent, say B^+, by

$$[LiL]^+ + B^+ \rightleftharpoons [BL]^+ + Li^+$$

will relate critically to the performance of an ionophore-based PVC electrode.

In attempts to establish a relationship between cavity sizes of crown-4 derivatives and lithium selectivities, Kitazawa et al. (58) examined the potentiometric behaviour of a series of fourteen highly lipophilic crown-4 compounds with 12-, 13-, 14-, 15- and 16-member rings respectively and the fixed compositions of each of the otherwise identical PVC sensor membranes based on (24–27) make for a facile comparison of their respective potentiometric behaviour (Table 3.15). In general, the 13-membered ring compounds are much less selective than 14- and 15-membered ring compounds. Benzo-13-crown-4 derivatives also showed poor lithium selectivities, and

Table 3.15 Characteristics of lithium ISEs based on crown compounds (55)

| Ionophore number | Membrane composition/mass% | | | Selectivity coefficient $k_{Li,B}^{pot}$ |
	Solvent mediator	PVC	Sensor	Mixed solution method
(24) $C_{12}H_{25}$	2-NPOE(70.2)	28.1	1.0 +0.7% KTpClPB	0.016
(25) $C_{12}H_{25}$ CH_3	2-NPOE(70.2)	28.1	1.0 +0.7% KTpClPB	6.6×10^{-3} 7.0×10^{-3} 2×10^{-3} For membrane with TOPO (1% m/m)
(26) $C_{12}H_{25}$ H CH_3 CH_3	2-NPOE(70.2)	28.1	1.0 +0.7% KTpClPB	6.2×10^{-3}
(27) $C_{12}H_{25}$ CH_3 CH_3 CH_3 CH_3	2-NPOE(70.2)	28.1	1.0 +0.7% KTpClPB +1% TOPO	7.2×10^{-3}

it was concluded that all the 12-crown-4, 13-crown-4 and benzo-13-crown-4 compounds preferred the other alkali metal ions to lithium. However, the lithium selectivity was dramatically enhanced in the case of the larger 14- and 15-crown-4 derivatives, but this desirable feature is again reversed in the largest ring compounds examined, namely 16-crown-4 derivatives.

The effect of the methyl substituent attached to the quaternary carbon of the crown-4 derivatives on the lithium selectivity is particularly interesting (58). Thus, the dodecyl methyl derivatives of both 14- and 15-membered ring molecules are superior to the corresponding dodecyl derivatives in terms of lithium selectivity relative to sodium and also potassium. This was attributed to some steric influence of the methyl group on the crown ring regarding complexation of cations, and the possible formation of sandwich-type 2:1 complexes with cations. This is supported by the dependence of the selectivity coefficients on the concentration of the crown ethers (24) and (25) in the PVC membrane. Thus, in the dodecyl methyl-14-crown-4 (25) sensor system, the selectivity coefficients for sodium and potassium were essentially unchanged on increasing sensor levels. On the other hand, the selectivity coefficients for these cations increased with increasing amounts of the sensor in the dodecyl-14-crown-4 (24) system.

It was concluded that dodecyl-14-crown-4 can complex Na^+ and K^+ ions on a 2:1 and 1:1 stoichiometric basis, whereas the methyl derivative has difficulty in forming sandwich complexes even at high crown ether levels. Thus, the methyl group in (25) is a steric barrier which offsets the access of two crown ether molecules to sandwich formation and so enhances lithium selectivity, i.e., the prevention of 2:1 crown ether–metal complexes with alkali metal cations, other than lithium, enhances the lithium selectivity of the original crown ether. Similarly, two methyl groups on opposite sides of the crown ether rings of (26) and (27) also improved the lithium selectivity (Table 3.15). Thus, sensors based on the three 14-crown-4 derivatives (25) to (27) portray sodium selectivities on a par with acyclic neutral carriers (Table 3.16). PVC membrane electrodes comprising dibenzo 14-crown-4 sensor with either 2-NPOE or dioctyl sebacate mediators showed quite poor lithium response in the presence of sodium. This feature was only marginally improved by incorporating KTpClPB in the PVC membranes (55).

To date, the best lithium electrode is realized with the dodecyl methyl-14-crown-4-based cocktail in Table 3.15 containing the additional component trioctyl phosphine oxide (TOPO), for which $k_{Li,Na}^{pot} = 2 \times 10^{-3}$ (see 55).

Acyclic quadridentate amido-ethers complex preferentially with lithium when their molecular structures are such that four oxygen atoms tend to form a cavity comparable in size with the rings of certain crown ethers (Table 3.15). A range of these (e.g. 28–32) has been synthesized and evaluated as lithium sensors with some well-established solvent mediators (Table 3.16). Three, namely (28), (30) and (32), are commercially available with appropriate contents of solvent/KTpClPB/PVC. The spectrum of $k_{Li,Na}^{pot}$ values ranges from poor (0.4) to 3.16×10^{-3} at best and the subtle influence of side chain

Table 3.16 Characteristics of lithium ISEs based on acyclic compounds (55).

Ionophore number	Membrane composition (mass %)			Selectivity coefficient $k_{Li,B}^{pot}$		
	Solvent mediator	PVC	Sensor		Separate solution method	Mixed solution method
(28)	TEHP(62.8)	31.4	5.8	Na	0.05	
				K	7.0×10^{-3}	
				NH$_4$	0.05	
				Ca	5×10^{-4}	
				Mg	1.99×10^{-4}	
	2-NPOE(65-66)	33	1-2	Na	0.08	
(29)	TEHP(66.7)	29.4	3.9	Na	0.051	0.063
				K	5.3×10^{-3}	0.104
				Ca	3.6×10^{-3}	7.1×10^{-3}
				Mg	4.0×10^{-4}	2.5×10^{-4}
(30)	2-NPOE(65-66)	33	1-2	Na	7.9×10^{-3}	
				K	6.3×10^{-3}	
				NH$_4$	7.9×10^{-3}	
				Ca	10^{-3}	
				Mg	6.3×10^{-3}	
	TEHP(64.6)	28.5	6.9	Na		0.15
(31)	2-NPOE(65-66)	33	1.2		0.4	
(32)	2-NPOE(65.6)	32.8	1.2	Na	5×10^{-3}	3.16×10^{-3}
			+0.4%TpClPB			3.55×10^{-3}
	BPA(65.6)	32.8	1.2			0.032
			+0.4%TpClPB			

structures alone on sodium selectivity is exemplified by ionophores (**30**) and (**31**), using the same 2-NPOE mediator.

An interesting selectivity study has been conducted with sensors obtained by replacing the dioxanonane backbone of ionophore (**29**) with a furan and a pyridine ring, respectively. The $k_{Li,Na}^{pot}$ value of 0.063 for (**29**) in a PVC/KTpClPB/2-NPOE matrix changed to 0.94 and 0.094 respectively. An extension of this work with sensors having electron-donating groups on their backbones is considered a worthwhile area of research (57). Despite considerable effort, the best acyclic sensor is still offered by ionophore (**32**) with 30 mass % KTpClPB relative to sensor and 2-NPOE rather than bis(1-pentyl adipate) (Table 3.16).

Poly(alkyleneoxy) compounds are characterized by a flexible array of donor oxygen atoms based on -CH(R)-CH$_2$O-units and are regarded as acyclic ethers, e.g. PPG1025, a poly(propylene glycol). The tetraphenylborate (TPB)-barium complex of PPG-1025 with di-octylphenyl phosphonate constitutes a useful PVC barium ISE (40) which becomes totally insensitive to barium even at 5 mM levels of lithium. This interesting feature has been exploited to develop a lithium sensor viable for clinical purposes which, although less selective than (25) or (32) (55, 59), is easier to prepare. Unfortunately the selectivity was not enhanced by incorporating KTpClPB.

3.2.9   Barium electrodes

Viable barium electrodes are based on acyclic ionophores (7, 60). The nonylphenoxypoly(ethyleneoxy)ethanol (Antarox CO-880) containing 12 ethylene oxide units (EOUs) and 2 moles of TPB per mole barium (12EOU·Ba·2TPB) is a classic example. The relevance of the viscosity and permittivity of mediators on its behaviour has been investigated with eleven aromatic solvents. The Nernstian slopes of PVC ISEs with 4-nitroethyl-benzene ($\eta = 1.99$ mN s m^{-2}, $\varepsilon = 24.1$) varied between 26.5 and 29.5 mV decade^{-1}, but fell dramatically after about two days, while response times increased. This behaviour was clearly associated with the loss of the low-viscosity solvent from the PVC matrix since the initial responses could be restored, albeit temporarily, by immersing the PVC sensor surface in fresh 4-nitroethyl benzene. PVC ISEs based on 2-ethyl and 3-ethylnitrobenzene, nitrobenzene, 2-nitro-3-methyl- and 3-nitro-2-methyltoluene, and 2-nitro-4-isopropyltoluene ($\eta = 1.68$–3.19 mN s m^{-2}, $\varepsilon = 17.7$–34.9) behaved similarly. During their brief operational lifetime, however, the calibration slopes were in the range 26–29 mV decade^{-1} (60).

Servicable barium ISEs resulted with incorporation of high-viscosity 2-nitrophenylphenyl ether ($\eta = 16.1$ mN s m^{-2}, $\varepsilon = 28.3$). They have a long linear calibration range, slopes of 28–29 mV decade^{-1} and lifetimes of about 30 days. The lifetimes of liquid membrane barium ISEs with all these nitroaromatic mediators in Orion 92-20 bodies were sufficiently long (20–30 days) so as to enable selectivity measurements. It is interesting that the variations in relative permittivities do not produce conclusive differences in selectivity within Group I or Group II cations. Clearly, solvent viscosity is more relevant to the design of long-lived ISEs than permittivity (60).

Barium ISEs find little application in terms of barium analysis but they are useful for the indirect titration of sulphate (60, 61) and especially for analysis of non-ionic surfactants and measuring of their critical micelle concentrations from the inflexions in emf v. log [surfactant] plots (62).

3.2.10   Uranyl electrodes

Organophosphoric acids and a variety of organophosphorus mediators have been examined as liquid ion-exchanger systems for PVC uranyl ISEs (63, 64) (Table 3.17). The sensors are readily made by multiple-stage extractions from

Table 3.17 Some characteristics of PVC uranyl ISEs (63, 65).

Sensor	Mediator	Slope (mV decade^{-1})	Selectivity coefficients, $K^{pot}_{UO_2,B}$ (separate solution method)			
			Na	Ca	Zn	Cu
Di(2-ethylhexyl) phosphoric acid	Dipentyl phosphonate	25 ± 2		2.7×10^{-2}	6.4×10^{-3}	9.3×10^{-3}
	Di (2-ethylhexyl)-ethyl phosphonate	26 ± 2		9×10^{-4}	1.7×10^{-3}	1.9×10^{-3}
	Tri (2-ethylbutyl) phosphonate	26 ± 2		1.4×10^{-3}	7×10^{-4}	5×10^{-4}
Ionophore (5)			50.1	0.5		
Ionophore (28)			0.16	0.79		
Ionophore (33)	1-chloronaphthalene	54.9 ± 0.9	3.98×10^{-2}	1.2×10^{-4}		
Ionophore (34)			1.99×10^{-3}	2.51×10^{-5}		
			2.5×10^{-4a}	7.9×10^{-5a}	$<10^{-4a}$	$<10^{-4a}$

[a] For PVC membranes with 2.8 mass% of ionophore (34)

aqueous uranyl nitrate with the free acid in chloroform:

$$UO_2(NO_3)_2(aq) + 2(RO_2)_2 POOH(chloroform) \rightarrow$$

$$UO_2[(RO_2)_2 PO(O)]_2(chloroform) + 2HNO_3(aq)$$

After removing chloroform, the residual uranyl salt is incorporated with mediator into PVC membranes. This simple technique has proved useful for preparing other sensor material, e.g. zinc and beryllium organophosphates.

The quest for improved uranyl sensors has rested with ionophores. Ionophores suitable for calcium (5) and lithium (28) are subject to sodium and calcium interferences (Table 3.17) whereas amongst newer ones, (34) particularly is considered to provide a UO_2^{2+} device of unrivalled specificity (65).

(33)   (34)

The hydrolysis of uranyl salts leads to four other uranium species, e.g., UO_2Cl^+, UO_2OH^+, UO_2OH^{3+} and $(UO_2)_2(OH)_2^{2+}$. From the standpoint of calibration, this problem is very readily accommodated by setting the pH of all standards at ~ 3 so as to largely maintain the uranium (VI) as the primary UO_2^{2+} species. The Nernstian slopes of this PVC-uranyl ISE in pH 3-uranium nitrate standards were 54.9 ± 0.9, 54.7 ± 1 and 55.3 ± 0.6 in mV decade^{-1} on days 6, 37 and 42, respectively, after assembly. These values correspond to permeation of a monovalent species, and the ionophore (34) evidently forms ion pairs, e.g. UO_2OH^+, in the PVC matrix. In effect, this enhances the sensitivity of a sensor specific for a formally divalent cation. The sodium selectivity is improved by increasing the amount of ionophore, whereas that for calcium is lowered (Table 3.17).

Calcium-phosphate-type sensors are essentially unaffected by large doses of ^{60}Co-gamma radiation (11) and thus the corresponding uranyl phosphate, and possibly neutral carrier types, should be immune to the very modest doses of internally generated uranium radiation.

3.3. Anion-selective electrodes

Many anion-selective ISEs are based on classic exchangers, complexes of 2-phenanthroline, and quaternary ammonium, phosphonium and arsonium salts. Not all of these possess the required selectivity, e.g. the electrodes based on dialkyltin(IV) compounds for HPO_4^{2-} and $H_2PO_4^-$.

Table 3.18 Some characteristics of PVC nitrate and chloride ISEs (66-68)

Property	Nitrate electrode (mass%)			Chloride electrode (mass%)	
	Corning cocktail(70) PVC(30)	Orion cocktail(70) PVC(30)	TOAN(6) DBP(65) PVC(30)	TDDM(6) 2-NPOE(65) PVC(29)	TDDM(6) 5-PP(65) PVC(29)
Selectivity coefficient: $k_{A,B}^{pot}$					
F	8.7×10^{-4}	7×10^{-4}			
Cl	5×10^{-3}	4×10^{-3}	5×10^{-3}	1.0	1.0
Br	0.16		0.13	10.0	5.01
I	17	16	14.1		
NO$_2$	0.066	0.06	0.07	3.98	2.51
NO$_3$	1.0	1.0	1.0	3.98	15.8
SO$_4$	$< 10^{-5}$	3×10^{-4}		0.08	0.25
ClO$_3$	1.66	1.66	3.01		
ClO$_4$	800	550	1.26×10^3	3.16×10^4	794
Nernstian slope (mV decade^{-1})	57	57	56.1	56.5 ± 2.4	55.9 ± 2.8

3.3.1 Nitrate electrodes

After the successful fabrication of PVC-calcium ISEs, viable PVC nitrate sensors were similarly constructed with commercial nitrate cocktails (66). Their selectivity for some parameters is quite similar, despite their different compositions—a tris(substituted 2-phenanthroline)nickel(II) exchanger in the Orion 92-07-02 cocktail and tridodecylhexadecylammonium nitrate sensor in the Corning 477316 cocktail (Table 3.18). Their respective lifetimes were ~ 3 and ~ 11 weeks (66).

Nielsen and Hansen have examined PVC electrodes based on five long-chain quaternary ammonium salts and dibutyl phthalate (DBP). Sensors were recrystallized until their constituted electrodes yielded a pH–nitrate response profile which was essentially pH-independent. However, this was not achieved with the tetratetradecylammonium sensor and its limit of detection was poorer than expected. In some cases purities, as in the tetraoctylammonium (TOAN) sensor, were verified by mass spectrometry (67).

Sensitivity and lifetime might be expected to improve on lengthening the alkyl chain of the sensor. However, in terms of selectivity, very little was gained by using sensors with alkyl chains $> C_{12}$ or C_{14}. Moreover, the synthesis and purification of the higher R_4N^+-sensor materials becomes disproportionately more difficult with increasing chain length, as exemplified by tetratetradecyl ammonium.

TOAN (6 mass %) in conjunction with DBP gave good-quality electrodes, with slopes of 56.1 mV decade^{-1} which fell to 52 mV decade^{-1} with a TOAN content of 1 mass %. The overall similarity between the TOAN- and Corning-type sensors is perhaps fortuitous, since the Corning sensor material is not only impure but the 2-NPOE mediator is different and their respective concentrations are unknown.

Measurements of nitrate in soils and waste waters using the TOAN-DBP electrode and the brucine method agreed closely. All these nitrate ISEs (66, 67) are subject to serious interference from iodide, chlorate and perchlorate. This feature can in turn be exploited, e.g., the Corning nitrate exchanger may be readily converted to a viable chlorate-sensing cocktail using an extraction technique as described for the uranyl phosphate sensor (section 3.2.10).

3.3.2   *Chloride electrodes*

The commercial product, Aliquat 336, also known as tricaprylammonium chloride, has been used as an anion-selective component in various ISE systems. However, it is a mixture of C_8- and C_{10}-side chain species, and evaluations of a PVC-Aliquat ISE may not readily relate to those for an ISE based on any one single component of Aliquat 336. A series of pure trialkyl methyl ammonium salts has been studied so as to assess that compound's most suitable component (68).

Selectivity, stability and lifetime are sensitive to the type and amount of quaternary salt and mediator (Table 3.18). The best model based on TDDM (tridodecylmethylammonium chloride) with 5-PP(5-phenylpentanol) was suitable for use in blood serum. Again, increasing the alkyl chain length results in improved Nernstian slopes and simultaneously enhances lipophilicity and, in turn, the lifetime of the sensor (68).

3.3.3   *Nitrite electrodes*

Until recently, nitrite could be assayed indirectly with ammonia gas or nitrate electrodes, respectively, following reduction to ammonium ion with nitrite reductase or oxidation to nitrate. A new PVC nitrite ISE, based on a lipophilic vitamin B_{12} derivative and bis(1-butylpentyl) adipate, permits nitrite assay in unmodified solutions of extracts from dried meat (69). The anion selectivity deviated from the classical selectivity sequence for anion exchangers, $ClO_4^- > SCN^- > NO_3^- > NO_2^- > Cl^-$; $k_{NO_2,NO_3}^{pot} = 6.3 \times 10^{-5}$ and $k_{NO_2,Cl}^{pot} = 1.58 \times 10^{-5}$. Nernstian behaviour was only achieved for sensor levels of ≈ 3 mass %, while lifetimes were reported to be at least 6 months (69).

3.4   The polymer matrix

The polymer matrix enables the entanglement of sensor, mediator and any further additives like KTpClPB. The maintenance of their intitial casting levels, and especially mass ratios, during use is critically related to the selectivity and lifetime of a sensor device.

3.4.1 Entangled poly (vinyl chloride) matrices

Considerable research effort has been devoted to the role of sensor and mediator, but relatively little information pertains to PVC save for occasional reference to its molecular mass and the almost obligatory ~ 30 mass % content needed for the mechanical strength of membranes.

Therefore, the behaviour of eight calcium ISEs with membranes comprising ionophore (4), DOPP and seven standard IUPAC PVCs plus the widely-used reference PVC, Breon III EP, have been examined (Table 3.19). All the master membranes were clear and flexible and (except for Nos. 2, 3, 4 and 6 with $\bar{M}_n$ values $< 40\,000$) non-sticky. In terms of the parameters listed in Table 3.19 there is little to recommend any one PVC type (70).

3.4.2 Alternative entangled polymer matrices

Despite the universal adoption of PVC, some investigations have centred on the prospect of alternative matrix materials. Except for poly(vinylisobutyl ether) (71), poly(methyl methyl acrylate) (72), poly(methyl acrylate) (39) and poly(vinyl chloride/alcohol) copolymer (73), practically no functional calcium ISEs were realized with poly(vinylidene chloride) (19), cellulose acetate, ethyl cellulose, collodion or pyroxylin (11). Photocured sensor membranes based on acrylates are considered in section 3.4.4.

Poly(urethane) can be substituted for PVC in nitrate sensor membranes (67). Nonetheless, such materials are unlikely to usurp the well-proven poly(vinyl chloride) matrix. It is essential to employ a proven PVC, e.g. Breon III EP, Flowell 470 or Fluka S 704, otherwise non-functional or poor-quality ISEs may be produced even with a well-tried cocktail. However, Silopren (a silicone rubber) offers a distinct advantage in that lipophilic anions do not interfere with its valinomycin electrode, and it is considered to have universal applicability for body fluids (41).

3.4.3 Polymer-bound sensor systems

Operative lifetimes of the PVC entangled sensor systems so far described are considerably shorter than their solid-state counterparts, e.g. the lanthanum fluoride model. The principal cause is the loss of active component(s) from the polymer matrix. Thus, the deterioration of PVC potassium electrodes relates to the leaching of potassium; membranes with an initial resistance of 5.8MΩ left in contact with deionized water for four days became paler and the resistance rose to 30MΩ compared with 50MΩ for a PVC-only membrane. Similar, but slower, detrimental processes arise with PVC nitrate electrodes (74).

The stiffening of calcium ISE membranes based on neutral carrier (4) and DOPP after use in 0.1M sodium dodecyl sulphate indicates a similar loss of

Table 3.19 Some characteristics of IUPAC PVCs and associated calcium ISEs (70)

PVC number	$\bar{M}_n$	$\bar{M}_w$	T_g (°C)	Resistance (MΩ)	Slope (mV decade^{-1})	Limit of detection (M)	Selectivity coefficient, $k_{Ca,B}^{pot}$ (separate solution method)		
							B = Mg	B = Na	B = K
1	47 150	175 300	102	2.02	32.0^a / 28.3^b	2.5 × 10^{-6a} / 1.0 × 10^{-5b}	1.28 × 10^{-3a} / 5.72 × 10^{-4b}	3.90 × 10^{-2a} / 3.12 × 10^{-2b}	4.83 × 10^{-2a} / 2.61 × 10^{-2b}
2	39 770	86 810	85	1.63	29.6 / 30.0	2.2 × 10^{-6} / 1.1 × 10^{-5}	9.81 × 10^{-4} / 6.31 × 10^{-4}	4.85 × 10^{-2} / 3.83 × 10^{-2}	3.53 × 10^{-2} / 3.68 × 10^{-2}
3	25 900	81 330	86	1.89	30.0 / 30.5	7.8 × 10^{-6} / 7.9 × 10^{-6}	3.50 × 10^{-3} / 8.98 × 10^{-4}	0.13 / 3.18 × 10^{-2}	0.11 / 5.90 × 10^{-2}
4	39 350	104 900	86	1.11	30.8 / 30.4	4.5 × 10^{-5} / 1.1 × 10^{-5}	1.04 × 10^{-3} / 8.20 × 10^{-4}	3.72 × 10^{-2} / 6.14 × 10^{-2}	7.59 × 10^{-2} / 4.01 × 10^{-2}
5	48 520	104 800	85	2.09	27.0 / 30.5	4.5 × 10^{-6} / 5.0 × 10^{-6}	1.28 × 10^{-3} / 1.30 × 10^{-3}	3.80 × 10^{-2} / 4.64 × 10^{-2}	4.74 × 10^{-2} / 5.93 × 10^{-2}
6	36 310	101 300	88	2.46	26.3 / 30.8	1.8 × 10^{-5} / 1.6 × 10^{-5}	1.31 × 10^{-3} / 2.74 × 10^{-3}	4.03 × 10^{-2} / 0.12	7.51 × 10^{-2} / 0.13
7	43 700	103 400	86	1.28	26.5 / 28.5	1.2 × 10^{-5} / 8.9 × 10^{-6}	2.22 × 10^{-3} / 1.86 × 10^{-3}	4.80 × 10^{-2} / 6.25 × 10^{-2}	9.99 × 10^{-2} / 0.12
8 (Breon 111EP)	77 000	^{c}	98	3.02	28.0 / 30.8	2.8 × 10^{-6} / 3.2 × 10^{-6}	2.18 × 10^{-3} / 1.47 × 10^{-3}	9.66 × 10^{-2} / 5.57 × 10^{-2}	1.03 × 10^{-2} / 6.46 × 10^{-2}

^{a,b} All first and second column parameters relate to measurements 1 and 12 weeks after fabricating each electrode
^{c} Unavailable

active components. X-ray fluorescence of their surfaces, and gas- and thin-layer chromatography of the concentrated aqueous leachates, indicate loss of calcium salt and DOPP. Under milder circumstances with $5 \times 10^{-4} M$ surfactant only DOPP could be detected as a leachate (19). Mediator, and perhaps ionophore, may also be lost by diffusion from the membrane into the PVC electrode support shaft (30). One-off loss into the small volume of internal reference solution may be neglected, and is non-existent in epoxy-based all-solid-state models which need no internal solution (38).

Models to predict operational lifetimes have been made on the kinetics of loss of mediator or sensor from the PVC into the sample solution. To ensure a continuous lifetime of one year, or more, the partition coefficient for both these components between sample and membrane should exceed 10^6 (30).

Various attempts to maintain the integrity of the initial membrane cocktail, and thus its selectivity performance, as well as to promote longer operational lifetime, relate to covalent bonding of the active components to an inert polymer (73, 75–77). Thus, for anionic surfactant ISEs a tertiary amine was grafted to the ends of PVC chains and then converted to a quaternary ammonium site with an alkyl bromide. Electrodes fabricated in the usual way were next conditioned in sodium dodecylsulphate (SDS) to exchange the bromide with surfactant anion (75). Polymerization of vinyl chloride monomer with SO_3^- radical anions gave a cationic surfactant ISE membrane (relative molecular mass $\sim 77\,000$) where about one-third of the polymer chains terminated in sulphate groups. Again, conditioning was affected by exchanging protons of the SO_3H group with the required surfactant cation (75).

Several grafted calcium ISEs have been reported (73, 76, 77). In one instance membranes were synthesized by cross-linking the styrene-b-butadiene-b-styrene (SBS) triblock elastomer with triallylphosphate, followed by alkaline hydrolysis of the resultant covalently bound trialkyl phosphate to give a pendant dialkyl phosphate. Good calcium electrodes with lifetimes greater than 6 months were obtained, but the limited selectivity for magnesium and sodium could relate to the absence of a solvent mediator (76).

An alternative approach involved combining monodecyl dihydrogen phosphate (73) or mono-[4-(1, 1, 3, 3-tetramethylbutyl) phenyl] dihydrogen-phosphate (77) with the hydroxyl groups of a partially hydrolysed vinyl choride–vinyl acetate copolymer (copolymer VAGH, relative molecular

$$\sim CH_2 - CH \sim \qquad \sim CH_2 - CH \sim$$
$$| \qquad\qquad\qquad |$$
$$O \qquad\qquad\qquad O$$
$$| \qquad\qquad\qquad |$$
$$RO - P = O \quad C_8H_{17}O - P = O$$
$$| \qquad\qquad\qquad |$$
$$OH \qquad\qquad\qquad C_6H_5$$

$$(35) \qquad\qquad\qquad (36)$$

mass $\sim 23\,000$). The resultant product (**35**), when incorporated in a membrane with DOPP mediator, provided a good-quality calcium ISE but without the expected longer lifetimes compared with the classical entangled PVC membranes (9, 11, 14, 16). This behaviour was due to leaching of DOPP as shown by TLC (77).

This grafting technique was extended to the mediator as a pendant via monoctylphenyl phosphonate (**36**). However, the slight increase in operational lifetime of electrodes comprising the double pendant sensor-mediator was quite marginal, particularly in view of the additional stages involved with the synthesis of grafted VAGHs (**35**) and (**36**).

3.4.4 Membrane-casting techniques

Functional PVC master membranes incorporating diverse cocktails have been successfully cast in the classic manner of Figure 3.2 (78). The liquid nature of PVC cocktails constitutes an additional asset in that they will conform to the shape of literally any surface. Consequently, on evaporation of tetrahydrofuran the PVC sensor film remains as a particular contour, which has resulted in important and improved designs of electrodes, e.g., coated wires/epoxy, tubular flow-through, micro- and all solid-state epoxy models.

The classic evaporation technique (9, 78) is nonetheless considered relatively slow, and only minimal control over the thickness and uniformity of the final membrane product is achieved (79).

A photocuring technique based on an acrylate matrix has thus been employed to fabricate calcium and potassium electrodes which dispenses with the volatile casting solvent, e.g. tetrahydrofuran (64, 79). Thus, the calcium salt of bis-[4-(1,1,3,3-tetramethylbutyl)phenyl]phosphoric acid (6 mass %); DOPP (18 mass %); Ebecryl 150-bisphenol A type epoxyacrylate (42 mass %); 1,6-hexanedioldiacrylate (26 mass %) and Uvecryl P36-copolymerizable benzophenone photoinitiator (8 mass %) were photolysed with a mercury lamp. Calcium ISEs with Nernstian slopes of 29.7 mV decade^{-1} were fabricated from the resultant clear, flexible, thin (0.1 mm) films (79). This procedure, which is unsuitable for a sensor or mediator absorbing in the ultraviolet, could find significant application in the realm of ISFET fabrication (64, 79).

The adhesion of classic PVC membranes to the active gate of an ISFET poses another problem. One recent innovation involves coupling the OH groups of a PVC–OH polymer ($\sim 1\%$ OH content) alone, or through the agency of silicon tetrachloride, to glass or silicon test surfaces (80). The resultant membranes had to be cut off the glass or natural SiO_2 surfaces, indicating the considerable measure of adhesion achieved. The performances of potassium ISEs made from these adhesive polymers and valinomycin–dioctyl adipate compared very favourably with the classic PVC model (section 3.2). This method, of course, will not necessarily conserve the initial loadings of sensor or mediator.

3.5 Conclusion

Literally thousands of sensor molecules are available, and the numbers are greatly compounded by the prospect of permutation with an equally large array of mediator solvents. A clearcut prediction of the ultimate behaviour of any sensor–solvent pair, let alone with variation in added KTpClPB, is still impossible, although useful ground rules are available. However, so many parameters contribute to the final package that the only meaningful approach is to 'try it and see', that is, to evaluate the ingredients as a PVC electrode arrangement and not as a solvent extraction exercise.

References

1. M.S. Frant and J.W. Ross, *Science* **154** (1966) 1553.
2. J.W. Ross, *Science* **156** (1966) 1378.
3. L.A.R. Pioda, V. Stankova and W. Simon, *Anal. Lett.* **2** (1969) 665.
4. *Orion Research Newsletter* **1** (1969) 13.
5. E. Pungor and K. Tóth, *Acta Chim. Acad. Sci. Hung.* **41** (1964) 239.
6. W. Simon, E. Pretsch, W.E. Morf, D. Ammann, U. Oesch and D. Dinten, *Analyst* **109** (1984) 207.
7. D. Ammann, W.E. Morf, P. Anker, P.C. Meier, E. Pretsch and W. Simon, *Ion-selective Electrode Revs.* **5** (1983) 3.
8. G.J. Moody and J.D.R. Thomas, *Ion-Selective Electrode Revs.* **3** (1981) 189.
9. G.J. Moody, R.B. Oke and J.D.R. Thomas, *Analyst* **95** (1970) 910.
10. G.J. Moody and J.D.R. Thomas, in *Ion-selective Electrodes in Analytical Chemistry*, H.Freiser (ed.), Vol. 1 Plenum, New York (1978) 287.
11. G.H. Griffiths, G.J. Moody and J.D.R. Thomas, *Analyst* **97** (1972) 420.
12. G.J. Moody and J.D.R. Thomas, *J. Power Sources* **9** (1983) 137.
13. K. Garbett and K. Torrance, *IUPAC Int. Symp. on Selective Ion-sensitive Electrodes*, UWIST, Cardiff, April 1973, Paper No. 12.
14. A. Craggs, L. Keil, G.J. Moody and J.D.R. Thomas, *Talanta* **22** (1975) 907.
15. A. Craggs, B. Doyle, S.K.A.G. Hassan, G.J. Moody and J.D.R. Thomas, *Talanta* **27** (1980) 277.
16. A. Craggs, G.J. Moody and J.D.R. Thomas, *Analyst* **104** (1979) 412.
17. G.J. Moody, N.S. Nassory and J.D.R. Thomas, *Analyst* **103** (1978) 68.
18. J. Růžička, E.H. Hansen and J.C. Tjell, *Anal. Chim. Acta* **67** (1973) 155.
19. A.J. Frend, G.J. Moody, J.D.R. Thomas and B.J. Birch, *Analyst* **108** (1983) 1072.
20. A.J. Frend, G.J. Moody, J.D.R. Thomas and B.J. Birch, *Analyst* **108** (1983) 1357.
21. S.A.H. Khalil, G.J. Moody and J.D.R. Thomas, *Analyst* **110** (1985) 353.
22. A. Craggs, P.G. Delduca, L. Keil, B. Key, G.J. Moody and J.D.R. Thomas, *J. Inorg. Nucl. Chem.* **40** (1978) 1483.
23. D. Ammann, M. Güggi, E. Pretsch and W. Simon, *Anal. Lett.* **8** (1975) 709.
24. A. Craggs, G.J. Moody, J.D.R. Thomas and A. Willcox, *Talanta* **23** (1976) 799.
25. G. Eisenman, *Anal. Chem.* **40** (1968) 310.
26. G.J. Moody and J.D.R. Thomas, *Chem. Ind.* (1974) 644.
27. G.J. Moody, *J. Biomed. Eng.* **7** (1985) 183.
28. U. Fiedler, *Anal. Chim. Acta* **89** (1977) 111.
29. D. Ammann, E. Pretsch and W. Simon, *Helv. Chim. Acta* **56** (1973) 1780.
30. U. Oesch and W. Simon, *Anal. Chem.* **52** (1980) 692.
31. U. Schefer, D. Ammann, E. Pretsch, U. Oesch and W. Simon, *Anal. Chem.* **58** (1986) 2282.
32. J. Petranék and O. Ryba, *Anal. Chim. Acta* **128** (1981) 129.
33. K. Kimura, K. Kumani, S. Kitazawa and T. Shono, *J. Chem Soc. Chem. Commun.* (1984) 442.
34. K. Kimura, K. Kumani, S. Kitazawa and T. Shono, *Anal. Chem.* **56** (1984) 2369.
35. M. Otto, P.M. May, K. Murray and J.D.R. Thomas, *Anal. Chem.* **57** (1985) 1511.
36. W.E. Morf, G. Kahr and W. Simon, *Anal. Lett.* **7** (1974) 9.
37. F. Lanter, D. Erne and W. Simon, *Anal. Chem.* **52** (1980) 2400.
38. S.A.H. Khalil, G.J. Moody, J.D.R. Thomas and J.L.F.C. Lima, *Analyst* **111** (1986) 611.
39. S.K.A.G. Hassan, G.J. Moody and J.D.R. Thomas, *Analyst* **105** (1980) 147.

40. A.M.Y. Jaber, G.J. Moody and J.D.R. Thomas, *Analyst* **102** (1977) 943.
41. P. Anker, H.-B. Jenny, U. Wuthier, R. Asper, D. Ammann and W. Simon, *Clin. Chem.* **29** (1983) 1447.
42. H. Tamura, K. Kimura and T. Shono, *Bull. Chem. Soc. Japan* **53** (1980) 547.
43. G.J. Moody, J.M. Slater and J.D.R. Thomas, *Anal. Proc.* **23** (1986) 287.
44. T.A. Nieman and G. Horvai, *Anal. Chim. Acta* **170** (1985) 359.
45. J. Petránek and O. Ryba, *Anal. Chim. Acta* **72** (1974) 375.
46. K. Kimura, T. Maeda, H. Tamura and T. Shono, *J. Electroanal. Chem.* **95** (1979) 91.
47. K. Kimura, A. Ishikawa, H. Tamura and T. Shono, *J. Chem. Soc. Perkin Trans. II* (1984) 447.
48. M.Y. Keating and G.A. Rechnitz, *Anal. Chem.* **56** (1984) 801.
49. M. Meyerhoff, *Anal. Chem.* **52** (1980) 1532.
50. O.G. Davies, G.J. Moody and J.D.R. Thomas, unpublished work.
51. P. Anker, H.-B. Jenny, U. Wuthier, R. Asper, D. Ammann and W. Simon, *Clin. Chem.* **29** (1983) 1508.
52. H. Tamura, K. Kimura and T. Shono, *Anal. Chem.* **54** (1982) 1225.
53. U. Oesch, Z. Brzózka, A. Xu, B. Rusterholz, G. Suter, H.V. Pham, D.H. Welti, D. Ammann, E. Pretsch and W. Simon, *Anal. Chem.* **58** (1986) 2285.
54. P. Schulthess, Y. Shijo, H.V. Pham, E. Pretsch, D. Ammann and W. Simon, *Anal. Chim. Acta* **131** (1981) 111.
55. G.D. Christian, V.P.Y. Gadzepko, G.J. Moody and J.D.R. Thomas, *Ion-selective Electrode Revs.* **8** (1986) 173.
56. E. Metzger, D. Ammann, R. Asper and W. Simon, *Anal. Chem.* **58** (1986) 132.
57. V.P.Y. Gadzepko, J.M. Hungerford, A.M. Kadry, Y.A. Ibrahim, R.Y. Xie and G.D. Christian, *Anal. Chem.* **58** (1986) 1948.
58. S. Kitazawa, K. Kimura, H. Yano and T. Shono, *J. Amer. Chem. Soc.* **106** (1984) 6978.
59. V.P.Y. Gadzepko, G.J. Moody and J.D.R. Thomas, *Analyst* **110** (1985) 1381.
60. A.M.Y. Jaber, G.J. Moody and J.D.R. Thomas, *Analyst* **101** (1976) 179.
61. D.L. Jones, G.J. Moody, J.D.R. Thomas and M. Hangos, *Analyst* **104** (1979) 973.
62. G.J. Moody and J.D.R. Thomas, in *Non-ionic Surfactants: Chemical Analysis*, Cross, J. (ed.) Marcel Dekker, New York (1986).
63. D.L. Manning, J.R. Stokely and D.W. Magouryk, *Anal. Chem.* **46** (1974) 1116.
64. G.J. Moody, J.M. Slater and J.D.R. Thomas, unpublished work.
65. J. Senkyr, D. Ammann, P.C. Meier, W.E. Morf, E. Pretsch and W. Simon, *Anal. Chem.* **51** (1979) 786.
66. J.E.W. Davies, G.J. Moody and J.D.R. Thomas, *Analyst* **97** (1972) 87.
67. H.J. Nielsen and E.H. Hansen, *Anal. Chim. Acta* **85** (1976) 1.
68. K. Hartman, S. Luterotti, H.F. Osswald, M. Oehme, P.C. Meier, D. Ammann and W. Simon, *Mikrochim. Acta II* (1978) 235.
69. P. Schulthess, D. Ammann, B. Kräutler, C. Caderas, R. Stepánek and W. Simon, *Anal. Chem.* **57** (1985) 1397.
70. G.J. Moody, B. Saad and J.D.R. Thomas, unpublished work.
71. O.F. Schäfer, *Anal. Chim. Acta* **87** (1976) 495.
72. R.W. Cattrall and H. Freiser, *Anal. Chem.* **43** (1971) 1905.
73. L. Keil, G.J. Moody and J.D.R. Thomas, *Analyst* **102** (1977) 274.
74. J.E.W. Davies, G.J. Moody, W.M. Price and J.D.R. Thomas, *Lab. Practice* **22** (1973) 20.
75. S.G. Cutler and P. Meares, *J.Electroanal. Chem.* **85** (1977) 145.
76. L. Ebdon, A.T. Ellis and G.C. Corfield, *Analyst* **104** (1979) 730.
77. P.C. Hobby, G.J. Moody and J.D.R. Thomas, *Analyst* **108** (1983) 581.
78. A. Craggs, G.J. Moody and J.D.R. Thomas, *J. Chem. Educ.* **51** (1974) 451.
79. R.W. Cattrall, P.J. Iles and I.C. Hamilton, *Anal. Chim. Acta* **169** (1985) 403.
80. T. Satschwill and D.J. Harrison, *J. Electroanal. Chem.* **202** (1986) 75.

4 Conducting polymers

W.J. FEAST

4.1 Introduction

Polymeric materials, both synthetic and natural, are widespread and consequently familiar. These familiar materials are rarely simple; indeed, they encompass an enormous structural variety and complexity. They may be organic, inorganic or organometallic; they may consist of an assembly of discrete linear or branched molecules with a distribution of molecular sizes; or, at the other extreme, they may be virtually infinite three-dimensional networks; to complicate the picture yet further, some materials are constructed from a mixture of several structural types. Another layer of complexity arises from the fact that such materials are frequently formulated with the aid of a range of additives: fillers, plasticizers, pigments, anti-oxidants, stabilizers, and so on. The detailed properties of a particular material depend in a complicated way on a set of interacting factors, including molecular structure; molecular size and size distribution; cross-link density and its distribution; the conformation and ordering of the various components of the material; the chemical and physical history of the sample under consideration; and often on the time scale of the observation. Polymeric materials are only rarely molecularly monodisperse, some of the important biopolymers providing the only really notable exceptions to this generalization (1).

The majority of polymeric materials do not conduct electricity to any appreciable extent; indeed, their frequently excellent insulating behaviour accounts for most of the applications of synthetic organic polymers in the electronics and electrical industries. Nevertheless there are a significant number of polymeric materials which do conduct electricity and these are the focus of our attention here. Other polymeric materials are already of importance in the sensors field and are discussed at various points throughout the book.

In practice, the term 'conducting polymers' includes a range of materials which display a wide variety of properties and consequently fit many diverse areas of established or potential application. The range of conductivity of interest is enclosed at one end of the scale by values associated with a typical good insulator, such as polyethylene, $c.\ 10^{-14}\ (\Omega\,cm)^{-1}$, and at the other end by those associated with a typical metallic conductor, such as copper, $c.\ 10^6$ $(\Omega\,cm)^{-1}$. Conductivity may be an intrinsic property of the material, for example poly(sulphurnitride) $(SN)_x$, the first 'non-metallic metal', has a room temperature conductivity of $c.\ 10^3\ (\Omega\,cm)^{-1}$; alternatively, it may result from adding a conductive filler to an insulating polymeric matrix. This additive approach has proved successful in producing practically useful materials, for example, conductive elastomers, fibres, plastics and paints are all produced by this technique. The added conductive particle is often a carbon black or chopped carbon fibre, but may also be a metal particle, fibre or flake; such materials find very widespread use in protection against build-up of static electricity and in shielding electronic devices against electromagnetic interference. The conductivities required are modest, typically between 10^{-7} and 1 $(\Omega\,cm)^{-1}$. The systems mentioned so far are all electronic conductors, but the addition of ionic species to an appropriate polymer can give ionically conducting materials; at present these are receiving a lot of attention as, *inter alia*, electrolytes for solid-state batteries.

A type of conducting polymer which has generated much research activity in the last fifteen to twenty years fits neither the intrinsic nor the additive conductor classification. These newer materials are generally insulators or poor conductors in their pristine state, but display marked increases in conductivity on exposure to oxidizing or reducing agents. Thus one may start with an insulating material and progressively oxidize or reduce it through the conductivity range of a semiconductor, *p*- or *n*-type respectively, into the metallic regime. The residue from the oxidizing or reducing reagent becomes part of the new material, which consequently has a structure and properties which are different from those of the starting material. The overall process, which is often referred to as doping, represents a large-scale modification of the starting material and should not be confused with the substitutive doping used in conventional silicon technology.

In this section our objective is to survey the kinds of conductive polymers which have been studied and to give an account of our present understanding of their properties. A large part of the subject of this section is in the process of rapid development at the time of writing and therefore our aim is to provide the reader with sufficient background to be able to read more detailed treatments and to evaluate new work as it appears. We shall give most attention to the newer kinds of material; and, since it is the paradigm for the field, we shall start by considering polyacetylene.

4.2 Polyacetylene

The electrical properties of an infinite polyene have intrigued theoretical chemists for more than fifty years. Shortly after Hückel introduced his π-electron theory for unsaturated systems in 1931, speculations began to appear about the carbon–carbon bond lengths to be expected in an infinite polyene. The question was interesting because, if the bonds were of equal length, π-electron theory predicted that the π-molecular orbitals would form a continuous band which would be half-filled, whereas alternating carbon–

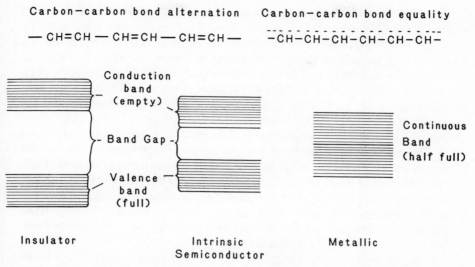

Figure 4.1 π-electronic band structure possibilities for pristine polyacetylene.

carbon bond lengths would lead to a π-electronic structure consisting of a filled band of bonding levels separated from an empty band by a gap. The first option is reminiscent of the bonding associated with metals, whereas the latter is like that of insulators or semiconductors, depending on the size of the gap; the idea is illustrated in Figure 4.1.

An experimental distinction between the various options seems trivially easy, but was delayed for a long time by the lack of a good synthesis of polyacetylene. Although the study of polyacetylene has, in recent years, proved very stimulating for the whole area of conducting polymers, it turns out that experimental work with this material is fraught with difficulties and pitfalls for the unwary. Early workers in the field discovered many of the difficulties and, although some progress was made, they were essentially unable to overcome them. Thus, although polyacetylene can be obtained via the obvious route, i.e. direct addition polymerization of acetylene, the polymer is insoluble, infusible and very sensitive to air oxidation. It is therefore difficult to handle and study, and most of the early reports refer to intractable powders. The breakthrough in the study of polyacetylene came when Shirakawa and co-workers discovered that exposure of the surface of a concentrated solution of a Ziegler–Natta catalyst to acetylene gas resulted in the formation of a self-supporting film of the polymer. This film form of the material, although it was not processable by any of the conventional solution or melt techniques of polymer science, could be washed and dried, and was suitable in the 'as-made' form or detailed study which revealed a most interesting set of properties.

To the naked eye the material produced by the Shirakawa technique appears to be a film, but electron microscopy reveals a fairly complex structure. The material consists of a mat of randomly oriented interconnected fibrils, the fibril diameter varying between 5 nm and 40 nm, being typically about 20 nm. Many of the structural details of these films can be varied as a function of the polymerization conditions, including the film thickness, the density, the distribution of fibril diameters, and the relative proportion of *cis* and *trans* vinylene units. The fibrillar regions of the material have crystalline order which can be studied by conventional scattering techniques, but the structure at the interfibrillar junctions is not well defined. It seems likely that virtually all samples of polyacetylene prepared by this route contain traces of catalyst residues, the concentrations and nature of such residues varying and depending on the details of synthesis, recovery and cleaning of the film. Trace contaminants of this kind may have an effect on the fine details of behaviour of the material but will not affect the general properties. A very great number of papers reporting the synthesis of polyacetylene have been published, but a particularly clear and detailed description of sound experimental procedures has been provided by Gibson and Pochan (2).

When acetylene is polymerized by the Shirakawa technique at low temperature, $c. - 70°C$, the product film has a yellow-gold appearance, it is an insulator with a conductivity of $c. 10^{-14}(\Omega \text{cm})^{-1}$, and it has a low free spin density. Diffraction and spectroscopic studies establish that there is regular single–double bond alternation and that the material is predominantly the *cis–cisoid* homopolymer. However, this material is stable only at low temperature, and when the polymerization is conducted at room temperature a different material with a silvery appearance and a predominantly *trans–transoid* structure is obtained; essentially the same material is produced when the *cis* polymer is allowed to warm to room temperature. This form of polyacetylene is a semiconductor with conductivity in the 10^{-8} to $10^{-5}(\Omega \text{cm})^{-1}$ range, and it has a spin density of about 10^{18} spins per g corresponding to about 1 per 3000 carbons. These structures are shown in Figure 4.2.

The proper description of the radical species shown as structure (*b*) in Figure 4.2, and the structure of the various electrically conducting forms of polyacetylene, has generated a lot of theoretical and experimental work and a great deal of discussion (3). From the viewpoint of basic organic chemistry we

a)

b)
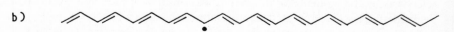

Figure 4.2 (*a*) *Cis-cisoid*- and (*b*) *trans-transoid* polyacetylene.

would expect the electron to be delocalized in a non-bonding molecular orbital, having coefficients only at alternate carbons. The resulting delocalized radical would have equal carbon–carbon bond lengths at the centre gradually changing to alternating single and double bonds in the adjacent pure *trans*-polyacetylene sequences, and the delocalization is believed to extend over about 13 to 15 carbons. The orbital defining the unpaired electron would also be expected to be located in the middle of the gap between the π-bonding and anti-bonding levels.

This picture describes a simplified version of the generally accepted view of the structure. The radical species is usually referred to as a soliton, and is predicted to have mobility along the chain direction. When such a material is oxidized, the electrons most easily removed will be those at mid-gap associated with the solitons, the next most easily removed being those at the top of the π-bonding level. Removal of the soliton electron creates a delocalized cation, or positive soliton, whereas removal of an electron from a pristine section of *trans*-polyacetylene creates a delocalized radical-cation, or positive polaron. If two electrons are removed from one section of pristine chain there may be a combination of spins and the resulting delocalized dication is known as a positive bipolaron; addition of electrons gives rise to the analogous negative species.

The presence of these species in a particular conducting polymer would be expected to be amenable to experimental test via magnetic, optical and vibrational spectroscopy; the experiments are often more complicated than might be expected but all the ground-state species listed above, together with some derived excited states, have been unambiguously identified in one or other of the conducting polymers which have been investigated. The overall picture and terminology are illustrated in Figure 4.3, using polyacetylene as the example; solitonic species are possible only in systems having the symmetry of polyacetylene, whereas polarons can be found in all types of conducting conjugated organic polymer—see (3), (4), and below.

These species do not, of course, exist on an isolated polymer chain in a vacuum but in close juxtaposition to each other and to the residues of the oxidizing or reducing agents used in their formation. Conductivity is a bulk phenomenon, and therefore if these kinds of species are involved in the process, they must have mobility between polymer chains as well as along them. The morphology of Shirakawa polyacetylene means that it has a high surface area, typically about $60 \, \text{m}^2\text{g}^{-1}$ for samples with a density of $0.4 \, \text{g cm}^{-3}$, which is advantageous for reversible oxidation and reduction, since the rate-limiting factor is the migration of the counter-ions into and out of the polymer. These processes are best accomplished electrochemically, and there was at one time hope that polyacetylene might be a suitable electrode material for lightweight high-energy density batteries. Although the feasibility of the concept has been demonstrated, this early optimism has not yet been justified and many workers in the field now feel it is unlikely

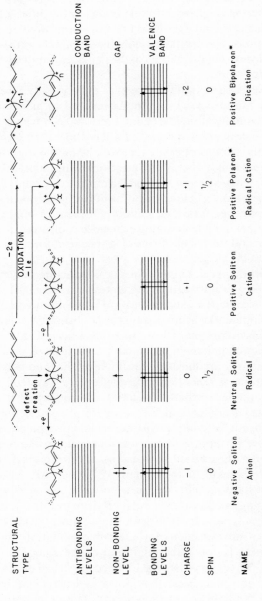

Figure 4.3 Band structure and electronic configuration for species implicated in conduction mechanisms in conjugated polymers, illustrated for polyacetylene. *Negative analogues via electron addition.

to be (5). It seems unlikely that polyacetylene will prove sufficiently stable over the periods required for this kind of application, even if all other technical problems can be solved.

One of the attractions of polymeric materials in general is the ease with which they may be fabricated into desired forms (fibres, films and mouldings). Unfortunately neither polyacetylene, nor many of the other conjugated polymers of interest in this field, offer this advantage, since they are insoluble and infusible and consequently very difficult to process. This general intractability of conjugated polymers also handicaps attempts to regulate their morphology, and since morphology often has an important effect on properties, this problem tends to delay detailed evaluation of materials and so inhibit the possible development of applications. Several attempts to regulate morphology have been described, including orientation via mechanical stretching, and polymerization under various ordering conditions, for example, in a shear field, at an ordered surface, using a liquid crystalline solvent, or in high magnetic fields, these various techniques being used either alone or in combination. Although some measure of success was obtained, no really satisfactory convenient processing for polyacetylene, and other conjugated polymers such as poly(paraphenylene) and poly(phenylene vinylene), has been developed.

The problem can, however, be circumvented by carrying out the processing and morphology regulation on a more amenable precursor material which is converted to the required conjugated polymer only at the last stage of the process. The development of an effective precursor route to polyacetylene has been recently reviewed and the advantages of the approach discussed in some detail (6). Effective routes to poly(paraphenylene) (7) and poly(paraphenylene vinylene) (8) have also been developed. This relatively new approach allows pure, highly anisotropic, solid continuous films of these polymers to be prepared, and the implications of these developments in synthesis for the physics and materials science of the polymers are only beginning to be explored. The idea is illustrated for the polyacetylene case in Figure 4.4. The tricyclic triene (1) is polymerized by ring opening metathesis at the cyclobutene double bond to give the precursor polymer (2), which is soluble and can be solvent cast to give orientable fibres or films which are thermally converted to polyacetylene as and when required.

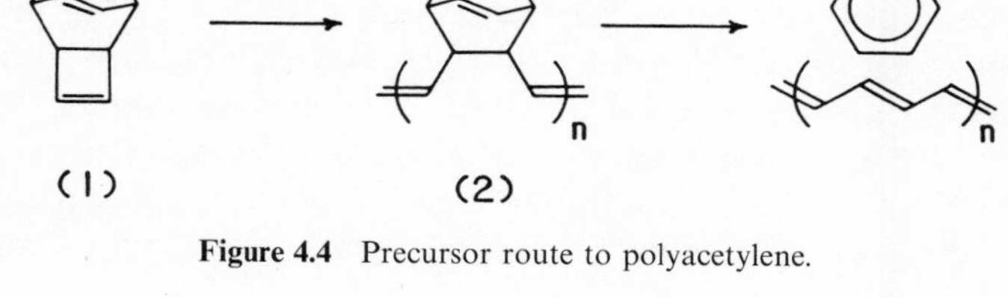

Figure 4.4 Precursor route to polyacetylene.

Another way of improving the processing properties of intractable polymers is to introduce substituents; this generally increases the disorder in the system and consequently lowers the glass transition temperature and melting point, and increases the solubility of the material. Several variants of this approach have been reported, including the synthesis of homopolymers of substituted alkynes, and the preparation of random, block, and graft copolymers. Blends of conducting polymers with conventional processable polymers have also been studied. These various approaches, while they have yielded some interesting results, all suffer from the same problem in that the further one moves away from the structure of the original conjugated conducting polymer, the bigger the gains in processability and losses of interesting electrical properties. All these approaches are discussed in reference (3), and in the remainder of this section some of the main trends will be outlined.

4.3 Other hydrocarbon polymers

The intensive investigation of polyacetylene's electrical, magnetic, and optical properties generated considerable enthusiasm for the field of electroactive polymers and, as the limitations of polyacetylene became clear, many research groups sought alternative materials which would combine interesting electrical properties with more acceptable stability and processing characteristics.

One might reasonably expect a conjugated aromatic polymer to have greater stability than a linear polyene; indeed, this supposition was the basis of many programmes of research generally pre-dating interest in conducting polymers and directed towards finding thermally and oxidatively stable polymeric materials. Expectation was justified in the event, but the improved thermal and oxidative stability was generally accompanied by less attractive electrical properties, and processing remained a major problem. Poly(paraphenylene) may be regarded as the parent of this family of materials and serves as a suitable example. A variety of attempts to synthesize well-defined samples of this polymer have been reported during the last fifty years; most have involved step-growth routes and have foundered as a consequence of the inherent insolubility and intractability of the product. Thus, direct oxidation of benzene, and many metal-catalysed couplings of aryldihalides, lead to the production of insoluble and oxidatively stable powders which are difficult to characterize with respect to molecular structure and size and are about as attractive as brick dust from the polymer processing point of view. Some ingenious approaches have been explored in the process of attempting to get round these difficulties, and two examples are given below.

Workers in the Allied Corporation have developed an approach in which the synthesis of the polymer and its oxidation into the conducting regime are carried out in one step and in solution. Thus, for example, p-terphenyl can be dissolved in AsF_3 containing AsF_5 which both oxidatively couples the

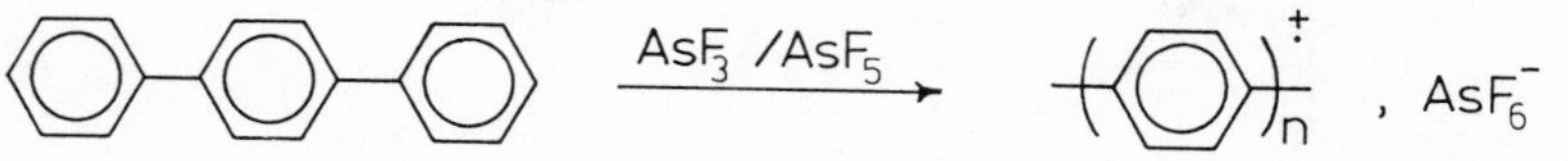

Figure 4.5 Simultaneous polymerization and oxidation of p-terphenyl.

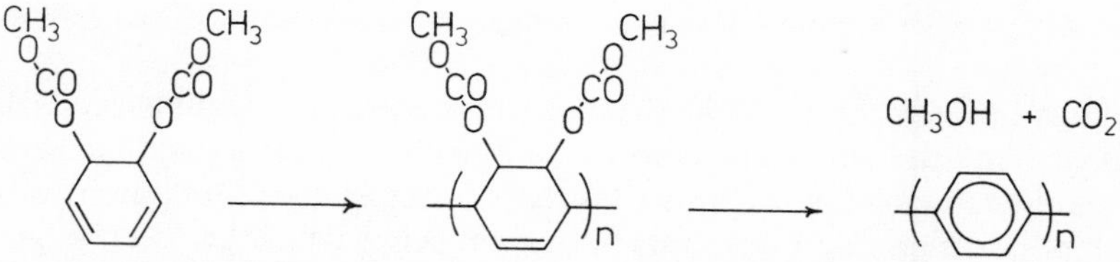

Figure 4.6 Synthesis of poly(paraphenylene) via a processable precursor polymer.

monomer and oxidizes the polymer to give an ionic solution in which the positive charges on the polymer are balanced by AsF_6^- anions (see Figure 4.5). The success of this method is dependent on the delocalization of the spin and charge on the polymer and solvation of the counter-ions by AsF_3. When such solutions are evaporated slowly, freestanding films of conducting poly(paraphenylene) are obtained which cannot be redissolved, but which have higher conductivity and better mechanical properties than analogous materials produced by direct solid phase oxidation of the pristine neutral polymer with AsF_5. The technique is applicable to a variety of other monomers, including heteroaromatics (9).

Perhaps the most elegant route to poly(paraphenylene) has been developed by workers at ICI, and involves a microbiological synthesis of a monomer which is used to prepare a processable precursor polymer. The starting material for this route is benzene, which is microbiologically converted to cis-5, 6-dihydroxycyclohexa-1, 3-diene; this diol is converted to the dimethyl carbonate shown as the monomer in Figure 4.6. Radical initiated polymerization gives a soluble and processable polymer which can be converted to poly(paraphenylene) as required (7).

The conductivity of pristine poly(paraphenylene) is about $10^{-12}(\Omega\,cm)^{-1}$, which makes it a respectable insulator; it has a bandgap of about 3.5 eV, which is considerably higher than that of polyacetylene, c. 1.5 eV, and accounts for its oxidative stability. Nevertheless, this stable fully aromatic insulating polymer may be converted to the metallic conductivity regime, $10^3\,(\Omega\,cm)^{-1}$ or greater, with strong oxidants such as AsF_5. The material can also be electrochemically oxidized and has been investigated as a potential battery electrode material. Although the possibility of higher operating potential, current density and stability exist, these apparent advantages over polyacetylene for example were outweighed by the lack of organic electrolytes which were stable under the required operating conditions. Both quinonoid and benzenoid forms of the rings can be drawn, but the quinonoid form is disfavoured by ortho–ortho interactions, and the planes of adjacent rings are inclined at about 20°.

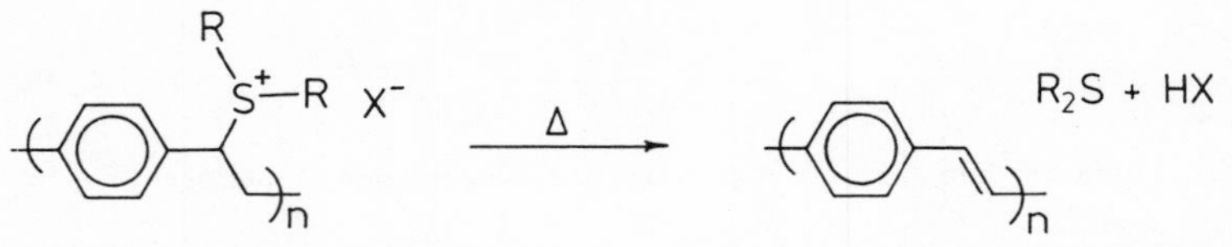

CHEMICAL SENSORS

Figure 4.7 Synthesis of poly(paraphenylene vinylene) via a processable polymeric electrolyte.

A fairly obvious variant of the structures discussed so far would be to include both vinylene and phenylene units in the polymer backbone, and a considerable number of such structures have been investigated. The parent of this group of polymers is poly(paraphenylene vinylene), which has been prepared by a number of different routes. Like polyacetylene and poly(paraphenylene), this polymer is insoluble and difficult to process; again the most promising synthesis appears to be that proceeding via a processable precursor (see Figure 4.7). In this case the precursor polymer is a polyelectrolyte which can be processed via standard solution techniques and converted to the required product as and when desired.

Many variations on the themes outlined above may be imagined, and many conjugated unsaturated polymers have been produced and studied. Those with interesting electrical properties generally turn out to be insoluble and infusible, but this problem can be overcome via careful regulation of the synthesis and doping, or by obtaining the required polymer via a processable precursor. The precursor route has a number of advantages including the opportunities presented for purification and morphological regulation, both of which can be expected to be important in respect of reproducibility and control of properties.

4.4 Pyrolysis routes to conducting polymers

It is common experience that high-temperature pyrolysis of organic compounds leaves a residual char or coke consisting predominantly of carbon. Controlled oxidative pyrolysis of organic materials has become an important technology and, despite the apparent crudeness of the technique, merits a mention here since it represents perhaps the most important practical source of conducting organics. It is, of course, the route by which carbon blacks are made, the materials which are the current-carrying components of the majority of conductive plastics. The technique also has an important place in the history of conducting polymers since, at the end of the last century, graphitization of natural organic fibres was the route by which both Swann and Edison produced the conducting incandescent carbon filaments for their electric light bulbs. Organic fibres of various kinds, predominantly poly-acrylonitrile, are still pyrolysed to produce carbon fibres which find use *inter alia* as a filler for conductive composites (10). This is a very active area of technological development and it has been argued that, when normalized

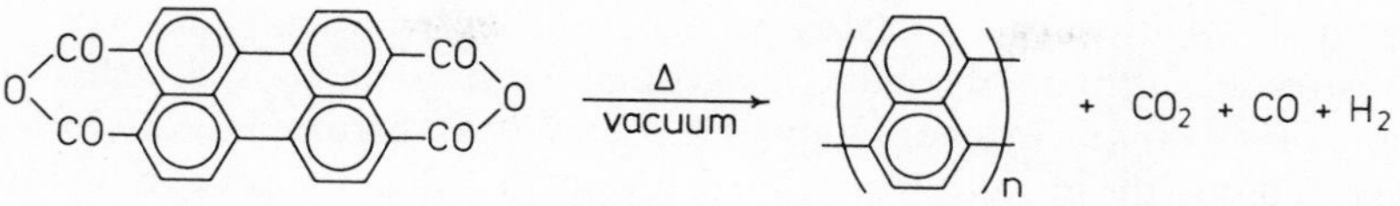

Figure 4.8 Attempted pyrolytic route to poly(perinaphthalene).

with respect to weight, some of the modified carbon fibres offer considerable advantages over conventional metallic conductors (11).

There have been many attempts to use pyrolysis in a controlled manner to produce molecularly well-defined materials; an example is shown in Figure 4.8. Pyrolysis of the dianhydride monomer shown in Figure 4.8 under conditions of high temperature and high vacuum leads to the deposition of thin, lustrous, and chemically inert films which display an undoped conductivity of 250 $(\Omega\,cm)^{-1}$. The expected structure was the ribbon polymer, poly(peri-napthalene), shown in Figure 4.8, but since hydrogen as well as carbon monoxide and carbon dioxide were eliminated this cannot be the real structure (12).

4.5 Conducting polymers having hetero-atoms in their backbone and/or substituents

The first, and still one of the very few intrinsically conducting, polymers is poly(sulphur nitride), $(SN)_x$, which, while it is undoubtedly a covalent polymer, behaves as a metal from the viewpoint of electrical conductivity and exhibits superconductivity below 0.3 K. Several synthetic methods are recorded for the preparation of this material, including polymerization of S_2N_2 in the solid state, decomposition of S_4N_4 while irradiating the decomposition products, and solution-phase preparations involving azides. Extensive development of this material has been inhibited by the knowledge that it detonates on percussion, as does one of the key intermediates in its synthesis, S_4N_4. Recent work by Banister and Hauptman has established less hazardous electrochemical routes to $(SN)_x$; for example, the electrolysis of cyclopentathiazenium chloride in liquid SO_2, or, more conveniently, the tetrafluoroborate in acetonitrile (see Figure 4.9).

This curious material has a high electrical conductivity and a high work function for electrons, and can consequently be coated on to electrodes in a reproducible way. It may therefore find use as an electrode for voltammetry or amperometry, and other thin film device applications such as light-emitting diodes (13).

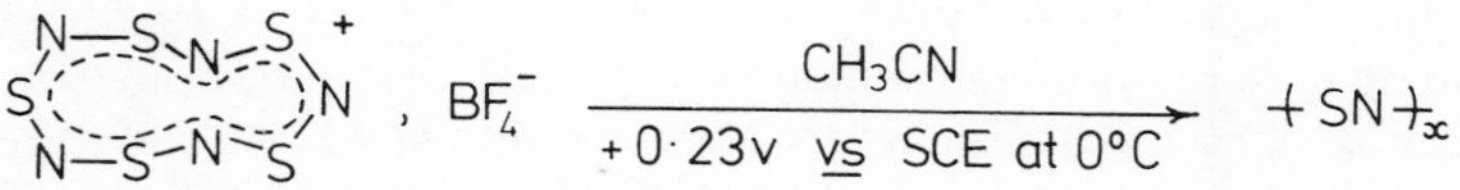

Figure 4.9 Electrochemical route to poly(sulphur nitride).

Several other polymers having hetero-atoms in the main chain have been investigated as potential conducting systems, perhaps the most detailed study being of poly (*p*-phenylene sulphide), a commercial thermoplastic, which can be doped into the metallic regime on exposure to AsF_5. The process of doping appears to involve irreversible chemical change, probably resulting in the formation of dibenzthiophene units in the main chain and some cross-linking.

Substituted conjugated polyenes are generally soluble and consequently are relatively easily manipulated and characterized. By and large they are not interesting from the point of view of their electrical properties, although those with aryl substituents appear to have semiconductor properties and interesting spectroscopic behaviour. As the bulk and the concentration of the substituent increases, the planarity of the polyene decreases. For example, poly(tert-butylacetylene), a genuine high polymer, is soluble and, since successive double bonds cannot assume a planar configuration with respect to each other, there is no conjugation and the material is white. One substituent which has attracted some theoretical speculation but relatively little experimental work is fluorine. Theoreticians have suggested that the fluorine atom is not big enough to cause severe steric perturbations, but may be expected to have beneficial effects on both the oxidative stability and the electrical properties of polyenes; there is some tentative evidence in patents and in the open literature which supports this hypothesis.

Another heterochain polymer which has received a lot of attention recently is polyaniline. This material is produced by chemical or electrochemical oxidation of aniline, and it seems likely that it is very similar to, if not always identical with, 'aniline black', an ill-defined material of some antiquity. The propensity of hetero-substituted aromatics to undergo coupling via intermediate radical cations is well established and, since a monosubstituted benzene is effectively a trifunctional monomer in this context, it seems likely that this product is a network or highly branched material, which would account for its unattractive handling properties.

4.6 Conducting polymers containing heterocyclic units

Electrically conductive films can be formed on electrode surfaces by electrochemical oxidation of solutions of heterocyclic compounds in electrolytes; similar materials can sometimes be obtained by the use of conventional chemical oxidizing agents. Most of the work carried out on such materials appears to have been concerned with their properties rather than their structure; consequently, there is often some uncertainty concerning exactly what these materials really are, a situation which leaves room for both imaginative speculation and argument.

The most likely mechanism for the electrochemical polymerizations seems to be an initial formation of radical cations which then undergo dimerization with the loss of two protons. This initial dimerization will occur at or near the

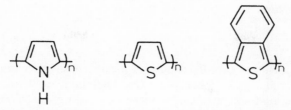

Figure 4.10 Poly(pyrrole), poly(thiophene), poly(isothionaphthene).

electrode and can presumably be repeated to build the polymer via a step-growth process. The resultant polymeric film is also electrochemically oxidized during the process, and the final product is a 'doped' electrically conductive material. The compounds which can be polymerized by this technique are required to undergo oxidation at relatively low anodic potentials, otherwise the solvent would degrade preferentially. Pyrrole, thiophene, carbazole, indole and so on are suitable candidates; some polycyclic aromatic hydrocarbons also meet this requirement, for example, azulene and pyrene. In the early stages of the process it is established that coupling takes place at the 2, 5-positions, and these materials are generally represented as shown in Figure 4.10.

The products may not be as structurally homogeneous as is implied by Figure 4.10, and the fine details of structure of the final products remains an area of dispute. Much of the earlier work in this field has been reviewed (3). The poly(isothionapthene) shown in Figure 4.10 was described relatively recently by Wudl and is perhaps the most interesting member of the group, since it displays the lowest bandgap so far recorded for a conducting polymer, introduction of the phenylene ring lowering the bandgap from 2.0 eV in polythiophene to 1.0 eV in poly(isothionapthene).

As is the case with other conjugated conductive polymers, most of these polymeric heterocyclics are insoluble and intractable; however, they are generally far more stable than polyacetylene and consequently considerable efforts have been made to overcome these processing penalties. Introducing pendant solubilizing groups has met with some success, and a series of 3-alkylthiophenes have been prepared and polymerized via oxidative coupling at the 2, 5-positions to give linear polymers which are stable and soluble in both their doped and undoped forms (14). Electrochemical polymerization of a variety of other substituted heterocyclic monomers has allowed the production of conductive films containing electroactive pendant groups; see, for example, the work of Parker and Munro. Developments of this kind may be significant for the sensors field (14).

4.7 Filled polymers

4.7.1 *Electronic conductors*

Carbon- or metal-particle-filled plastics are the basis of a number of well-established technologies; they are mentioned here only for the sake of completeness, and the reader who wishes to explore this subject further is referred to the literature (10, 15).

4.7.2 *Ionic conductors*

It is now well established that solvent-free films can be cast from solutions of polyethers (such as poly(ethylene oxide)) and alkali metal salts, and that these films can display high ionic conductivity. Most of the effort devoted to this field has been based on the potential of such materials as solid-state electrolytes for battery applications. In this context, from viewpoints of both ionic mobility and weight, lithium salts in PEO have attracted the most intensive research and appear to offer the most promise; such materials are discussed elsewhere. The preparation of materials displaying both electronic and ionic conductivity raises interesting possibilities both in the field of batteries and sensors and is beginning to attract attention (16).

It is to be hoped that this brief summary of the active and constantly changing field of research generally labelled 'conducting polymers' will be of use in introducing readers of this specialist text to a possibly relevant topic, and that it will allow access to the literature and an easier understanding for those who delve further.

References
1. There are many good textbooks on polymer science; a useful general introduction is provided in *Polymers: Chemistry and Physics of Modern Materials*, J.M.G. Cowie, Intertext Books [Blackie Publishing Group] (1973).
2. H.W. Gibson and J.M. Pochan, in *Encyclopedia of Polymer Science and Engineering*, vol. 1, 2nd edn., John Wiley, New York (1984) 87; H. W. Gibson, in *Quasi-One-Dimensional Organics*, E.M. Conwell (ed.), Academic Press, New York.
3. A selection of articles on this and various other aspects of conducting polymers is to be found in *Handbook of Conducting Polymers*, T.A. Skotheim (ed.), Marel Dekker, New York (1986). An alternative view, probably more likely to appeal to chemists, has been presented by Wegner; see, for example, G. Wegner, in *Contemporary Topics in Polymer Science*, E.J. Vandenberg (ed.), **5** (1984) 281; and *Angew. Makromol. Chem.* **145/146** (1986) 181, and references therein.
4. J.L. Bredas and G.B. Street, *Acc. Chem. Res.* **18** (1985) 309.
5. H. Munstedt, in *Electronic Properties of Polymers and Related Compounds*, H. Kuzmany, M. Mehring, and S. Roth (eds.), Springer Series in Solid State Sciences **63** (1985) 8.
6. D.C. Bott, C.S. Brown, C.K. Chai, N.S. Walker, W.J. Feast, P.J.S. Foot, P.D. Calvert, N.C. Billingham and R.H. Friend, *Synthet. Met.* **14** (1986) 245.
7. D.G.H. Ballard, A. Courtis, I.M. Shirley and S.C. Taylor, *J.C.S. Chem. Commun.* (1983) 954; and European Patent Application, 0 076 605; 13/04/1983 to ICI plc.
8. M. Kanabe and M. Okawara, *J. Polym. Sci. A1* **6** (1968) 1058; and F.E. Karasz, R.W. Lenz *et al.*, *Polymer Bull.* **12** (1984) 293 and **15** (1986) 181.

9. L.W. Shacklette, H. Eckhardt, R.R. Chance, G.G. Miller, D.M. Ivory and R.H. Baughman, *J. Chem. Phys.* **73** (1980) 4098; and J.E. Frommer, *Acc. Chem. Res.* **19** (1986) 2.
10. Interesting and useful discussions of these topics will be found in sources such as *The Kirk–Othmer Encyclopedia of Chemical Technology*, 3rd edn., John Wiley, New York (1982).
11. J.S. Murray, D.D. Dominguez, J.A. Moran, W.D. Lee and R. Eaton, *Synthet. Met.* **9** (1984) 397.
12. M.L. Kaplan, P.H. Schmidt, C-H. Chen and W.M. Walsh, *Appl. Phys. Lett.* **36** (1980) 867.
13. A.J. Banister, Z.V. Hauptman, A.G. Kendrick, U.K. Patent, GB 2 147 889 B, published 10/12/1986; A.E. Thomas, J. Woods and Z.V. Hauptman, *J. Phys. D: Appl. Phys.* **16** (1983) 1123; A.J. Banister, Z.V. Hauptman, A.G. Kendrick and W.H. Small, *J. Chem. Soc., Dalton Trans* (1987) in press.
14. R.L. Elsenbaumer *et al.*, *A.C.S. Polymeric Mater. Sci. and Eng.* **53** (1985) 79; J.G. Eaves, H.S. Munro and D. Parker, *J. Chem. Soc. Chem. Commun.* (1985) 684; *idem, Synthet. Met.* **16** (1986) 123.
15. This topic has generated a considerable volume of technical and patent literature. Recent discussions of the scientific basis of these systems will be found, for example, in E.K. Sichel *et al.*, *Phys. Rev. B.* **18** (1978) 5712 and T.A. Ezquerra *et al.*, *J. Mater. Sci. Lett.* **5** (1986) 1065 and references therein.
16. See, for example M. Armand *et al.*, *J. Chem. Soc. Chem. Commun.*, (1986) 1636.

5 Chemically modified electrodes

G.G. WALLACE

5.1 Introduction

The electrochemical reduction of silver ions and the understanding of limiting currents by Solomen (1) was perhaps the first indication that voltammetry could be analytically useful. However, it was not until the introduction of a practical, reproducible working electrode, the dropping mercury electrode (DME), that the analytical utility of the technique was realised. The DME alleviated problems associated with surface phenomena which complicated the electrochemistry of solid electrodes previously employed. These same surface phenomena are now much more clearly understood and the chemistry associated with them may be used in appropriate analyses to enhance analytical responses.

Such is the area of chemically modified electrodes (CMEs). What may have been considered a contaminated or fouled electrode 20 years ago may now be considered a CME. The difference now is that the modification is purposefully carried out and carefully controlled to produce a desired result.

Several excellent reviews on CMEs already exist (2–4). The purpose of this chapter is to highlight the considerations necessary when these devices are to be used as chemical sensors and to illustrate some useful examples. The design, preparation, characterization and application of CMEs as sensors will be discussed. Challenges which need to be overcome to expand the range of applications will be highlighted.

Even in the early days of CMEs the ability to function as chemical sensors was realised. For example, Cheek and Nelson (5) determined silver ions using an electrode modified to contain ethylenediamine. Using this electrode, silver could be determined at the 10^{-11} M level. This example illustrates a property of CMEs which makes them attractive for use as chemical sensors. Many voltammetric methods require derivatization of the analyte or matrix modification of some kind for optimum performance. Since voltammetry is a surface technique, it is obvious that there are benfits to be obtained from derivatizing/modifying where it matters most—*at the electrode surface*. Using CMEs this is easily achieved, since the derivatizing agents or matrix modifiers can be attached to the electrode surface. This should result in the development of sensors capable of performing with minimum external sample pretreatment.

Following breakthroughs in electrochemical theory in the 1960s and in instrumentation in the 1970s, the future of electroanalytical chemistry now surely lies in the development of new useful working electrodes. Increasing the range of sensors available can only expand the capabilities of voltammetry. Chemical sensing using CMEs can be improved by occluding interferents from the electrode surface, using the principles of electrocatalysis to enhance signals or designing electrodes capable of preconcentrating species in a chemical environment which will enhance sensitivity. Ideally, chemical sensors which perform optimally with *no* sample preparation are required. In designing a CME for chemical sensing, this requirement should be kept in mind.

The ideal CME for chemical sensing should have the following properties.

(i) *Easily prepared*: Raw materials should be inexpensive and the preparation procedure should be rapid.

(ii) *Mechanically and chemically stable*: it should be able to withstand reasonable laboratory treatment and storage procedures. Mechanical stability becomes more important in flow-through or on-site applications. Chemical stability is also important; CMEs should be as inert as possible to matrix effects. The chemical form of the CME should be independent of the matrix and the CME should not be liable to chemical oxidation or reduction.

(iii) *Electrochemically inert*: the electrochemical potential range available should be suitable for the analytical problem at hand.

(iv) *Ability to discriminate against matrix components*: the components of the matrix should not produce analytical signals nor interfere with the signal(s) of interest.

(v) *Ability to resolve analyte signals*: most analytical problems call for the analysis of a small number of major components. Given the resolution capabilities of modern voltammetric methods then it should be feasible to design CMEs capable of simultaneous analysis of 5 or 6 species.

(vi) *High sensitivity*: this can be achieved via either preconcentration or electrocatalysis. Since analysis time is of extreme importance in considering a new analytical procedure then sensitivity (response per unit concentration) per unit time is perhaps a more relevant parameter.

(vii) *High capacity*: important to enable analysis of high concentration solutions.

(viii) *Reusable*: a finite number of analyses should be attainable before cleaning or regeneration of the CME is required.

(ix) *Cleaning/regeneration*: this should be easily and reproducibly achieved.

(x) *Electrochemical processes*: should be well understood. In general the electrode processes at modified electrodes involve more than just electron transfer at a well-defined boundary. To understand the limitations of a modified electrode and to maximize the benefits, the mechanism of performance should be clear. The electrode process should be induced at low anodic potentials to ensure optimum performance by keeping residual current values to a minimum. All

CMEs will have limitations in one or more of the above areas, and it is therefore important that a range of electrodes be developed to fill the gaps in voltammetric analysis.

5.2 Designing a CME

When designing a CME for a particular purpose, both the chemistry and electron transfer process must be optimized. CMEs are generally employed to induce a desired chemical or electrochemical effect during analysis.

5.2.1 *Basis of operation*

Usually, CMEs which have been designed to act as chemical sensors employ chemical procedures used previously in solution analyses. CMEs which involve (i) ionic interaction (positive and negative); (ii) complexation; (iii) electrocatalysis including bioelectrocatalysis; or (iv) size exclusion have been previously employed.

The first two types may achieve extra sensitivity and selectivity by electrostatic attraction or complexation of the analyte on to the electrode surface, achieving a preconcentration effect similar to stripping voltammetry (Figure 5.1) at a mercury-based electrode. Using charged modifiers to preconcentrate on the basis of electrostatic attraction will discriminate against species of similar charge to the modifier. Employing complexation to preconcentrate has additional benefits. The analyte can be trapped in a chemical environment which can be used to influence the nature of the electrode process, and also a selective complexing agent can be used to provide a selective sensor. The catalytic processes achieve extra sensitivity and selectivity by enhancing the electron transfer process for a particular analyte.

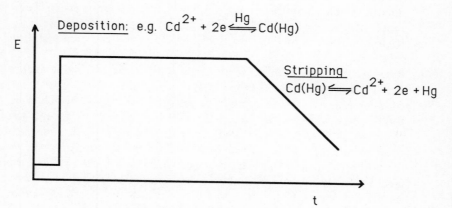

Figure 5.1 Potential-time profile for stripping voltammetry.

5.2.2 *Preparation of the modified electrode*

Having decided on what chemical or physical principles the sensor is to operate, a means of electrode preparation must be devised. This section can be divided into several areas: choice of substrate, substrate derivatization, and electrode characterization.

Choice of substrate. If the coating (modifier) to be placed on the electrode is totally impermeable, then the choice of substrate will be governed mainly by the requirements of the preparative stage. If the coating is permeable, then the electrochemical behaviour of the substrate will be a more important consideration. The ideal substrate should be:

  (i) Stable (mechanically, chemically and electrochemically)
 (ii) Chemically inert to species other than the modifier
(iii) Easy to manipulate and machine
(iv) Amenable to derivatization or the coating procedure to be employed
  (v) Inexpensive
(vi) Transparent (in some instances transparent electrodes can be most useful).

The surfaces of most solid electrodes are difficult to maintain in an oxygen-free state. Mercury drop or mercury film electrodes may be employed, but these have mechanical limitations as substrates for CMEs as well as a limited anodic potential range. Other solid electrodes, such as gold, platinum or carbon (particularly basal plane pyrolitic graphite) can be rendered oxygen-free by abrasion in an oxygen-free environment or by treatment in a reductive rf plasma (6). CME synthesis on such substrates is difficult, since all operations must be carried out in an oxygen-free environment such as a nitrogen glove box. The oxygen content of substrate surfaces can be reduced using various other methods; for example, heat treatment under vacuum (7) or irradiation have been employed to clean electrode surfaces (8).

Gold, platinum and carbon (particularly edge plane pyrolitic graphite and glassy carbon) electrodes are known to have oxygen-containing groups on the surface. The oxygen content of these surfaces can be increased using chemical oxidation with suitable oxidizing agents. However, if oxidizing conditions are too strong, for example, if concentrated H_2SO_4 or $K_2Cr_2O_7$ is used, non-conducting graphite oxide layers can be formed on the surface of some carbon electrode materials (9). Electrochemical pretreatment (10), simple mechanical polishing in an oxygen environment (11) or treatment in an oxidative rf plasma (12) have been shown to result in increased oxygen content on electrode surfaces. Of the methods available for controlling oxygen content on the electrode surface, electrochemical methods are probably the most convenient to apply. The electrolysis time along with the applied potential can be used to control the extent of oxygenation. Other metal oxides have been employed as electrode substrates. Electrodes such as RuO_2, PbO_2, SnO_2, In_2O_3 and MnO_2 have been employed in the area of electrocatalysis and energy

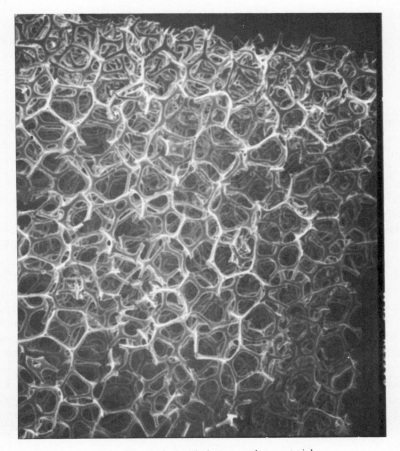

Figure 5.2 Reticulated vitreous carbon material.

conversion (13). Preparation of these substrates is relatively straightforward and is outlined in reference (13). RuO_2 and SnO_2 are the most popular of the metal oxides encountered in the field of chemically modified electrodes.

Along with the more conventional substrates discussed above, more unusual electrode materials have been investigated in recent years. Carbon-based materials such as reticulated vitreous carbon (RVC) (14), carbon cloth (15), and carbon fibres (16) have been employed. RVC is somewhat like glassy carbon in appearance; however, it has an open porous structure as depicted in Figure 5.2. It has similar electrochemical characteristics to glassy carbon, but the open structure as well as the large surface area makes it ideal for flow-through applications. RVC has been chemically modified by various workers (17, 18). Carbon cloth has a texture similar to graphite but no details are available on the surface composition. The porous nature plus the high surface area makes it ideal for flow-through applications.

Carbon fibres also have similar electrochemical properties to glassy carbon, but once again the physical form differs greatly. They are fibres with a small diameter (typically $< 10\,\mu m$). These small electrode substrates have some advantages over more conventionally-sized electrodes. The small currents drawn result in very little IR drop, and so often much simpler instrumentation can be employed. Furthermore they can be employed in highly resistive media or at fast scan rates, and their small physical size makes them ideal for *in-vitro* or *in-vivo* applications. Other more unusual substrates include amalgam electrodes (19), and plastic materials impregnated with SnO_2, In_2O_3 or gold to make them conductive (20).

Substrate derivatization. This is undoubtedly the aspect of CMEs which has received most attention in the last ten years, and justifiably so. Without the development of convenient, reproducible preparation methods the study of CMEs will remain somewhat academic. Fortunately, it appears this is not to be the case, and feasible, practical electrode preparation methods appear regularly in the literature. As pointed out previously, the electrode should be easily and inexpensively prepared. It should also exhibit the desired chemical and physical properties, have an appropriate capacity and an adequate electron transfer rate constant for the analyte electrode process.

It is important at this stage to define first-order and second-order derivatization. First-order CMEs are prepared by adding the chemically active group directly to an electrode substrate. There is a limited number of instances where this approach is appropriate. Second-order derivatization, which involves adding the chemically active group to a previously modified substrate, greatly expands the number of different CMEs which can be prepared:

Bare substrate	First order	Second order
	CM_1	$CM_1 - CM_2$
	CM_1	$CM_1 - CM_2$
	CM_1	$CM_1 - CM_2$
	CM_1	$CM_1 - CM_2$

At present not many modified electrodes are derivatized beyond second order, although it is obvious that to do so would again expand the range of electrodes which could be prepared.

Most CME preparation methods fall into one of four major categories:

(i) Composite electrode preparation
(ii) Chemisorption
(iii) Covalent bonding
(iv) *In-situ* polymerization.

All involve addition of a modifier, or chemically active constituent to what would normally be a chemically inactive substrate.

CMEs which have the modifier physically incorporated (mixed in) with the electrode material fall into category (i). The most common example involves carbon paste electrodes which have a modifier, usually less than 5% (w/w), added to the graphite/Nujol mixture during electrode preparation (21–23). Such electrodes are very simple and inexpensive to prepare. They are easily regenerated by simply exposing a fresh electrode surface. The capacity of such electrodes is relatively high, in that 'surface' coverages of 5×10^{-8} mol cm^{-2} have been reported. This is certainly greater than a monolayer coverage. High capacities would be expected, given the porosity of carbon paste electrodes. The greatest problems with composite CMEs of this type are associated with leaching of the modifier from the electrode surface. It is difficult to envisage that any modifier would be totally insoluble in common solvents. Whilst most composite CMEs described to date simply involve physical mixing of the modifier, some workers (5) have actually employed covalent bonding of the modifier (discussed in more detail later) via silanization to the graphite particles before mixing. This approach should overcome the modifier leaching problem.

The chemisorption approach (ii) is most useful for modifiers which are insoluble in the analyte solvent. The modifier is usually dissolved in a volatile solvent. It is then pipetted on to the electrode surface (Figure 5.3) and the solvent allowed to evaporate, leaving the modifier on the surface. It is imperative that the solvent/modifier completely covers the electrode surface to provide a homogeneous covering. To achieve this, an intermediary wetting solvent may be applied to the substrate before depositing the modifier.

The chemisorption approach has been widely used for coating electrode surfaces with polymers (24–26) as well as various other organic (27) and inorganic (28, 29) compounds. Uniform coatings varying in thickness from 10^{-10}–10^{-7} mol cm^{-2} have been obtained using this procedure, which is extremely easy to implement. For stable uniform coatings the electrode substrate must be meticulously prepared and often several hours are required for drying. Electrodes left to dry slowly appear to have better mechanical properties than those dried quickly. Electrodes prepared in this way suffer the same instability problems as composite electrodes in that the modifier gradually leaches from the substrate (24). On the other hand, it has been shown that at least some chemisorbed materials demonstrate long-term stability (27). The chemical and electrochemical properties of these electrodes depend on the substrate and on the modifier employed. The capacity of the electrode is determined by the thickness of the modifier. For example, coverages of 1.7×10^{-7} mol cm^{-2} quinhydrone (27) have been reported. Dip coating or spin coating can also be used to prepare chemisorbed modified electrodes.

If one considers the evolution of stationary phase materials for chromatography then the advent of *bonded-phase electrodes* (iii) appears to be a

Figure 5.3 Preparation of chemisorbed chemically modified electrodes.

natural progression in the evolution of chemically modified electrodes.

Both composite electrodes and those prepared using chemisorption suffer from instability due to leaching of the modifier from the substrate. Furthermore, such coating procedures are relatively difficult to control and hence to reproduce from one electrode to another. Using bonded-phase electrodes these problems are alleviated. Some of the more common methods used for bonding modifiers to electrode substrates are discussed below.

Direct bonding: using this procedure, modifiers are bound directly to the substrate with no intermediary chemistry. The procedure is therefore relatively simple to apply. However, it is limited in that the substrate and modifier must be amenable to direct bonding. Two examples follow. More details are available in reference 30.

Silanization: In general, this class of reaction may be represented as

$$\text{〉-OH} + \text{XSiX}_2\text{R} \rightarrow \text{〉-O-SiX}_2\text{R} + \text{HX, (X = Cl for example)}$$

The silanizing agent ($XSiX_2R$) reacts with oxygen-containing groups on the electrode substrate, to produce a silanized surface. This surface can then be used as it is, or derivatized.

This chemistry relies heavily on the presence of oxygen on the electrode material (see choice of substrate, above), and so substrate selection and preparation are important. Silanization procedures have been used to bind modifiers to substrates such as platinum (31), RuO_2 (32), carbon (33) and SnO_2 (34). This method is extremely versatile since various R groups (chains of different lengths) and X groups (X being either chemically or electrochemically active) can be chosen in designing a particular sensor. However, these preparation procedures are more difficult to implement than those previously discussed. Polymer formation can only be avoided by using scrupulously dry reagents and removing excess organosilane from the surface before exposing the electrode to the atmosphere. In general, monolayer coverages of $\sim 10^{-10}\,\text{mol cm}^{-2}$ are obtained using this procedure, giving well-defined but low-capacity electrodes. The electrodes formed are highly stable. In fact, in some instances the modifiers are more stable on the electrode surface than they would be in solution (31).

Cyanic chloride reactions once again involve reactions of an intermediary with oxygen-containing groups on the electrode surface:

This procedure has been widely applied to carbon materials (35, 36). Cyanic chloride has proven extremely useful as a coupling agent for enzymes, since the extremely labile chlorine atoms undergo rapid reactions with nucleophiles such as amines, amino acids and hydroxy groups.

Carbodiimide reactions have been used to attach enzymes (E) to electrode surfaces (35). Activation is usually carried out in aqueous media and bonding in acid media. Attachment can be achieved through a variety of amino acids.

The thionyl chloride reaction (37) is once again a convenient method for attaching chemically and electrochemically active groups to electrode surfaces.

R may be an electrochemically reactive moiety such as a tetraphenyl porphyrin ring. It has been shown (38) that the thionyl chloride can be replaced by acetyl chloride.

The Williamson reaction (39): following/reaction ion exchange during which sodium ions replace hydrogen on the substrate the Williamson reaction can be used to attach a range of species to the surface.

The final category of CME preparation is (iv) *in-situ* polymerization. In general, a monomer in solution undergoes a reaction which produces a polymer coating on the electrode surface. Chemical, photo-induced, plasma-induced or electrochemical polymerization may be employed. Chemical polymerization has been discussed in detail elsewhere (40) as has photopolymerization (41). Neither of these have been applied extensively in the preparation of modified electrodes. Glow discharge (42, 43) or plasma (44) polymerization have proven popular in recent years. Using this technique, stable, reproducible polymer layers can be deposited. Copolymerization is readily achieved (42) and the degree of cross-linking can be controlled by varying the plasma temperature. The polymer thickness is controlled by deposition time. A major drawback of this technique is that it requires rather sophisticated instrumentation.

In-situ electrochemical polymerization requires somewhat simpler instrumentation—a device capable of applying either a constant current or a constant potential. A soluble monomer is usually oxidized or reduced in solution to form an insoluble polymer. The working electrode—the substrate—is subsequently coated with the polymer.

TEAP = tetraethyl ammonium perchlorate

In the author's experience, more even coatings are obtained employing galvanostatic as opposed to potentiostatic deposition. Thickness is controlled by deposition time. Various deposition parameters can be used to influence the morphology of the polymer. For example, the applied potential or the current density can influence the porosity, as can the choice of deposition solvent. Also, the nature and concentration of the supporting electrolyte can affect the conductivity and the stability of the polymer. It is possible to coat electrodes with either non-conducting (45–47) or conducting (48–52) polymers using the above approach. With electrochemical methods, deposition of non-conducting polymers is limited to relatively thin coatings without the addition of a charge carrier to increase the conductivity.

With *in-situ* polymerization it is relatively easy to incorporate chemically active species into the polymer using codeposition methods (48, 49) or by incorporating the electroactive species as a counter-ion in a conducting polymer. Using the latter method, species such as $IrCl_6^{2-}$ and $Fe(CN)_6^{4-}$ have been incorporated into polypyrrole (52).

Functional groups may also be incorporated by derivatizing the polymer using conventional derivatization procedures such as covalent bonding.

Characterizing modified electrode surfaces. Various surface analysis techniques have been used to characterize modified electrodes. Efforts should be made to characterize the CME in conditions similar to those it will be subjected to as a sensor. For example, if the CME is designed for solution analysis (the most common application), the electrochemical characterization methods which can be performed in solution are most relevant.

Electrochemical techniques are particularly amenable to solution analysis; however, the information obtained is usually limited and this necessitates the use of other analytical methods to gather more detailed, quantitative data. Both the physical properties and the chemical composition of polymers are important.

Cyclic and differential pulse voltammetry are the most commonly used electrochemical methods for characterizing electrode surfaces (53). Cyclic voltammetry requires relatively simple and inexpensive instrumentation and the method is rapid (for further details see Chapter 8). Readouts, as shown for the example in Figure 5.4, are usually obtained.

The theory is governed by diffusion in a confined space, which is a thin layer, and so the peak current (i_p) is given by

$$i_p = -\frac{n^2 F^2 \tau v}{4RT}$$

where τ is surface concentration and v is scan rate.

Cyclic voltammograms for CMEs which contain electroactive species are usually symmetrical and, as the above equation shows, i_p is directly proportional to scan rate. In the ideal case the peak width at half height is given by $90.6/n$ mV. However, due to the proximity of redox sites on the electrode surface, there may be some interaction between species. It is therefore usual to employ surface activity rather than surface concentration on the electrode surface for detailed theoretical treatments.

This behaviour has analytical implications, since the degree of interaction will vary with surface concentration, leading to distortion of the analytical response. Attractive interaction serves to decrease peak widths while repulsive

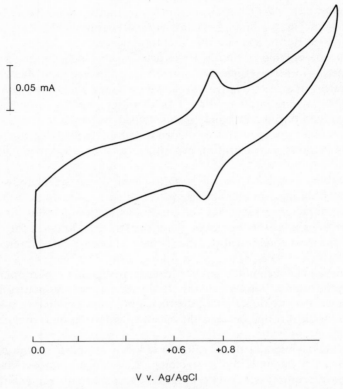

0.05 mA

0.0   +0.6   +0.8

V v. Ag/AgCl

Figure 5.4 Cyclic voltammogram of $Ru(bpy)_3^{2+}$-containing polymer attached to a glassy carbon
electrode surface.

interaction increases peak widths beyond what would be expected for an ideal
system.

Coverages as low as 10^{-10} mol cm^{-2} can be detected by cyclic voltammetry,
making it a very sensitive method for surface-confined species. The technique is
semi-qualitative in that redox potentials may be used to verify the presence of a
particular species.

Differential pulse votammetry can be even more sensitive than cyclic
voltammetry. However, due to the rapid decay of the current signal with
surface controlled responses, an external resistence in series with the working
electrode may be required to enable a well-defined signal to be obtained (53).
More recently, square wave voltammetry has been used to investigate CME
surfaces (54).

Chronoamperometry (55) and rotating disk electrode (56) studies are
primarily used to obtain information on the electrode process of interest, in
particular the effective diffusion coefficient which is important in optimizing
sensor performance. The effective diffusion coefficient accounts for the
diffusion of charge which may occur in the modifier. This information is

important, as it is necessary to maximize the diffusion coefficient for the analyte while minimizing those for potential interferents.

As well as the above techniques, dry electrical conductivity measurements using the four-point probe technique may be useful. However, it is important to remember that this is likely to be much different from solution conductivity.

In the developmental stage of new CME electrodes, information obtained from other analytical techniques is invaluable. Information on modifier thickness, uniformity, permeability and chemical composition can be most helpful in evaluating the usefulness of a new CME as a chemical sensor. Some of this information can be obtained using electrochemical methods, but other techniques are also useful (57). Scanning electron microscopy (SEM) can be used to obtain information on the polymer morphology and thickness; SEM is particularly useful for conducting modifiers since complications, due to surface charging, arise when viewing non-conducting materials. Ellipsometry may be used to obtain films from sub-monolayer to several microns thick. To obtain the weight of a modifier and substrate, a quartz crystal microbalance may be used.

Several analytical techniques which can be used to obtain information on the chemical composition of modified surfaces are available (58, 59). For example, x-ray photo electron spectroscopy (XPS) can be useful for analysis of thin layers (to depths of 20 Å) on substrates. XPS can provide both qualitative and quantitative information on the elements present as well as on their oxidation state, organic structure and bonding information. Auger electron spectroscopy (AES) is a similar technique, but offers only marginal information on the chemical environment of the elements. As for XPS, AES is a highly surface-sensitive technique. It is usually the outermost 2–6 atomic layers which are analysed. These surface-sensitive techniques are very prone to interference from absorbed contaminants. Careful handling of the sample between preparation in the electrochemical cell and the characterization experiment is therefore most important. AES is quantitative only to $\pm 50\%$ (60). Electron microprobe analysis (EPMA) provides much more accurate quantitative data.

In these laboratories we have recently employed Fast Atom Bombardment Mass Spectrometry for analysis of CMEs. Depth penetrations of 1 nm per minute can be achieved, depending on beam intensity. The low penetration rate allows for excellent depth profiling studies to be carried out. Examples of FAB mass spectra obtained from a modified electrode surface are shown in Figures 5.5–5.6. This technique holds much promise for the future, since a mass spectrum of the electrode surface can be obtained.

Infrared spectroscopy is useful in obtaining information on the chemical composition and the chemical structure on the electrode surface. Previous workers (61) have used this technique to elucidate the method of binding of some films to an electrode surface. Infrared spectroscopy can be used either in the transmission mode or the reflectance mode. In the transmission mode the

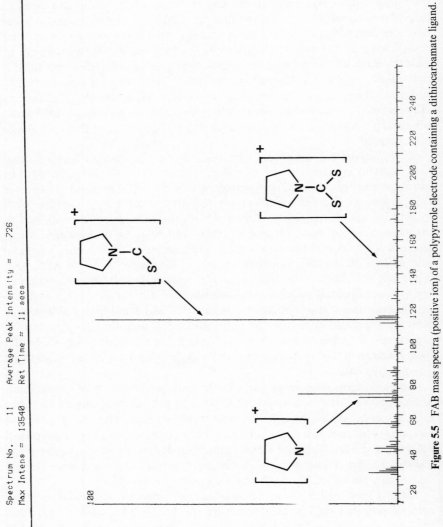

Figure 5.5 FAB mass spectra (positive ion) of a polypyrrole electrode containing a dithiocarbamate ligand.

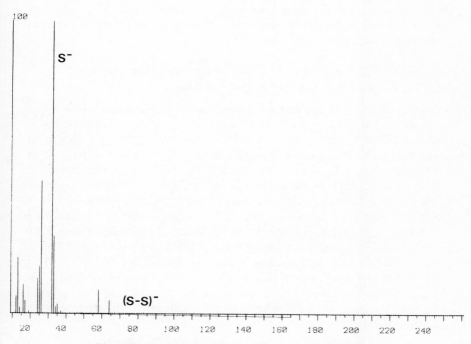

Figure 5.6 FAB mass spectra (negative ion) as for Figure 5.6.

modifier must be placed on an IR transparent substrate such as impregnated NaCl plates (62) or it must be removed from the substrate and prepared in a Nujol mull or a KBr disc form. Using reflectance ir spectroscopy, the analysis can be performed directly on the electrode surface. Both techniques suffer from a lack of sensitivity; however, sensitivity has been improved with the introduction of Fourier Transform Infra Red (FTIR) spectrometry. FTIR has also introduced the possibility of performing spectroscopic analysis of the electrode surface in solution.

5.2.3 *Basis of signal generation*

To be employed as a chemical sensor the CME must be capable of producing an electrical signal which is related to the concentration of the analyte. (An excellent study of the mechanism of electrode transport in modified electrodes

F

can be found in reference 63.) The signal may be due to a change in resistance
of the CME, a change in the potential difference across the CME (potentiometric methods), or to generation of a current arising from oxidation or
reduction of the analyte (voltammetric/amperometric methods). Voltammetric methods give rise to the most sensitive techniques with the most rapid
response times, and consequently they have been used in the majority of
applications to date. We shall discuss these in detail.

Voltammetric signal generation at a conventional clean metal electrode
surface depends on

 (i) Transport of the species to the electron transfer boundary
(ii) The rate of electron transfer at that boundary.

In general, optimum responses are obtained for species which are at a high
concentration at the boundary and where electron transfer is rapid (reversible
systems). This will result in well-defined responses with intensity proportional
to analyte concentration (see Chapter 8). With CMEs similar processes occur,
in that the analyte must be transported to the electron transfer boundary
before the redox process occurs. However, the electron transfer boundary is
not so clearly defined on CMEs as it is on conventional metal electrodes. If the
modified sensor is coated with a material to occlude other species, then the
movement of the analyte (Y) may also be hindered, as depicted in Figure 5.7(i).
However, with such sensors electron transfer will usually occur at the metal
substrate surface with the usual rate of electron transfer. If a catalytic CME is
employed then a process as depicted in Figure 5.7(ii) may be observed. The
analyte migrates to the electrode surface, in this case the modifier–solution
boundary. The analyte may then be oxidized or reduced by the electrocatalyst
according to:

$$Ru^{II} \rightarrow Ru^{III} + e \qquad (1)$$

(substrate/modifier interface)

$$Ru^{III} + Ru^{II} \rightarrow Ru + Ru^{III} \qquad (2)$$

(throughout modifier to the modifier/solution interface)

$$Ru^{III} + Fe^{II} \rightarrow Re^{III} + Ru^{II} \qquad (3)$$

(at modifier/solution interface)

The overall electron transfer process is much more complicated that at a
clean metal surface. Following electron exchange at the surface, the charge
must be transported in some way from the substrate surface to the modified
boundary. This may occur via a charge hopping process (self-exchange
reactions) or some limited diffusion procedure. The process may also rely on
diffusion of a counter-ion in the opposite direction to balance the charge.
Obviously then the properties of the counter-ion, in particular the mobility,

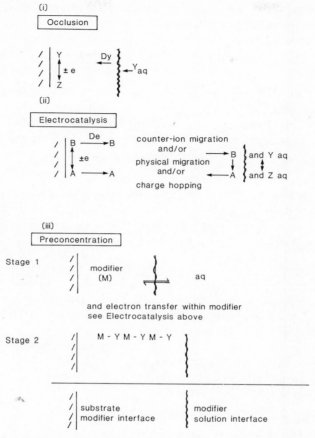

Figure 5.7   Signal generation at a chemically modified electrode.

would be extremely important. The high concentration of counter-ion in the modifier can also cause chemical complications, such as precipitation of analytes. When catalytic CMEs are used, the response is improved because the mediator improves the nature and rate of the electron transfer process. Electrocatalytic electrodes may also be used to improve responses by providing reactants such as oxygen from the electrode surface or by catalysing the electrochemical reaction at less extreme potentials.

Responses obtained at mediator modified electrodes can be complicated if some of the analyte diffuses through to the bare electrode surface. Conventional voltammetric stripping procedures introduce another step into conventional voltammetry, namely, preconcentration of the analyte on or into the electrode surface. Similarly, with preconcentrating modified electrodes (Figure 5.7, iii) the analytical performance depends on:

(i) The ability to preconcentrate-both the rate constant (k) and the equilibrium constant (K) are important, since

$$A \underset{k}{\overset{K}{\rightleftharpoons}} A$$

(solution)   (modifier)

(ii) The nature of the electron transfer process, which, as discussed above, will be more complicated than at bare metal substrates.

Electron transfer may occur in much the same way as in a redox polymer with electron hopping from site to site. Alternatively, if the analyte species are not rigidly held in the modifier, they may diffuse to the electrode surface.

Whether electron hopping or molecular diffusion takes place will depend on which of these processes is faster. In some cases the centres may diffuse short distances to allow electron hopping to occur. This argument was substantiated by previous workers (65) who measured $d-d$ distances of 2.5–3.4 nm for $Fe(CN)_6^{3-}$. For electron self-exchange to occur, distances of 0.92 nm are required. Where the response is limited by the rate of charge transfer, a small charge carrier additive may prove useful. For example, a ferro-ethylenediaminetetra-acetic acid additive has been used (66). It is most likely that the mechanism of transport will vary with analyte concentration and also with the morphology of the polymer (67).

For rapid electron transport, a high density of electroactive groups is required. For rapid counter-ion transport, a high free volume content is required. Both will influence the electrochemical behaviour of the polymer. Polymer morphology is influenced by:

(i) The nature of the electroactive pendant group
(ii) The extent of backbone functionalization
(iii) The size and structure of the side chain linking the electroactive group to the backbone
(iv) The polymer backbone
(v) The extent of ionization
(vi) The presence of monomeric dopants
(vii) The conditions of film preparation.

Yet another mechanism of electron transfer at preconcentrated electrodes can be envisaged. This mechanism involves dissolution/retrapping of the species, according to

$$A_{trapped} \underset{trapping}{\overset{dissol}{\rightleftharpoons}} A_{soln} \underset{process}{\overset{electrode}{\rightleftharpoons}} B_{soln} \underset{soln}{\overset{trap}{\rightleftharpoons}} B_{solid}$$

In this case all the electrochemistry actually takes place in solution. This is analogous to a mechanism proposed for reduction of an organic crystalline material on an electrode (68).

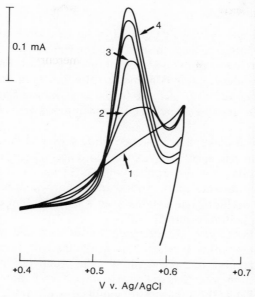

Figure 5.8 The 'break-in-process'-signal due to oxidation of mercury ions trapped on a polypyrrole-dithiocarbamate electrode (1–4, scans 1–4 respectively).

After preparation of a CME, electroactive sites on the surface may display behaviour as depicted in Figure 5.8. Similar behaviour may be observed if a CME is used to preconcentrate an electroactive analyte. This break-in process is attributed to the establishment of conductive chains brought about by reorganization of the polymer (69).

For optimum sensor performance, the electron transfer process should be rapid and chemically reversible. With conventional metal electrodes it is unlikely that reversible electron transfer processes will change with analyte concentration. However, with CMEs it is likely this will occur. If so, then an extra degree of complexity is added to the interpretation of the CME sensor response in that non-linear calibration curves as well as non-linear standard addition curves will be obtained. Furthermore, the electron transfer process is also dependent on the chemical environment. For example, the nature and concentration of the counter-ion in conductive polymers have a significant influence. In such cases, matrix matching between standards and samples must be carried out.

Chemical reactions, such as precipitation in the polymer subsequent to the electron transfer process as well as during preconcentration, are also important. This re-emphasizes the fact that attention must be paid to the proper choice of a supporting electrolyte (counter-ion). The electro-active zone must be taken into account since it is necessary to electrolyse 100% of the trapped analyte for increased sensitivity.

5.3 Applications

Given the challenges previously outlined in the areas of electrode characterization and in defining the mechanism of electron transfer, it is not surprising to find that the number of analytical applications of CMEs is limited at present. As the level of understanding continues to rise, however, the application of CMEs as sensors will also increase in the near future. Some examples of applications using the three types of CMEs mentioned in the introduction are the following.

Examples which make use of the occluding power of modified electrodes may be found in references (70–72). Cellulose acetate is the most common coating used to date for this application, and it has been used to occlude species from the substrate on the basis of molecular size. It has proved useful, for example, in occluding surfactants from a mercury film electrode (70) or proteins from a platinum surface (72).

Electrocatalytic CMEs are designed to increase the sensitivity for a particular element, either by mediating in the electron transfer process, by providing reactants or regenerating the analyte, or by shifting oxidation/reduction potentials to less extreme values. Examples can be found in references (73)–(77). A substrate modified with $IrCl_6^{2-}$ was used to determine nitrite without nitrate interference, the modifier acting as an electron transfer mediator (73). Platinum substrates modified with iodide have been employed for the electrocatalytic determination of metals (74) and the determination of dopamine in brain tissue. Electrodes capable of catalysis via oxygen transfer have been employed for determination of both organic and inorganic species (75). Bioelectrocatalysis has been achieved by attaching enzymes to electrode surfaces, e.g. for determination of xanthine (76) and for H_2O_2 determination (77).

Examples of applications involving preconcentration may be found in references (79–85). Modified electrodes capable of preconcentrating the analyte onto the electrode surface have been employed to enhance the analytical sensitivity and selectivity. The most common applications appear to be in the area of metal analysis. Preconcentration has been achieved on the basis of electrostatic attraction for analysis of chromium as $Cr_2O_7^{2-}$ (78). For more selective preconcentration, however, complexation has been employed (79–81). Dimethylglyoxime has been employed for nickel determinations (81), and dithiocarbamates (79, 80, 82, 83) for copper determination. Dithiocarbamate modified electrodes have been used for mercury analysis (84, 85).

Complexation prior to the electron transfer step is particularly attractive, since species may be trapped in a chemical environment which is conducive to rapid electron transfer and facilitates more unusual electron transfer mechanisms (e.g. stabilization of unusual oxidation states). Conducting polymers have been employed for determination of electro-inactive species on the basis of counter-ion replacement (86).

5.4 Future trends

The development of simple manufacturing and characterization procedures should allow the design and implementation of very useful sensors in the near future. Drawing on the experience of other workers who have been involved with the fundamental development of CMEs, the purposeful design of new chemical sensors will become easier to achieve. The number of applications continues to expand, as is evident from the increasing number of publications in the literature.

References

1. E. Solomen, *Z. Physik. Chem.* **24** (1897) 55.
2. A.J. Bard, *J. Chem. Educ.* **60** (1983) 302–304.
3. R.W. Murray, *Acc. Chem. Res.* **13** (1980) 135–141.
4. L.R. Faulkner, *Chem. Eng. News.* 1984, Feb. 28–45.
5. G.T. Cheek and R.F. Nelson, *Anal. Lett.* **5** (1978) 393–402.
6. W.E. van der Linden and J.W. Dieker, *Anal. Chim. Acta* **119** (1980) 1–24.
7. D.T. Fagan, I.F. Hu and T. Kuwana, *Anal. Chem.* **57** (1985) 2759–2763.
8. E. Hershenhart, R.L. McGreery and R.D. Knight, *Anal. Chem.* **56** (1984) 2256–2257.
9. E. Theordoridou, J.O. Besenhand and H.P. Fritz, *J. Electroanal. Chem.* **922** (1981) 67–71.
10. R.C. Engstrom and V.A. Strasser, *Anal. Chem.* **56** (1984) 136–141.
11. J.F. Rusling, G.N. Kaman and W.S. Willis, *Anal. Chem.* **57** (1985) 545–551.
12. J.F. Evans and T. Kuwana, *Anal. Chem.* **49** (1977) 1632.
13. S. Trasatti, *Electrodes of Conductive Metallic Oxides.* Elsevier, Amsterdam (1980).
14. J. Wang, *Electrochimica Acta* **26** (1981) 1721–1726.
15. D. Yaniv and M. Ariel, *J. Electronal. Chem.* **129** (1981) 301–313.
16. T.E. Edmonds, *Anal. Chim. Acta* **175** (1985) 1–22.
17. H.J. Wieck, C. Shea and A.M. Yacynych, *Anal. Chim. Acta* **142** (1982) 277–279.
18. C.W. Anderson and K.R. Lung, *J. Electrochem. Soc.* **129** (1982) 2505–2508.
19. R. Bilewicz and Z. Kublik, *Anal. Chim. Acta* **171** (1985) 205–213.
20. N.R. Armstrong and J.R. White, *J. Electroanal. Chem.* **131** (1982) 121–136.
21. M.K. Halbert and R.P. Baldwin, *Anal. Chem.* **57** (1985) 591–595.
22. R.P. Baldwin, J.K. Christensen and L. Kryger, *Anal. Chem.* **58** (1986) 1790–1798.
23. P.W. Geno, K. Ravichandran and R.P. Baldwin, *J. Electroanal. Chem.* **183** (1985) 155–166.
24. C.A. Koval and F.C. Anson, *Anal. Chem.* **50** (1978) 223–229.
25. L.L. Miller and M.R. van de Mark, *J. Amer. Chem. Soc.* **100**(2) (1978) 639–640.
26. N. Oyama and F.C. Anson, *J. Amer. Chem. Soc.* **101**(3) (1979) 739–741.
27. H. Gomathi and G. Prabhakara Rao, *J. Electroanal. Chem.* **190** (1985) 85–94.
28. L.R. Taylor and D.C. Johnson, *Anal. Chem.* **46** (1974) 262–266.
29. J.H. Larochelle and D.C. Johnson, *Anal. Chem.* **50** (1978) 240–243.
30. R. Nowak, F.A. Schultz, M. Umana, H. Abruna and R.W. Murray, *J. Electroanal. Chem.* **94** (1978) 219–225.
31. H.D. Abruna, T.J. Meyer and R.W. Murray, *Inorg. Chem.* **18** (1979) 3233–3240.
32. P.R. Moses and R.W. Murray, *J. Electroanal. Chem.* **77** (1977) 393–399.
33. L.M. Elliott and R.W. Murray, *Anal. Chem.* **48** (1976) 1247–1254.
34. P.R. Moses, L. Wier and R.W. Murray, *Anal. Chem.* **47** (1975) 1882.
35. J.A. Osborn, R.M. Ianniello, H.J. Wieck, T.F. Decker, S.L. Gordon and A.M. Yacynych, *Biotechnol. and Bioeng.* **24** (1982) 1653–1664.
36. H.J. Wieck, R.M. Ianiello, J.A. Osborn and A.M. Yacynych, *Anal. Chim. Acta* **140** (1982) 19–27.
37. J.C. Lennox and R.W. Murray, *J. Electroanal. Chem.* **78** (1977) 395–401.
38. R.D. Rocklin and R.W. Murray, *J. Electroanal. Chem.* **100** (1979) 271.
39. N. Jannakoudakis, N. Missaelidis and E. Theodoridou, *Synthet. Met.* **11** (1985) 101–108.

40. H.R. Allcock and F.W. Lampe, *Contemporary Polymer Chemistry*, Prentice-Hall, New Jersey (1981).
41. S.S. Labana (ed.), Ultraviolet light induced reaction in polymers, *ACS Symp. Ser.* **25** (1976).
42. K. Doblhofer, W. Durr and M. Jauch, *Electrochim. Acta* **27** (1982) 677–682.
43. K. Doblhofer, W. Durr, *J. Electrochem. Soc.* **127** (1980) 1041–1044.
44. P. Daum, J.R. Lenhard, D. Rolison and R.W. Murray, *J. Amer. Chem. Soc.* **102** (1980) 4649–4653.
45. M.C. Pham, P.C. Lacaze and J.E. Dubois, *J. Electrochem. Soc.* **131** (1984) 777–784.
46. J.E. Dubois, P.C. Lacaze and M.C. Pham, *J. Electroanal. Chem.* **117** (1981) 233–241.
47. M.C. Pham and J.E. Dubois, *J. Electroanal. Chem.* **199** (1986) 153–164.
48. L. Roullier and E. Waldner, *J. Electroanal. Chem.* **187** (1985) 97–107.
49. S.E. Lindsey and G.B. Street, *Synthet. Met.* **10** (1984) 67–69.
50. P. Pfluger and G.B. Street, *J. Chem. Phys.* **80** (1984) 544–553.
51. G. Tourillan, E. Dartyge, H. Dexpert, A. Fontaine, A. Jucha, P. Lagarde and D.E. Sayers, *J. Electroanal. Chem.* **178** (1984) 357–366.
52. R. Noufi, *J. Electrochem. Soc.* **130** (1983) 2120–2128.
53. A.P. Brown and F.C. Anson, *Anal. Chem.* **49** (1977) 1589–1595.
54. E.S. Takeachi and J. Osteryoung, *Anal. Chem.* **57** (1985) 1770.
55. T. Ikeda, R. Schmehl, P. Denisevich, K. Willman and R.W. Murray, *J. Amer. Chem. Soc.* **104** (1982) 2683–2691.
56. K. Shigehara, N. Oyama and F.C. Anson, *J. Amer. Chem. Soc.* **103** (1981) 2552–2558.
57. H.G. Tompkins, *Thin Solid Films* **119** (1984) 337–348.
58. D.M. Hercules and S.H. Hercules, *J. Chem. Educ.* **61** (1984) 403–408.
59. D.M. Hercules and S.H. Hercules, *J. Chem. Educ.* **61** (1984) 483–489.
60. B.G. Baker, *Trends in Electrochemistry*, 4th Austr. Electrochem. Conference (1976).
61. N. Oyama, T. Oshaka and T. Shimizu, *Anal. Chem.* **57** (1985) 1526–1532.
62. N.A. Surridge and J.J. Meyer, *Anal. Chem.* **58** (1986) 1576–1578.
63. R.M. Ianiello, H.J. Weick and A.M. Yacynych, *Anal. Chem.* **55** (1983) 2067–2070.
64. W.J. Albery and A.P. Hillman, *J. Electroanal. Chem.* **175** (1984) 27–49.
65. K.-N. Kuo and R.W. Murray, *J. Electroanal. Chem.* **131** (1982) 37–60.
66. T. Shimomura, N. Oyama and F.C. Anson, *J. Electroanal. Chem.* **112** (1980) 265–270.
67. A.H. Schroeder and F.B. Kaufman, *J. Electroanal. Chem.* **113** (1980) 209–224.
68. L. Roullier, E. Waldner and E. Laviron, *J. Electrochem. Soc.* **132** (1985) 1121–1125.
69. D.A. Buttry and F.C. Anson, *J. Electroanal. Chem.* **130** (1981) 333–338.
70. J. Wang and L.D. Hutchins-Kumar, *Anal. Chem.* **58** (1986) 402–407.
71. K.A. Robinson, T.W. Gilbert and H.B. Mark, *Anal. Chem.* **52** (1980) 1549–1551.
72. G. Gittampalam and G.S. Wilson, *Anal. Chem.* **55** (1983) 1608–1610.
73. J.A. Cox and P.J. Kulesza, *J. Electroanal. Chem.* **175** (1984) 105–118.
74. D.C. Johnson and E.W. Resnick, *Anal. Chem.* **49** (1977) 1918–1924.
75. D.S. Austin, J.A. Polta, T.Z. Polta, A.D. Tang, T.D. Cabelka and D.C. Johnson, *J. Electroanal. Chem.* **168** (1984) 227–248.
76. R.M. Ianniello, T.J. Lindsay and A.M. Yacynych, *Anal. Chem.* **54** (1982) 1980–1982.
77. R.M. Ianniello and A.M. Yacynych, *Anal. Chem.* **53** (1981) 2090.
78. J.A. Cox and P.J. Kulesza, *J. Electroanal. Chem.* **159** (1983) 337–346.
79. L.M. Wier, A.R. Guadalupa and H.D. Abruna, *Anal. Chem.* **57** (1985) 2009–2011.
80. A.R. Guadalupa and H.D. Abruna, *Anal. Chem.* **57** (1985) 142–149.
81. R.P. Baldwin, J.K. Christensen and L. Kryger, *Anal. Chem.* **58** (1986) 1790–1798.
82. D.M.T. O'Riordan and G.G. Wallace, *Anal. Chem.* **58** (1986) 128.
83. D.M.T. O'Riordan and G.G. Wallace, *Anal. Proc.* **22** (1985) 199.
84. D.M.T. O'Riordan and G.G. Wallace, *Anal. Proc.* **23** (1986) 14.
85. M.D. Imisides, D.M.T. O'Riordan and G.G. Wallace, *Proc. Int. Conf. Chem. Sensors*, Elsevier, Amsterdam (1986) 293.
86. Y. Horiyama and W.R. Heineman, *Anal. Chem.* **58** (1986) 1803–1806.

6 Immunoassay techniques

N.J. SEARE

6.1 Principles of immunoassay

The basic principles of immunoassay were first reported by Berson and Yalow (1) and since then it has become an extremely powerful technique in the determination of a broad spectrum of compounds. The power of the technique lies chiefly in the areas of specificity and versatility.

The principles of immunoassay are basically straightforward. If the substance of analytical interest is foreign to an animal, typically a rabbit, sheep or goat, injection of that substance into the animal will cause the production of a glycoprotein, known as an antibody (Ab). The antibody produced will have a specificity for the substance that initiated its production, the antigen (Ag). Antigens are generally naturally-occurring macromolecules, such as proteins, polysaccharides, nucleic acids, etc. Smaller molecules, such as drugs, hormones, peptides, etc., do not themselves initiate antibody production, but when coupled to a macromolecular carrier, such as a protein or a synthetic polypeptide, antibody production may be initiated. The resultant antibodies will react with the carrier-linked molecule and also the small molecule alone. A small molecule of this type is known as a hapten.

Immunoassay is based on the antigen (or hapten)–antibody binding reaction which is both reversible and non-covalent:

$$Ag + Ab \rightleftharpoons Ag\text{--}Ab$$

If a label is covalently attached to the antigen in such a way that it does not block the reaction site for the antibody, the presence of the label will not significantly affect the binding reaction. Thus, in a situation in which a mixture of labelled (Ag^*) and unlabelled (Ag) antigen react with antibody, competition for the antibody binding site occurs:

$$Ag^* + Ag + Ab \rightleftharpoons Ag^*Ab + AgAb$$

Using labelled antigen (Ag^*) it is possible to determine an unknown amount of antigen in a given solution. If a competitive binding reaction between Ag and Ag^* is set up and allowed to reach equilibrium, the amount of labelled antigen–antibody complex will decrease for a fixed amount of labelled antigen with increasing unlabelled antigen concentration in the reaction mixture. The measurement of some property proportional to the amount of antigen–

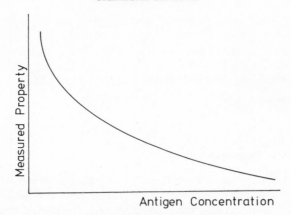

Figure 6.1   A typical immunoassay calibration curve.

antibody is used to construct a curve where the property measured is plotted as a function of unlabelled antigen concentration. A typical example of this type of curve is shown in Figure 6.1. The unknown antigen concentration in a given sample may thus be determined from a competitive binding reaction with known amounts of labelled antigen and antibody, the antigen content subsequently being determined from the standard curve.

An alternative approach to immunoassay was described by Miles and Hales (2) and termed immunoradiometric assay (IRMA), since it initially used isotopically labelled antibodies. In this type of assay the labelled component is a specific antibody (Ab*), the process again being based on the antigen–antibody binding reaction:

$$Ag + Ab + Ab^* \rightleftharpoons Ag{-}Ab + Ag{-}Ab^*$$

This method may now become more important with the introduction of monoclonal antibodies (3); see also Chapter 1 of this volume. The major disadvantage of monoclonal antibodies, however, lies in the increased cost of the product.

Immunoassays may generally be divided into two distinct types: heterogeneous assays and homogeneous assays. In a heterogeneous immunoassay procedure a separation step is required to separate bound and free labelled antigen prior to measurement of one of the fractions to give an assay value. In the case of a homogeneous procedures, no such separation step is required, as the properties of the label change on binding antibody such that they may be determined independently of the original labelled antigen.

The major advantages of immunoassay are its specificity, versatility and sensitivity. Specificity and versatility may be attributed to the selective and specific nature of the antigen–antibody binding reaction and the large number of molecules of interest that themselves, or coupled to a carrier (i.e., haptens), may be used to initiate antibody production. The sensitivity of an immunoassay procedure is dependent upon the choice of label and its limit of detection in the matrix of interest. Typical radioimmunoassay procedures have limits of detection in the pmol L^{-1} range, while an enzymatic immunoassay for macromolecular analytes reported by Ishikawa and Kato (4) has a reported sensitivity of 10^{-18} moles.

6.2 Labels in immunoassay

In the original immunoassay of Yalow and Berson (5), the label of choice was a radioisotope (^{125}I), and today radioisotopes are still widely used as labels (see The Immunoassay Index I, *Lab. Pract.* **32**(8) (1983) 21), ^{125}I being the most commonly used. Radioimmunoassay (RIA) has been used extensively in the determination of steroid and peptide hormones, drugs (both for therapeutic monitoring and for the detection of drugs of abuse) and various macromolecules of clinical interest.

The principal advantages of radioimmunoassay are the selectivity of the assay, conferred as in all immunoassays by the specificity of immunological reactions, and the sensitivity of radioisotope determination. The latter is the factor that has led to RIA becoming the most extensively used form of immunoassay, as background interference is not a problem.

However, RIA does have several major disadvantages which have led to extensive research into the use of alternative labels. The most important of these include: (i) the inherent instability of radiolabelled compounds; (ii) the need for specialized equipment and associated safety precautions; (iii) the cost of preparing radiolabelled compounds; (iv) the heterogeneity of all RIAs, because the properties of the label do not change on Ag–Ab binding, hence automation on a small scale is difficult; (v) disruption of the reaction by radioactivity.

When chosing a non-isotopic label for immunoassay perhaps the most important factor to be considered is the level at which it can be determined in the matrix of analytical interest. Other factor include: (i) freedom from hazards; (ii) ease of introduction into the sample molecule; (iii) cost and stability; (iv) suitability for homogeneous assays and automation; (v) possibility of combination with another label to develop a multicomponent analysis; (vi) application of the method in sensor development.

Alternative labels that have been investigated are reviewed in Table 6.1. As can be seen from the table, certain of these methods along with RIA are unsuitable for application to sensor development and will not be mentioned further. Other methods, particularly fluorescence immunoassay (FIA), will be discussed in more detail later with particular reference to their application in sensors.

Table 6.1 A review of non-isotopic immunoassay procedures

Immunoassay type	Label(s)	Detection method	Comments	Reviews and examples
Enzyme immunoassay	Enzymes	(i) Spectrometry (ii) Fluorimetry	Homogeneous methods, commercially available methods; cost may be a disadvantage; enzymes are labile	(12), (13), (14)
Co-enzyme immunoassay	NAD ATP	Spectrophotometry	Complicated labelling; simple, but often imprecise detection	(15), (16)
Chemiluminescence and bioluminescence immunoassay	Chemiluminescent groups	Luminometry	Complicated labelling; detection simple, but time dependent; low detection limits possible, homogeneous methods	(17), (18), (19), (20) (21), (22)
Fluorescence immunoassay (FIA)	Fluorophores	Fluorimetry	Simple labelling; homogeneous methods possible; background effects possible	(23), (24), (25), (26), (27)
Metallo immunoassay	Metal atoms	(i) Colorimetry (ii) Fluorimetry	Complicated labels	(7), (28), (29)
Sol-particle immunoassay (SPIA)	Metal atoms	(i) Visible response (ii) Colorimetry	Simple labelling	(30)
Particle counting immunoassay	(i) Latexes (ii) Particles	(i) Visible response (ii) Photometry	May be difficult to obtain viability of label	(31)
Viroimmunoassay	(i) Virus (ii) Bacteriophage	Bacterial culture	Complicated method, difficult to retain viability of label	(32), (33)
Free-radical immunoassay	Free radicals	ESR	Complex labelling; complex instrumentation	(34)
Voltammetric immunoassay (VIA)	(i) Isoenzyme electrodes (ii) Metal chelates	Voltammetry	Experimental methods may be quite complicated	(35)

6.3 Fluorescence immunoassay (FIA)

FIA in principle needs only simple equipment and, with pg mL^{-1} detection limits in pure solution, has the potential sensitivity required of an alternative label; however, background interferences associated with the sample matrix may cause certain problems. Background effects, for example with blood serum samples, include native fluorescence of the sample (chiefly associated with the fluorescent bands of the proteins and protein-bound materials such as bilirubin), scattering, high sample absorbance and quenching or enhancement effects. These problems have been overcome in several ways including careful choice of label, sample matrix preparation, chemical treatments both of the label and within the method and by the use of various spectroscopic methods. The use of these techniques in homogeneous and heterogeneous FIA are discussed below.

6.3.1 *Heterogeneous FIA*

In heterogeneous FIA the separation step generally serves two functions: firstly, the antibody-bound and free fractions are separated (a requirement of the assay), and secondly, interfering substances present in the sample are removed, resulting in an increase in sensitivity in many cases provided that adequate washing procedures are employed. Heterogeneous FIA techniques have been widely based on solid-phase systems; commonly used solid-phase FIA techniques include (i) competitive, (ii) sandwich, and (iii) fluoroimmunometric assays. Typical supports that may well also find application in sensor development are polyacrylamide beads, magnetizable particles and polystyrene tubes (6).

Heterogeneous FIA methods, although more complicated than homogeneous methods, generally tend to be more sensitive.

6.3.2 *Homogeneous FIA*

Homogeneous FIA methods are made possible by changes in the fluorescent properties of the label resulting from the binding of antibody. These changes may include enhancement or quenching of the fluorescence, energy transfer effects, changes in fluorescent wavelength, a change in fluorescence polarization, etc., or a combination of these effects. The major advantages of homogeneous assays lie in the speed of operation, convenience and the potential ease of automation or incorporation into a sensor.

Homogeneous FIA methods with potential application in sensor development and the spectroscopic techniques used in these assays are given in Table 6.2. The assay techniques are described below.

6.3.3 *Polarization FIA*

The fluorescence associated with a small molecule is normally unpolarized, or very nearly so, because the molecule undergoes many random changes in

Table 6.2 A review of fluorescent techniques used in immunoassay

Spectroscopic technique	General comments	Reviews and examples
Quenching and enhancement FIA	Homogeneous assays, possible background serum (urine) interference	(36), (37)
Fluorescence polarization (polarization FIA)	Homogeneous assays, rapid and precise specialized equipment sensitivity often low	(38)
Energy transfer	Homgeneous assay, two labelling procedures is a disadvantage	(23)
Fluorescence protection	Homogeneous assays, two anti-bodies required	(39), (40)
Fluorescence lifetime (time-resolved FIA)	Serum background overcome	(41)
Fluorescence internal reflectance	Many practical disadvantages	(42)

orientation during the lifetime of the singlet state. When bound to macro-molecules, small molecules may, however, show appreciably polarized fluorescence, since the molecular motions are much slower in this case. The increase in fluorescence polarization on antibody binding to a fluorescent-labelled small molecule may thus be used in the development of a homo-geneous immunoassay.

Fluorescence polarization as a method of FIA is quick and precise; however, it does have disadvantages. The equipment for this type of assay is specialized and the polarization change tends to be small, resulting in assays with sensitivity limited to the μ mol L^{-1} to the upper μ mol L^{-1} range.

6.3.4 *Quenching and enhancement*

An increase or decrease in the fluorescence of a labelled antigen on binding antibody may be used as the basis of a homogeneous FIA. Although not all the mechanisms of these effects are fully understood, quenching or enhancement assays are simple and do not require specialized equipment. A possible disadvantage of this type of method is that the degree of quenching or enhancement may vary with different batches of antibody. Occasionally the quenching or enhancement is also accompanied by a shift in fluorescent wavelength which effectively reduces the background.

6.3.5 *Fluorescence internal reflectance*

This technique uses a surface to which antibodies are bound (an immuno-

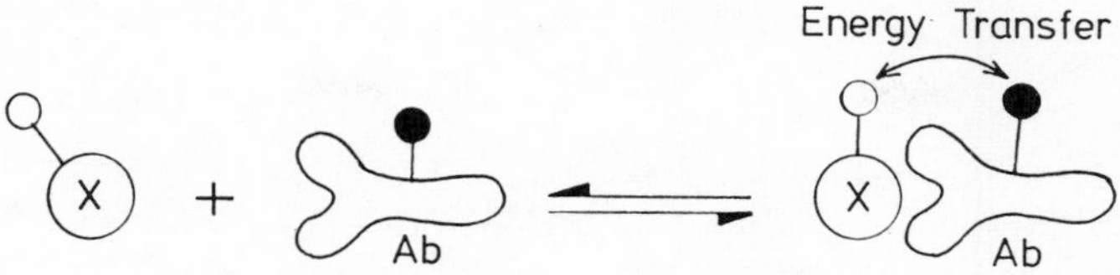

Figure 6.2 Energy transfer immunoassay.

logically reactive quartz plate) and fluorescent-labelled and unlabelled antigen are allowed to compete for the antibody binding sites as in any solid phase assay. Total internal reflection of light then excites the fluorophores bound to the surface and not those in the solution. The method has many practical disadvantages (e.g. scattered light) and has found little application.

6.3.6 *Energy transfer immunoassay*

This method involves the labelling of both antibody and antigen and is illustrated in Figure 6.2. The fluorescent labels are chosen such that on antibody–antigen binding non-radiative energy transfer occurs from one label to the other, the fluorescence of the donor being quenched and the acceptor possibly enhanced. Typical donor and acceptor pairs are fluorescein–rhodamine and fluorescamine–fluorescein. The major disadvantage of this technique is the need for two labelling procedures.

6.3.7 *Fluorescence protection immunoassay*

In its simplest form, the fluorescence protection assay consists of a fluorescent-labelled antigen, an antibody to that antigen (first antibody) and an antibody to the fluorescent label (second antibody) (Figure 6.3). In the protected assay the fluorescent-labelled antigen binds to the first antibody, and the subsequent binding to the fluorescent label of the second antibody is sterically prevented by the immune complex. Quenching occurs when there is sufficient unlabelled antigen to occupy the binding sites of the first antibody, allowing the second antibody to bind the fluorescent label.

    This method is reported to overcome some of the problems of various energy transfer assays but does still have the disadvantage that it requires two antibodies. The method has also been referred to in the literature as double antibody FIA, double-receptor FIA, indirect quenching FIA, and FIA using mixed binding reagent.

6.3.8 *Fluorescence lifetime immunoassay*

Although not strictly a homogeneous method, fluorescence lifetime immunoassay is a useful technique with particular applications in the clinical

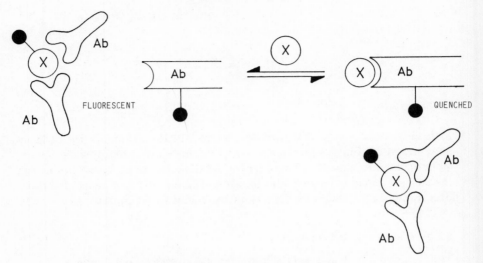

Figure 6.3   Fluorescence protection immunoassay.

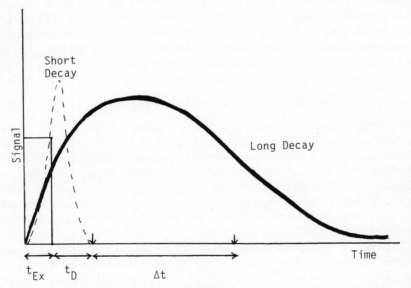

Figure 6.4   Time-resolved fluorescence measurements. t_{Ex}, excitation pulse (<1 ns); t_D, delay time; Δt, counting time.

field. Figure 6.4 shows the principles of lifetime immunoassay and how it may be used to remove background fluorescence with short decay times.

The fluorescence associated with the proteins of blood serum typically has a decay time of 10 ns. Accordingly, this emission may be allowed to decay prior to the measurement of the longer-lifetime fluorescence from the label, effectively removing the high background associated with such samples. Examples of labels suitable for this type of assay are europium and terbium chelates (with various α-diketones and salicylates) and a number of organometallic complexes (7). This type of assay requires a pulsed light source; however, pulsed xenon lamps are now available and have simplified such measurements.

A commercial immunoassay kit (the Fluoromatic System, Biorad, Richmond, California, USA) also utilizes a method with potential applications in immunoassay development using sensors. In this assay polyacrylamide microbeads having a refractive index equal to the assay medium, making them transparent in the assay medium, are used. The method uses a dedicated automatic fluorimeter with photon counting techniques for increased sensitivity (8).

6.3.9 Fluorescent labels

The choice of label is critical in designing an FIA method. The label should be stable, the fluorescence emission should be clearly distinguishable from the background, the quantum yield should be high and ideally it should have a high Stokes shift (difference between excitation and emission wavelength). A suitable method of linking the label to the antigen is also necessary and the antigen–antibody binding reaction should also be unaffected.

Typical labels that have been used in FIA are given in Table 6.3, and

Table 6.3 Fluorescence labels

FITC	Fluorescein isothiocyanate
RBITC	Rhodamine-β-isothiocyanate
RB200SC	Lissamine rhodamine-B200 sulphonyl chloride
DNS-CL	Dimethylaminonaphthalene-5-sulphonyl chloride (dansyl chloride)
Fluorescamine	4-Phenylspiro [furan 2(3H), 1'-phthalan]-3, 3'-dione
MDPF	2-Methoxy-2, 4-diphenyl-3(2H)-furanone
QM	9-(4-diethylamino-1-methylbutylamino)-6-chloro-2-methoxy acridine mustard
ANS	8-Anilino-1-naphthalenesulphonic acid
EITC	Eosin isothiocyanate
Erythrosin ITC	Erythrosin isothiocyanate
SITS	4-Acetamido-4'-isothiocyanatostilbene-2, 2'-disulphonic acid
Anthracene ITC	Anthracene-2-isothiocyanate
NPM	N-(3-pyrene) maleimide
NBD-CL	7-Chloro-4-nitrobenzo-2-oxa-1, 3-diazole
SBD-CL	7-Chloro-4-sulphonbenzofurazan
Lucifer Yellow CH	3, 6-disulphonate-4-aminonaphthalimide
BMC	4-bromomethyl-7-methoxy coumarin
Tb-Transferrin	Complex of terbium with transferrin
ADAM	9-diazomethylanthracene
9, 10 DAP	9, 10-diaminophenanthrene
HMA	9-hydroxymethylanthracene

Table 6.4 Properties of some fluorescence labels

Label	λ_{Ex}	λ_{Fl}	Quantum yield ϕ_f	Lifetime, τ(ns)
FITC	492	~ 520	0.85	4.5
RBITC	550	585	0.70	3
RB200SC	~ 565	~ 590	0.04	~ 1
DNS-CL	340	480	0.30	14
		~ 520		
Fluorescamine	~ 390	~ 490	0.10	7
MDPF	~ 390	~ 490	0.10	
QM	~ 350	~ 520	0.1–03	~ 10
ANS	385	471	0.80	16
EITC	~ 520	~ 550	0.19	0.9
				1 ms[†]
Erythrosin	~ 530	~ 560	0.02	0.1
ITC				0.5 ms[†]
SITS	~ 320	430		~ 1
Anthracene	~ 355	460	0.6	
ITC				
NPM	340	~ 380		100
NBD-CL	~ 470	530		
SBD-CL	~ 390	510		
Lucifer Yellow CH	~ 430	540	0.25	
(derivatives)				
Tb-transferrin	295	550		1.24 ms[†]

[†] Delayed fluorescence
λ_{Ex} Excitation wavelength
λ_{Fl} Fluorescence wavelength

Table 6.4 gives the corresponding excitation and emission wavelengths, quantum yields and lifetimes.

6.3.10 *Instrumentation*

The basic components of a fluorimeter are a light source, excitation and emission wavelength selectors, a sample holder and a detector. Which of these components is used depends upon the application and the required sensitivity.

Light sources. Light sources commonly used in fluorimeters are quartz halogen, tungsten and xenon lamps. The wavelength ranges of these sources are listed in Table 6.5. Alternative sources that may be considered are light-emitting diodes (LEDs) and lasers.

In terms of sensor development several of these sources have distinct advantages, notably LEDs, pulsed xenon lamps and lasers. LEDs have the advantage that they are inexpensive and simple, although not very intense. However, they may find application in the development of simple purpose-built equipment. Pulsed xenon sources when coupled to a gated detector are particularly useful because a probe based on this type of system may be used in an open beaker without interference from ambient light. Laser sources have the advantages of high power and excellent collimation. The laser sources

Table 6.5 Wavelength ranges for various sources

Source	Wavelength range (nm)
Xenon	200–800
Quartz–halogen	330–800
Tungsten	330–800
LEDs	Blue Yellow Green Red

potentially useful in fluorescence excitation are the argon ion laser with a tuneable range 450–570 nm, the nitrogen laser at 337 nm, and certain dye lasers.

Wavelength selection. Filters or monochromators may be used for wavelength selection. In commercial instrumentation the choice is usually reflected in the cost and sophistication of the instrument and may be a combination of the two types. The choice then is again dependent upon the application.

Detectors. Depending on the wavelength of interest and the sensitivity required for a given application, the detector may be chosen from photodiodes, photomultiplier tubes and photon-counting devices. In certain cases a diode array detector may also find application.

It is important that an instrument with a good optical system is used to develop methods before they are applied to simpler instruments or fiber optic sensors. Fluorescence instrumentation was recently reviewed by Wehry (9).

6.4 Multicomponent assays

Multicomponent analysis in immunoassay has recently become an important technique and in principle this type of assay should be possible in FIA. The labels for multicomponent analysis must be chosen to have excitation and emission characteristics which enable them to be determined independently in the same solution. The assay may take the form where the excitation of the two overlap and the emission bands are separated, and in this case it may be possible to excite the molecules simultaneously and split the emission to two detectors giving simultaneous measurements. Alternatively, the labels may be measured independently. In certain cases derivative or synchronous spectra may be used to further resolve the labels.

Multicomponent FIA methods may be carried out in solution or they may be solid-phase. In the case of solid-phase assays, the antibodies may be immobilized randomly on the solid phase or in sections which may be viewed selectively.

6.5 Alternative immunoassay methods with potential application in sensor development

Other methods with potential application in sensor development include certain enzyme immunoassays and sol-particle immunoassay. The enzyme methods that are potentially useful are those where the detection of the enzyme is carried out with a substrate that produces colour that may be determined colorimetrically, or a substrate that produces fluorescence.

The method of sol-particle immunoassay uses colloidal metal particles as the label. Typically these labels have been colloidal gold and colloidal silver. Detection in this type of assay is carried out visually, colorimetrically or by atomic absorption spectroscopy with electrothermal atomization. The colorimetric methods obviously offer potential in sensor development and the detection limits in the pmol l^{-1} range are useful.

Although this section has discussed immunoassay methods with potential for the development of sensors based on fibre optics, electrochemical immunoassay methods (10) may also find application in sensors development. Of particular interest is the use of piezoelectric crystals in immunoassay (11).

References
1. S.A. Berson and R.S. Yalow, *J. Clin. Invest.* **38** (1959) 1996.
2. L.E.M. Miles and C.N. Hales, *Nature (London)*, **219** (1968) 186.
3. A. Albertini and R. Ekins, in *Monoclonal Antibodies and Developments in Immunoassay*, Elsevier North Holland Biomedical Press, Amsterdam (1981) 3.
4. E. Ishikawa and K. Kato, *Scand. J. Immunol.* **8** (1978) 63.
5. R.S. Yalow and S.A. Berson, *J. Clin. Invest.* **38** (1959) 1996.
6. J.N. Miller, *Pure Appl. Chem.* **57**(3) (1985) 515–522.
7. M. Cais, US Patent 4 205 952 (1980).
8. M.G. Simonsen, *Clin. Biochem. Anal.* **10** (1981) 97.
9. E.L. Wehry, *Anal. Rev. Anal. Chem.* **58** (1986) 13R–33R.
10. W.R. Heineman and H.B. Halsall, *Anal. Chem.* **57**(14) (1985) 1312A.
11. M. Thompson, C.L. Arthur and G.K. Dhaliwal, *Anal. Chem.* **58** (1986) 1206–1209.
12. A.H.W.M. Schuurs and B.K. van Weeman, *J. Immunoassay* **1** (1980) 229.
13. E. Ishikawa, T. Kawai and K. Miyai, *Enzyme Immunoassay*, Igaku-Shoin Ltd., Tokyo (1981).
14. M. Oellerich, *J. Clin. Chem. Clin. Biochem.* **22** (12) (1984) 883–894.
15. R.J. Carrico, J.E. Christner, R.C. Boguslaski and K.K. Yeung, *Anal. Biochem.* **72** (1976) 271.
16. R.J. Carrico, K.K. Yeung, H.R. Schroeder, R.C. Boguslaski, R.T. Buckler and J.E. Christner, *Anal. Biochem.* **76** (1976) 95.
17. T.P. Whitehead, L.J. Kricka, T.J.N. Carter and G.H.G. Thorpe, *Clin. Chem.* **25** (1979) 153.
18. M.E. Wilson, in *Immunoassays, Clinical Laboratory Techniques for the 1980s*, R.M. Nakamura, W.R. Dito and E.S. Tucker (eds.), Alan R. Liss, New York (1980) 4.
19. M.A. DeLuca and W.D. McElroy (eds.), *Bioluminescence and Chemiluminescence*, Academic Press, New York (1920).
20. T. Olsson and A. Thore, in *Immunoassays 80s, Proc. Conf.* 1980, A. Voller, A. Bartlett and D. Bidwel (eds.), University Park Press, Baltimore (1981) 113.
21. W.R. Seitz, *CRC Crit. Rev., Anal. Chem.* **13** (1981) 1.

22. W.G. Wood, *J. Clin. Chem. Biochem.* **22**(12) (1984) 905–918.
23. E.F. Ullman and P.L. Khanna, in *Methods in Enzymology*, (1981), 74C.
24. E.F. Ullman, M. Schwarzberg and K.E. Rubenstein, *J. Biol. Chem.* **251** (1976) 4127.
25. P.S.C. van der Plas, F.A. Huf and H.J. De Jong, *Pharm. Weekbl.* **116** (1981) 1341 (*Anal. Abs.* **42**(5) (1982), Abstract No. 5D5).
26. G.C. Visor and S.G. Schulman, *J. Pharm. Sci.* **70** (1981) 467.
27. J.M. Hicks, *Human Pathol.* **15**(2) (1984) 112.
28. M. Cais, *Nature (London)* **270** (1977) 534.
29. N.J. Wilmott, J.N. Miller and J.F. Tyson, *Analyst*, **109**(3) (1984) 343.
30. J.H.W. Leuvering, P.J.H.M. Thal, M. van der Waart and A.H.W.M. Schuurs, *J. Immunoassay* **1** (1980) 77.
31. C.L. Cambiasio, A.E. Leek, F. De Steenwinkel, J. Billen and P.L. Masson, *J. Immunol. Meth.* **18** (1977) 33.
32. J. Haimovich and M. Sela, *J. Immunol.* **97** (1966) 338.
33. O. Makela, *Immunology* **10** (1966) 81.
34. R.K. Leute, E.F. Ullman, A. Goldstein and L.A. Harzenberg, *Nature, New Biol.* **236** (1972) 93.
35. C. Yuan, S. Kaun and G.G. Guilbault, *Anal. Chem.* **53** (1980) 190.
36. E.J. Shaw, R.A.A. Watson, J. Landon and D.S. Smith, *J. Clin. Pathol.* **30** (1977) 526.
37. C.S. Lim, Ph.D. Dissertation, Loughborough University (1980).
38. W.B. Dandliker, J. Kelly, J. Dandliker, J. Farquhar and J. Levin, *Immunochemistry* **10** (1973) 219.
39. R.F. Zuk, I. Gibbons, G.L. Rowley and E.F. Ullman, US Patent No. 4,281,061 (1980).
40. R.F. Zuk, G.L. Rowley and E.F. Ullman, *Clin. Chem.* **25** (1979) 1554.
41. T. Jackson, S. Dakubu and R.P. Ekvis, *Med. Lab. World* **23** (1983) 25–26, 78, 75.
42. M.N. Kronick and W.A. Little, *J. Immunol. Meth.* **8** (1975) 235.

7 Selective chemical transduction based on chemoreceptive control of membrane ion permeability

M. THOMPSON and W.H. DORN

7.1 Introduction

Biological organisms respond to chemical messages internally in the form of various transmitters and hormones, and externally via a host of sensory receptor cells (1). Examples of the former are the postsynaptic membrane in neurons which is influenced by the neurotransmitter acetylcholine (2), and the protein insulin which interacts with receptors for control of the glucose level. External sensory processes of great significance are the senses of smell (olfaction, Figure 7.1) and taste (gustation) (3). Despite the enormous differences in the engineering of these senses across the spectrum of biological species, a common feature is the use of receptor cells which respond to chemical stimuli in the outside environment. Chemical sensitivity is employed in the search for food, declaration of alarm, trail marking, aggregation and dispersion, mating and general communication. In this respect, the role of the pheromone–insect antenna interaction has often served as a model for the study of olfaction because of the relatively simple array of chemical moieties detected by the species.

Although the cells which respond to neurotransmitters and the olfactory neurons have, of course, completely different electrophysiological functions, the recognition and transduction mechanisms apparently share some common ground. Molecular recognition is effected by lipid membrane-embedded proteins, which govern the degree of chemical stimulant binding by shape and electronic distribution criteria. This binding event is translated into a sudden change in permeability of the membrane to inorganic ions (opening of the ion channel). At a certain level of occupation of protein binding sites the membrane undergoes depolarization. It is fair to say that the detailed link between binding of stimulant and electrophysiological event is far from understood at the present time. In contrast with the postsynaptic acetylcholine receptor-containing cell, the olfactory epithelium is made up of 10–20 million cells which can apparently differentiate between a number of odorant types. Here, it is certain that molecular identification involves pattern recognition across the two-dimensional array set up by the epithelial structure. One further point worth noting is that the natural sensory process clearly involves out-of-equilibrium digital processes (i.e. firing of neurons).

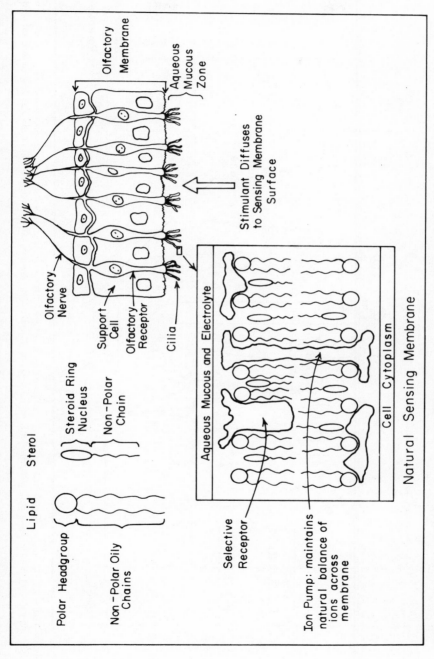

Figure 7.1 Schematic of olfactory neurons and simplified version of receptor-containing membrane. Stimulant diffuses through mucous layer to bind to molecular receptor. Threshold occupation of molecular receptors on cilia of neuron result in membrane depolarization. (Reprinted by kind permission of Elsevier Science Publishers, B.V., Amsterdam).

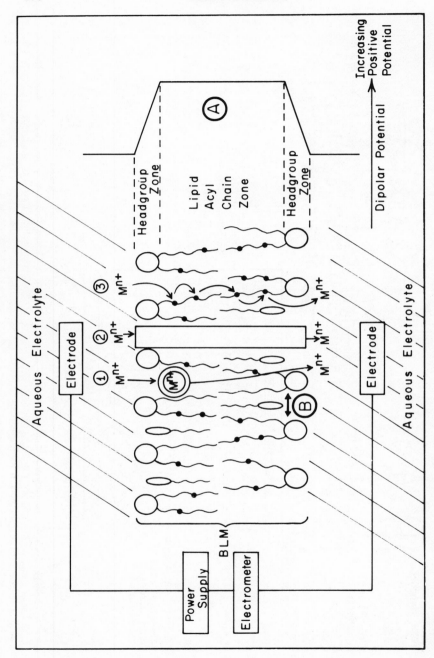

Figure 7.2 Representation of BLM structure configured to operate in electro-analytical operation. Ion transport processes are hydrophobic carriers (1), pores (2) and passive electrodiffusion via polar binding sites (3), (Reprinted by kind permission of Elsevier Science Publishers, B.V., Amsterdam).

Table 7.1 Developing technologies based on biological function

Technology	Biological process
Computer science; parallel processing	Brain, neural networks
Shape recognition, object identity in robotics	Vision
Tactile systems, robotics	Sense of touch
Hydrodynamics of ship motion	Surface free energy of proteins employed in skin of aquatic species
Molecular electronics	Organized membrane assemblies
Chemical and biosensors	Sense of smell, electrophysiology hormone stimulation, chemoreception

This type of mechanism has been virtually ignored by the chemical sensor community which has relied totally on 'analogue' signals from optical, electrical, acoustic wave, etc., devices (4).

The membrane electrochemistry alluded to above can be simulated to some degree by conducting experiments with artificially constructed bilayer lipid membranes (BLM). With regard to chemical sensor technology, Del Castillo and co-workers (5) were the first to point out that the BLM could be the basis of a membrane-based device. The term 'lipid–membrane conductimetry' was coined as a result of this suggestion. This concept was resuscitated by Thompson *et al.* (6–8) in an examination of the technique as applied to the detection of antibiotics. Basically, the operation of the system relies on the alteration of membrane parameters which control a small but finite ion flux through the membrane by a selective complexation at the membrane surface (Figure 7.2). This perturbation of membrane physical chemistry changes the conductivity of the membrane to ions when a specific driving force for transmembrane ion movement is available. Such a perturbation can involve a number of different scenarios which will be discussed later. In essence the process can be associated with gating of ion channels, 'pumping' of ions, hydrophobic ion chemistry, or straightforward perturbation of membrane parameters such as surface dipolar potential, fluidity or Gouy–Chapman potential.

In summary, the attempt is made to copy at least an element of the natural sensor technology, viz. electrophysiology, although, of course, we are very far from being able to achieve this aim at the present time (4). Despite the fundamental and technical difficulties, the model is an attractive one from several angles. Furthermore, the direct 'mimicking' of biological processes for technological development is a rapidly burgeoning area, as can be seen in Table 7.1.

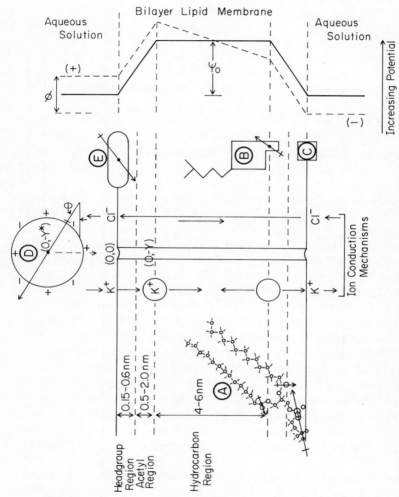

Figure 7.3 Origin of membrane dipole potential. Key. A–phosphatidyl choline and dipole structure, B—location of cholesterol, C—ion adsorption, D—long-range dipole interaction involving macromolecule, E—dipole contribution of membrane embedded species. See Figure 7.2 for description of modes of ion transport (centre of diagram). Right of figure depicts dipole potential. (Reprinted by kind permission of Elsevier Science Publishers, B.V., Amsterdam).

7.2 Theoretical aspects of ion permeation in unmodified membranes

In this section, we will be concerned with a concise review of the physical chemistry of inorganic ion transport in hypothetical unmodified BLM, that is, those membranes which do not contain protein or other mediating species. We shall not be concerned with hydrophobic or carrier transport or with channel or pore effects which have been reviewed extensively elsewhere (9, 10). For the interested reader, much of the appropriate physics in this area has been reviewed by Lakshminarayanaiah (11), and the electrostatic aspects have been critically evaluated by Dorn (12).

We begin with a brief description of the structure of a hypothetical BLM made from phosphatidyl choline (PC) which is a major constituent of many cell membranes. Many of the salient features are depicted in Figure 7.3. The membrane is of 6–8 nm thickness and is composed of a double layer of lipid molecules, that is, two planar and adjacent monolayers oriented in a symmetrical but opposing manner, with the polar groups making up the outer surface of the membrane. This structure has a central non-polar hydrocarbon core region bounded on both sides by polar sheets of lipid headgroups which are hydrated in supporting aqueous electrolyte. The headgroup of PC is zwitterionic with the centre of positive charge being located on the nitrogen atom, and the negative charge being spread between the phosphorus and oxygen atoms. This configuration yields a dipole moment of some 20 Debye units, often termed the P–N vector. Dipoles also originate with the two carbonyl functionalities and water structure at the headgroup, resulting in an overall net perpendicular dipole. The effect of this close-packed and aligned array of molecular dipoles is to produce a positive transmembrane potential, the so-called surface dipolar potential. This electrostatic field forms part of the barrier to cation transport shown schematically at the right of Figure 7.3.

Models for the transport of ions across BLMs are usually based upon the fundamental laws of electrostatics and electrodiffusion. Traditionally, these treatments consider the membrane to be a uniform macroscopic phase, devoid of chemistry (such as incorporation of cholesterol shown in Figure 7.3). This is a very serious fault, but at least the simple derived expressions provide a qualitative picture of the problem. In broad terms, it is possible to consider two separate factors which determine the mechanism of charge transport, electrostatic and steric (lipid fluidity), although there may well be a synergism between the two influences. The potential energy barriers to introduction of ions into the membrane can then be included into generalized expressions for electrodiffusion.

The Nernst–Planck approach to ion conduction, which is the model most frequently used, treats the membrane as a homogeneous, macroscopic dielectric of specified thickness. The ions have infinite mobility in the aqueous phase, activity coefficients are unities, there are no bound charges and the ion concentrations at large distances from the membrane are constant. For convenience, anionic and cationic valences are usually taken as -1 and $+1$, respectively. The flux J of an ion is taken as the gradient of the electrochemical potential μ of that ion:

$$J = -uC\frac{d\mu}{dx} = -uC\frac{d(\mu_0 + kT\ln C + zF\phi)}{dx}$$

when C is its concentration, x is the distance along the line normal to the surface of the bilayer, ϕ is the potential at some point along x and u is the mobility of the ion. The term μ_0 is the standard chemical potential, equal to the work W required to transfer the ion from infinity to a point x in the membrane plus an additive constant:

$$\mu_0(x) = W(x) + \text{const.}$$

This potential energy has also been called the intrinsic potential, and it is important to note that it contains the surface dipolar potential discussed above. It is now necessary to make one or more of the following assumptions:

(i) Electroneutrality holds in the membrane
(ii) The electrostatic field in the membrane is constant
(iii) The ionic concentrations at the interface are equal.

In the constant field assumption the potential drops off linearly across the membrane:

$$\phi = Vx/d$$

where V is the applied voltage, and ϕ is the potential at point x in a membrane of thickness d. If it is assumed that the potential drops off very quickly outside the membrane, that is

$$W(x=0) \sim 0, \quad W(x=d) \sim 0$$

a solution can be found using standard methods for solving linear first-order differential equations:

$$J = -D\frac{C''\exp(V) - C'}{\displaystyle\int_0^d \exp(W+\phi)}$$

and

$$C(x) = \exp-(W+\phi)\left[C' + (C''\exp(V) - C')\frac{\displaystyle\int_0^x \exp(W+\phi)\,dx}{\displaystyle\int_0^d \exp(W+\phi)\,dx}\right]$$

where D is a diffusion coefficient, and C' and C'' are the ion concentrations at $x=0$ and $x=d$ respectively. The flux can then be determined for various values of the intrinsic barrier, $W(x)$. The barrier 'shapes' that have been examined are shown in Figure 7.4. The results of these calculations are then generally compared with experimental I-V curves. Interestingly, if the single central barrier is sufficiently narrow and high, then the conductance is

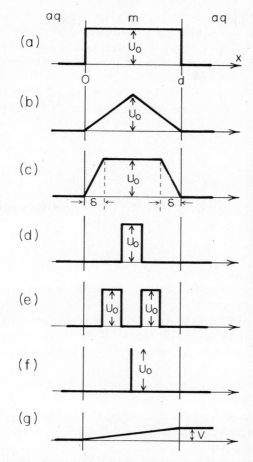

Figure 7.4 Typical potential energy ion barriers postulated for membrane ion transport.

proportional to $\sin h\,(zV/2)/V$, as one would obtain from Eyring rate theory. The rate constant (K_i) is then

$$K_i = \frac{kT}{h}\exp\left(-(W_0 + zV/2)\right)$$

Now we need to consider on a theoretical basis the total potential energy (U) for an ion in the membrane:

$$U = U_{Born} + U_{Image} + U_{Dipole}$$

The Born or 'self' energy is the energy gained by the ion due to polarization of the medium surrounding it. The energy difference between aqueous and membrane phases is

$$U_{Born} = \frac{q_0^2}{8\pi\varepsilon_r}\left(\frac{1}{\varepsilon_m} - \frac{1}{\varepsilon_{aq}}\right)$$

where r is the radius of the ion, ε_0 is the permittivity of free space, q_0 is the charge on the ion, and ε_m and ε_{aq} are the dielectric constants for the membrane (~ 2) and water (80), respectively. The image potential is the contribution from polarization of the aqueous medium in the vicinity of the membrane surface, caused by the ion in the membrane. Finally, we must add the dipolar potential outlined above. For two dipolar layers of finite size (radius R), whose dipoles are pointing vertically, the electric potential energy profile (V_v) is given by

$$V_v = -\frac{P}{2\varepsilon A}\left\{2 - \frac{x}{(R^2 + x^2)^{1/2}} - \frac{d - x}{[R^2 + (d - x)^2]^{1/2}}\right\}$$

where P is the dipole moment and A is the membrane area. In the case where the dipoles point at some angle to the plane, the horizontal field components will cancel, leaving only the vertical components to contribute to the potential (V_a). For the aligned dipole area:

$$V_a = \frac{P}{4\varepsilon A}\left[\ln\frac{R^2 + x^2}{x^2} + \ln\frac{R^2 + (d - z^2)^2}{(d - x)^2}\right]$$

A further crucial consideration is the possible, but unsubstantiated, effect of dipolar reorientation. Dorn (12) has suggested that the polar head group could reorient under the field of an ion. Such a process would yield a decrease of the overall barrier to ion transport. If the dipole is completely free to rotate under the influence of temperature, the interaction energy would be

$$U_{ion-dipole} = -\frac{1.27 \times 10^4 z^2 p^2}{r^4 \varepsilon_r^2 T} kJ/mole$$

where z is the valence of the ion, and the dipole moment (p) is given in Debyes, r is the separation between the ion and part dipole in Å, and ε_r is the relative dielectric constant ($\varepsilon/\varepsilon_0$).

As a final remark in this section, we should emphasize that perturbation of the low currents involved in BLM offers both a sensitive and generic mechanism for sensor technology. The key, of course, from a fundamental standpoint lies in an understanding of alteration of the energy barriers discussed above by complexation of strategically placed receptors by stimulants (13). Clearly, one possibility is given by electrostatic effects. The membrane provides a low dielectric medium in which electrostatic interactions can be very large. One can expect the electric potential and field magnitudes to be about forty times higher than those found in the aqueous environment. Accordingly, receptor 'conformational' change on stimulant binding may be important in controlling membrane ion permeability.

7.3 Molecular receptors and possibilities for signal generation

Receptorology has become a very important field in biochemical research.

Here, we shall only be concerned with a brief treatment of the area in the context of sensor development. For an excellent review of the field, the reader should consult the text provided by Levitzki (14). Receptors are specific cellular components that interact with certain ligands such as hormones, neurotransmitters, antibodies, growth factors etc. The binding of the ligand results in a biological effect through the effector system to which the receptor is coupled. The effector system may be an enzyme such as adenylate cyclase or an ion channel. An agonist triggers a response, whereas an antagonist binds to the receptor, but does not activate the appropriate biochemical process.

It is apparent that the biological receptor–effector process incorporates much of the transduction physical chemistry required for the design of biosensors. The dynamic system involves occupancy of the receptor by an agonist, coupling of the biochemical process with the receptor, instigation of a series of intracellular events (the overall cell response) and desensitization. Table 7.2, provided by Levitzki, classifies the specific intracellular biochemical processes elicited by receptor–ligand interactions. Important features of receptor function are ligand selectivity, the nature of the effector system (which can be quite complex), and the overall time-scale of the event. With respect to the latter, the nicotinic receptor is activated within 1 millisecond by acetylcholine, whereas the β-adrenergic system takes several seconds after ligand occupancy. Despite this enormous range in time-scale, the process is generally much faster and, therefore more attractive, than existing biosensor technology, which can involve response times of minutes (viz. enzyme electrodes).

Now we turn to some brief remarks concerning aspects of ligand–receptor binding, energetics and transduction. With respect to the former, the

Table 7.2 Categories of chemoreceptive processes at lipid membranes (14)

Biochemical process	Receptor

1. Ion flux
    (a) Influx of Na^+ during
        depolarization event
	Nicotinic

(b) Influx of Ca^{2+}	Muscarinic (type 1), α_1-adrenergic, H$_1$ histaminergic
(c) Efflux of K^+	Cardiac muscarinic
(d) Influx of Cl^-	γ-Aminobutyric acid ergic, glycine
2. Activation of adenylate cyclase	Glucagon, adrenocorticotrophical hormone, β-adrenergic, vasointestinal peptide, adenosine (A2)
3. Inhibition of adenylate cyclase	Muscarinic (type 2), α_2-adrenergic, adenosine (A1), μ-type morphine
4. Phosphorylation of a receptor or receptor-linked component	Epidermal growth factor, insulin platelet derived growth factor

distinction between affinity of binding and effector potency must be emphasized. An antagonist may well bind to a receptor site more 'tightly' than an agonist, but in fact, it can result in the *blocking* of biological response. In this respect it is often the case that antagonists are poisonous, an example being the neurotoxin, α-bungarotoxin, that binds to the nicotinic receptor. Controlled introduction of such 'poisons' often constitutes drug therapy! Returning to the affinity theme, as with any conventional chemical interaction the binding event involves a change of system free energy:

$$A_o + B_i \rightleftharpoons AB_i \quad K_{oi} = \exp(-G_{oi}/RT)$$

where an effector A_o will bind with ligands B_i, but only B_o or B_i will cause a biochemical response. Among an array of interactive forces contributing to the change in free energy will be entropy, dipole–dipole, dipole–induced dipole and hyrophobic interaction, hydrogen bonds and change transfer. In the case of the immune complex where great specificity is required (recognition by A_o, the antibody, must occur for B_o, the antigen, from over 10^5 self-antigenic determinants) several of these forces are involved, in a concerted manner, to maximize the binding energy. However, this is not necessarily the case for the receptor–agonist interaction. A picture is emerging which shows that receptor binding of relatively small ligands involves a low number of weak forces, viz. hydrogen bonds in a defined spatial arrangement. One example is shown in Figure 7.5. Dopamine apparently binds to one of the receptor sites in a tetrahedral configuration (15). This binding 'structure' is currently being examined by computer graphic techniques in our laboratory. In summary, it appears that a relatively low energy of binding might confer some reversibility on the system, without total loss of selectivity.

With respect to receptor energetics it appears to be an accepted dogma that agonist binding induces a conformational change in the receptor macro-

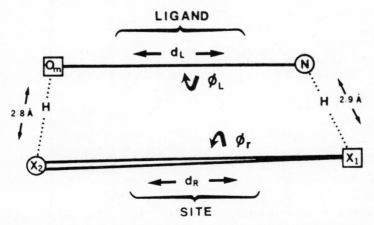

Figure 7.5 Binding characteristics of dopamine to putative receptor sites (15).

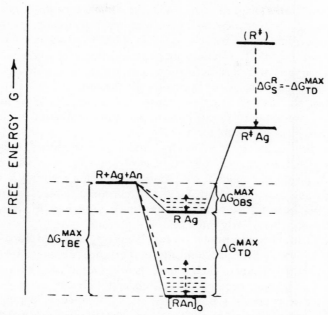

Figure 7.6 Proposed free energy description for receptor binding to agonist and antagonist. (Reprinted by kind permission of Imprimerie Biscaye, Bordeaux).

molecule. Such a perturbation is considered to be crucial in eliciting the overall biochemical response. Despite this acceptance there is at the present time no comprehensive theory about the process, nor, indeed, for that matter, any detailed experimental evidence of conformational changes. Based on the arguments of Jencks (16) concerning enzymes, Franklin (17) and Tedesco *et al.* (18) have argued that prior complexing of receptor to ligand provides the energy required for conformational changes (Figure 7.6). Indeed, the measured binding energy is quite different from the intrinsic energy of binding mentioned above in that some energy has been used in receptor complex 'destabilization.' Turning to transduction, the state change referred to above probably occurs during the lifetime of the agonist–receptor complex. Thus, it may proceed either concurrent with the initial complex formation (producing adsorption-rate-dependent kinetics for biochemical response) or during the course of the complex's lifetime (yielding occupation-dependent kinetics) or concurrent with dissociation of the complex (desorption-rate-dependent kinetics). Apparently there are two major categories of membrane transduction mechanisms set up by these state changes: (i) the perturbation of ion permeability, and (ii) the so-called second messenger system. Here, cyclic AMP is involved, for example, to act as a messenger.

One final remark should be made here regarding the concept of desensitization of receptors. From the sensor point of view, one would argue that this effect constitutes 'dead time' in which the structure cannot respond. To the biochemist, however, desensitization can mean a number of things. It can apply to loss in a receptor's affinity for an agonist, or to a loss of receptors (or at least sites), or both. In addition, there is always the distinction to be made between receptor desensitization and transduction desensitization which is associated with loss of biochemical response.

What does all this mean for the sensor specialist who is interested in mediation of membrane ion currents by receptor–stimulant interactions? There are two clear philosophies here.

1. *Direct* use can be made of electroactive receptor species, such as the acetylcholine receptor extracted from the electroplax organ of fish. The idea would be to purify the material for reconstitution in a membrane which is immobilized on a sensor structure. Such a system could be used as a 'negative' sensor for neurotoxins. Naturally, the military branches of a number of governments are particularly interested in this route.

2. Through study of receptor physical chemistry we would hope to learn enough about the salient mechanisms to persuade designed membrane/synthetic receptor systems to elicit electrochemical responses. It is important to note here that, although second-messenger biological receptor systems exhibit no electrochemical (viz. channel) effects of obvious significance, in an *artificial* configuration such changes may indeed be observed. This could also be the case for some generic bimolecular interactions as we shall see.

With respect to point 2, taking into consideration the theoretical discussion of the previous section, we can summarize the following possibilities for perturbation of membrane ion currents in a non-biochemical fashion (19):

(i) Receptor–ligand complexation could result in an alteration of the overall surface dipolar potential of the sensor membrane. Such a perturbation could be associated with disturbance of the ligand headgroup structure itself and/or a realignment of dipoles in the receptor–stimulant complex itself.

(ii) Modification of membrane fluidity and molecular packing. Membrane ion conduction is sensitive to the viscosity and availability of interstitial spaces in the hydrocarbon zone (the 'steric' effect alluded to above). Receptor–ligand interaction can alter adjacent intermolecular interactions (effect of conformational changes?) resulting in changes in the intrinsic properties of the membrane.

(iii) A synthetic pore or channel system could be 'gated' by conformational receptor-based processes.

7.4 Experimental BLM systems

The basic experiment is shown schematically in Figure 7.7. Membranes can be formed in a relatively large aperture (0.5 mm radius) punched into a sheet of

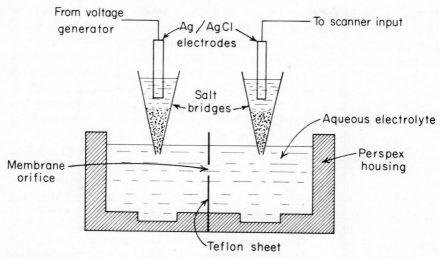

Figure 7.7 Basic electrochemical cell used for the study of BLM. (Reprinted by kind permission of Elsevier Science Publishers, B.V., Amsterdam).

Teflon, or as arrays of 'mini-BLM' in the microstructure of various porous substrates. (Planar solid-phase support of BLM for sensor development will be discussed later.) A Teflon sheet is incorporated into a membrane housing involving separation of two compartments which contain electrolyte. Typically, a membrane-forming solution of approximately 2% weight for volume solution of both phosphatidylcholine and cholesterol in n-decane is applied to the orifice by brush or syringe. Within a few minutes the lipid film thins to bilayer thickness, and usually the film is observed during this process by light microscope to detect appropriate interference patterns. This procedure requires considerable experimental skill, as one would expect with a membrane of approximately 8 nm thickness. The thinning process can also be monitored by an electrometer. Transmembrane conductance is measured with single-junction silver–silver chloride electrodes placed in each electrolyte compartment in conjunction with a microprocessor-controlled electrometer. The cell is usually mounted on a vibration-damping platform which is enclosed in a well-grounded Faraday cage.

Typical BLM currents are of the order of 10–100 pA for applied dc voltages of about 20 mV. Given that the membrane covers an area of approximately 0.0075 cm^2 and is of the order of 5 nm thick, the following data are typical:

Resistance $\sim 10^9\,\Omega$
Area-specific resistance $\sim 7.5 \times 10^6\,\Omega\,cm^2$
Area-specific conductance $\sim 1.3 \times 10^{-7}\,S\,cm^{-2}$
Resistivity $\sim 6.7 \times 10^4\,\Omega\,cm^{-1}$
Permeability $\sim 3.48 \times 10^{-10}\,cm\,s^{-1}$

The membrane and electrodes can be represented by a parallel combination of a resistor with a capacitor. The electrodes and aqueous electrolyte solution have resistances which are orders of magnitude lower than that of the membrane and can generally be ignored. The electrometer, operating in the voltmeter mode, must, on the other hand, have a very high resistance ($10^{15}\,\Omega$ input impedance). With the voltmeter operating in series it is possible to observe the discharging of the membrane when the electrometer is switched from current to voltage mode. From this type of experiment the membrane capacitance, about 4 nF, can be obtained.

The technique described above results in bilayers containing trapped solvent which supposedly leads to poor reproducibility in electrochemical measurements. Accordingly, a number of designs have been generated in recent years to produce 'solvent-free' bilayers. However, if measurement trends are all that is required the above will suffice.

When the membrane is completely thinned and stable, various stimulants can be introduced into the electrolyte in one compartment with stirring, and the conductance can be monitored over a suitable period of time. One interesting variation of this experiment was the incorporation of a special flow-through cell to allow the system to be included in a flow injection apparatus (20). In this way, the conductance can be measured as plugs of stimulant flow past a bilayer membrane surface.

In view of the difficulties usually experienced in producing BLM of area $\sim 0.008\,\mathrm{cm}^2$, porous structures have been used to form mini-membranes. Two examples of this are nylon polymer films grown at an interface and polycarbonate films normally employed in organic chemistry for the purpose of filtration (21, 22). In the former case, a nylon film was synthesized at the interface between water and an organic solvent. Electron microscopy showed that the polymer structure was close-packed on one with a 'Swiss-cheese' structure on the other. Treatment of the films with lipid solution resulted in functional mini-membranes. Polycarbonate films can be employed in a similar manner. Although such systems are relatively simple to work with, one large imponderable is the overall area of membrane produced for study.

7.5 Examples of membrane conductance mediated by biomolecular/receptor interactions

7.5.1 *The nicotinic acetylcholine receptor*

This system has attracted enormous attention over recent years. Stimulation of the presynaptic membrane of a neuron results in the release of acetylcholine which then diffuses across the synapse. Binding occurs on the receptor (AchR) located in the postsynaptic membrane, producing the opening of an ion

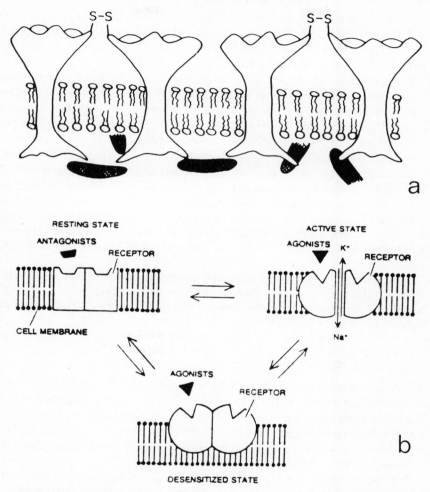

Figure 7.8 Schematic representation of acetylcholine receptor in BLM; cross-hatched area depicts associated protein (*a*); proposed AchR states (*b*).

channel and causing a small Na^+ flux into the neuron. Nearly all research involving this receptor has been conducted on material extracted from the electroplax organ of electric eels and rays. The AchR is composed of four discrete subunits of approximate molecular weights 40K (α subunit), 50 K (α), 60 K (γ) and 65 K about 4–7% by weight, with 25% of the AchR having helical secondary structure. Electron microscopy and neutron diffraction has shown that the AchR is 11 nm long, protruding 5 nm long and 9 nm wide into the synapse, and 1.5 nm long and 6 nm wide into the cytoplasm (Figure 7.8*a*). Apparently, there is a central hole which is 2.5 nm wide at the synapse side and 0.65 nm at the narrowest point. It has been proposed that the receptor complex involves three conformation states (Figure 7.8*b*); a resting state, active state and desensitized state (23). These states have different affinities for the agonist. The active state, which is ion-conductive, is in rapid equilibrium with the low-affinity ground or resting state.

Reconstitution of the receptor in planar BLM allows electrochemical studies of ion-channel physics to be performed (24). Acetylcholine binding results in the opening of large groups of ion channels and single ion-channel events. Currents are in the range of pA (a net flow of about 12 000 ions per millisecond). The channel behaviour appears to be sensitive to the nature of surrounding lipid and to the type of extracted protein material that is used. Interestingly, the channels open without the presence of agonist, and noise is observed for open-channel events. Finally, it should be mentioned that the link between agonist binding and electrophysiology has not yet been identified, as is the case for suggested conformational changes. The opinion is very widely held that the ions permeate down the centre of a hydrophilic channel formed by the protein, but as yet there is only circumstantial evidence that this is the case.

This receptor system is particularly important for the sensor specialist since it fulfils both the applicational criteria specified above, that is, it could be utilized directly, and it offers promise as a role model for selective sensing technology.

7.5.2 *The plant auxin receptor*

Maize coleoptile membranes contain high-affinity auxin-binding sites that have attracted significant research attention in view of their importance with respect to plant growth biochemistry. Reported dissociation constants for the auxin naphthalene-1-acetic acid (NAA) are in the range $0.1-1\ \mu M$. There is a suggestion that auxin binding to the receptor induces activation of a proton-pumping ATPase in the cell membrane.

Purified receptor together with ATPase was reconstituted in BLM by Thompson *et al.* (25) for electrochemical experiments as outlined above. Figure 7.9 shows the electrochemical response of a BLM to fractions of receptor protein, ATP and naphthalene-1-acetic acid in various time sequences under an applied dc voltage of 15 mV. Only when all three components are present (in any order) is a substantial increase in current observed. Adjustment of hydronium ion concentration in solution resulted in a correlation of current at various pH values with the known receptor biochemical activity at the same values. This behaviour indicates that previous suggestions that an ATPase pump acts to translocate protons are indeed correct. From the sensor point of view, it is interesting to note that the signals exhibited a limit of detection of about 10^{-7} M NAA and were quite selective for the hormone.

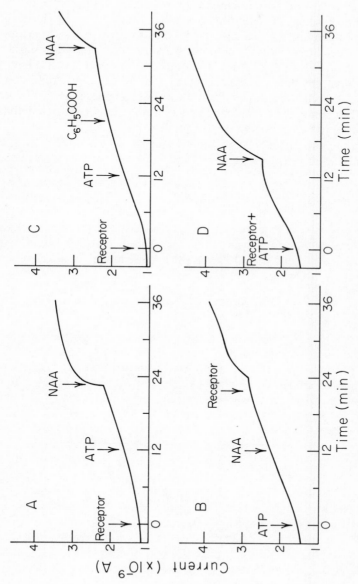

Figure 7.9 BLM response to auxin receptor, naphthalene 1-acetic acid (NAA), ATP and C_6H_5COOH in various combinations. (Reprinted by kind permission of Academic Press, Orlando, Fl).

7.5.3 *The lectin–polysaccharide and antibody–antigen combinations*

The lectin concanavalin A is a globulin capable of binding certain saccharide residues in a fashion that resembles an antibody–antigen interaction. The protein has a particular conformation that is induced by the binding of a transition metal ion, for example Mn^{2+}, and calcium ion. Its structure has been elucidated by x-ray diffraction at 0.2 nm resolution, indicating dimensions of about $4.2 \times 4.0 \times 3.9$ nm. The double ion site is located 2.3 nm from the carbohydrate binding site, implying that large-range conformational effects must occur. An independent hydrophobic pocket is available and suitable for non-specific binding to membranes. It is considered that the protein does not span membrane structures.

Membrane experiments were carried out by Thompson *et al.* (26) by allowing BLM to interact with Mn^{2+} and Ca^{2+}, followed by introduction of

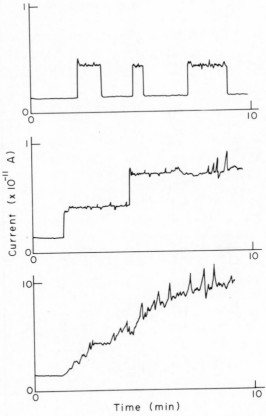

Figure 7.10   BLM responses to concanavalin A–Mn^{2+}, Ca^{2+} – polysaccharide combinations. (Reprinted by kind permission of Elsevier Science Publishers, B.V., Amsterdam).

polysaccharides such as glycogen. The protein itself caused little or no change in the transmembrane current, but in the presence of the sugar highly characterizable step-changes in current were observed (Figure 7.10). In view of the fact that the lectin system has often been used as a model for immunochemical interactions, analogous experiments were performed with rabbit IgG antiserum and rabbit IgG immunoglobulin. Similar results were obtained to those found for the concanavalin A experiments. Previously, such electrochemical behaviour has been interpreted in terms of the formation of transient lectin–sugar 'pores'. In reality, a completely different structure must apply. The membrane step-current responses are clearly caused by aggregate formation involving the complexes at the membrane surface. In some way the membrane is being perturbed by stochastic processes probably involving two-dimensional nucleation and aggregate growth. The change in current must be caused by the aggregate dipole contribution to the surface of the BLM and/or lipid headgroup dipole reorientation, as discussed above. Strong support for this argument comes from experiments conducted with immune complex system in the subphase of a Langmuir–Blodgett trough where transient changes in measured surface dipole potential are observed (27). Finally, from an analytical point of view the responses described above are highly selective and originate from $\sim 10^{-8}$ M concentrations of sugar or antiserum.

7.5.4 *Olfactory epithelial protein*

In a particularly interesting experiment, Vodyanoy and Murphy (28) pro-cessed samples of olfactory epithelial tissue obtained from rats to obtain a homogenate containing vesicles. Introduction of the homogenate to one side of a BLM prepared in a somewhat similar manner to that described above resulted in the observation of single-channel conductance fluctuations. Apparently these died to zero after about thirty minutes, but addition of a solution of diethyl sulphide (an odorant?) to the same side of the membrane produced a new set of channel openings. The mean time in the open state and average conductance for spontaneous channels was 29.3 ± 7.8 seconds and 58 ± 7 pS, respectively. In the presence of the sulphide the analogous values were 42.3 ± 10.00 seconds and 56 ± 10 pS, respectively. The authors concluded that the data suggested they had been able to generate discrete chemosensitive ion channels in their artificial BLM in association with the initial events of chemoreception.

7.6 Towards a practical chemoreceptive BLM sensor

The route to a practical sensor based on the physical chemistry offered by the BLM depends essentially on the solution of three different technical problems, viz. the production of stable, substrate-supported membranes, the incorporation of natural or synthetic receptors into these structures, and the fabrication of a real device. Notwithstanding research efforts in these areas, work will, of course, continue into the understanding of potential of chemoreceptive mechanisms.

The practical use of the BLM is severely handicapped by the delicate nature of the structure which is entirely dependent for its stability on hydrogen bonding and Van der Waals' forces. In order to approach a solution to this difficulty, several methods have been developed to transfer the intact structure to a solid support. These are:

(i) The use of Langmuir–Blodgett film technology
(ii) The production of lipid species capable of being cross-linked between hydrophobic chains or elsewhere in the molecule-polymerizable lipids—this technique can be associated with the Langmuir–Blodgett procedure
(iii) The synthesis of lipid species capable of being covalently bound to a substrate surface.

In the method developed by Langmuir and Blodgett in the 1920s, a lipid (surfactant) monolayer is compressed on a trough of water. The organized assembly of lipid molecules can then be transferred to a solid substrate by dipping the latter through the film to produce a monolayer, or, by repeated dipping, multilayers. Recent times have seen a veritable renaissance in this technology with emphasis being placed on thin-film applications in drug encapsulation, semiconductor fabrication, sensors and molecular electronics. In our own work we were able to produce a simple device by depositing bilayers of lipid films on to polyacrylamide gel (29). The device was prepared by production of a Ag/AgCl film on a glass wafer, followed by introduction of a well using polymer resin. The well was filled with gel and the structure included in a Langmuir–Blodgett/electrochemical station as shown schematically in Figure 7.11. The configuration described above could not be removed from the trough, but known BLM perturbants were added to the subphase electrolyte to examine the nature of the film deposited on the gel. In a number of experiments significant increases in dc current were observed for introduction of valinomycin (a hydrophobic ion carrier). Despite this encouraging beginning, significant difficulty was experienced in reproducibility, probably caused by differences in gel cross-linking, hydration character and surface morphology.

A second possibility is provided by phospholipid species containing diacetylenic groups which can be polymerized by introduction of light. In an elegant experiment, Dalziel et al. (30) described the use of polymerized lipid films produced in association with Langmuir–Blodgett technology. The intact films were transferred to quartz substrates for further interaction, with films generated from lipid vesicles containing extracted AchR (see above). From this work it was not clear whether functional receptor could be incorporated into a bilayer film in view of the fact that the molecule, which is expected to span the bilayer, would not only have to penetrate the polymer layer, but also be introduced in the correct orientation.

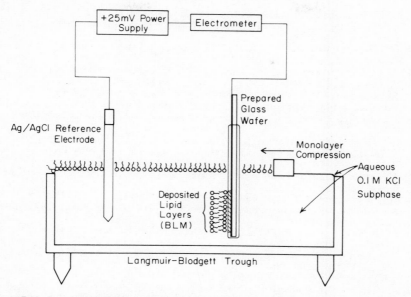

Figure 7.11 Langmuir–Blodgett configuration used for *in situ* deposition and electrochemical perturbation of device-BLM structure. (Reprinted by kind permission of Elsevier Science Publishers, B.V., Amsterdam).

It is expected that a significant component of lipid membrane physical chemistry could be retained if the molecules were attached to a surface by bondable groups located at the end of the acyl chain. In this respect Kallury *et al.* (31) from the Chemical Sensors Group at the University of Toronto has devised a route to yield silanized lipid. Studies of binding of this moiety to model silicon surfaces have so far proved to be particularly encouraging.

The Group mentioned above is collaborating with the University of Utah in the development of a surface-supported ac impedance-membrane based sensor. The structure will incorporate AchR to generate a negative gas sensor for nerve toxins. In this effort, Langmuir–Blodgett depositions on various surface morphologies of silver chloride (electrodic substrate) and design of ac circuitry are being developed. Eventually miniaturization of the whole structure is planned.

Acknowledgements

The Chemical Sensors Group at the University of Toronto is grateful to the Natural Sciences and Engineering Research Council of Canada, The Ministry of the Environment, Province of Ontario, Defence Advanced Research Project Agency and Allied-Signal, Canada for support of their work.

References

1. W. F. Ganong, *The Nervous System* (2nd edn.), Lange Medical Publications, Los Altos, California (1979).

2. B.M. Conti-Tronconi and M.A. Raftery, *Ann. Rev. Biochem.* **51** (1982) 491.
3. E.C. Carterette and M.P. Friedmann (eds.) *Handbook of Perception*, Vol. VI A, Academic Press, New York (1978).
4. M. Thompson, W.H. Dorn, U.J. Krull, J.S. Tauskela, E.T. Vandenberg and H.E. Wong, *Anal. Chim. Acta* **180** (1986) 251.
5. J. de Castillo, A. Rodriguez, C.A. Romero and V. Sanchez, *Science* **153** (1966) 185.
6. M. Thompson, P.J. Worsfold, J.M. Holuk and E.A. Stubley, *Anal. Chim. Acta* **104** (1979) 195.
7. M. Thompson, U.J. Krull and P.J. Worsfold, *Anal. Chim. Acta* **117** (1980) 133.
8. U.J. Krull and M. Thompson, *IEEE Trans. Elect. Devices* **ED-32** (1985) 1180.
9. P. Lauger and B. Neumcke, in *Membranes*, G. Eisenmann (ed.), Marcel Dekker, New York, Vol. 1 (1972).
10. D.A. Haydon and S.B. Hladky, *Quant. Rev. Biophys.* **5** (1972) 876.
11. N. Lakshminarayanaiah, *Equations of Membrane Biophysics*, Academic Press, New York (1984).
12. W.H. Dorn, 'The Dipole Potential in Monolayers, Bilayers and Biological Membranes', M.Sc. Thesis, University of Toronto (1985).
13. M. Thompson and U.J. Krull, *Anal. Chim. Acta* **147** (1983) 1.
14. A. Levitzki, *Receptors, a Quantitative Approach*, Benjamin/Cummings Publishing Co., Menlo Park, California (1984).
15. J.L. Tedesco, 'High Affinity Binding Sites for Dopamine', Ph.D. Thesis, University of Toronto (1985).
16. W.P. Jencks, in *Chemical Recognition in Biology*, F. Cahpeville and A.-L. Haenni (eds.), Springer Verlag, Berlin (1980) 3.
17. T.J. Franklin, *Trends Pharm. Sci.* (Nov. 1980) 430.
18. J.L. Tedesco, U.J. Krull and M. Thompson, pers. comm.
19. U.J. Krull, M. Thompson and H.E. Wong, *Chemical Sensors*, D. Schuetzle and R. Hammerle (eds.), *ACS Symp. Ser.* **309** (1986) 351.
20. M. Thompson and U.J. Krull, *Anal. Chim. Acta* **142** (1982) 207.
21. M. Thompson, U.J. Krull and P.J. Worsfold, *Anal. Chim. Acta* **117** (1980) 188.
22. M. Thompson, R.B. Lennox and R.A. McClelland, *Anal. Chem.* **54** (1982) 76.
23. H.A. Lester, *Sci. Amer.* (March 1977) 107.
24. D.P. Corey, *Neuroscience Comment.* **1** (1983) 111.
25. M. Thompson, U.J. Krull and M.A. Venis, *Biochem. Biophys. Res. Commun.* **110** (1983) 300.
26. M. Thompson, U.J. Krull and L.I. Bendell–Young, *Bioelectrochem. Bioenerg.* **13** (1984) 255.
27. H.E. Wong. 'Langmuir-Blodgett Film Technology in the Development of Selective Chemical Sensors', Ph.D. Thesis, University of Toronto (1987).
28. V. Vodyanoy and R.B. Murphy, *Science* **220** (1983) 717.
29. A. Arya, U.J. Krull, M. Thompson and H.E. Wong, *Anal. Chim. Acta* **173** (1985) 331.
30. A.W. Dalziel, J. Georger, R.R. Price, A. Singh and P. Yager, in *Proc. Membrane Proteins Symp.*, San Diego, California, 1986, L.M. Hjelmeland, R. Gennis, M.G. McNamee and S.C. Goheen (eds.), Bio-Rad Publishing Co., in press (1987).
31. R.K.M.R. Kallury, U.J. Krull and M. Thompson, *J. Org. Chem.*, in press (1987).

4

ELECTROCHEMICAL TRANSDUCTION

8 Voltammetric and amperometric transducers

T.E. EDMONDS

8.1 Introduction

A voltammetric measurement is made when the potential difference across an electrochemical cell is scanned from one preset value to another, and the cell current is recorded as a function of the applied potential. The curve so generated is known as a voltammogram. Amperometric measurements are made by recording the current flow in the cell at a single applied potential. At first sight transducers based on voltammetric and amperometric principles appear to be near-ideal devices for probing the selective chemistry of a sensor. In both cases, the essential operational feature of such a transducer is the transfer of an electron or electrons to or from the probed chemistry. It is the flow of these electrons that constitutes the output signal of the transducer. Additionally, voltammetric and amperometric devices are capable of conferring a degree of selectivity to the overall sensing process. This arises from the dependence of the measured current on the applied potential, a dependence that, to some extent, is a function of the standard potential of the redox couple under study. Voltammetric and amperometric transducers are also relatively simple: in the most elementary case, the transducer (electrochemical cell) consists of two electrodes immersed in a suitable electrolyte. A more complex arrangement involves the use of a three-electrode cell, one of the electrodes providing a reference potential.

In order to probe a chemical system with a voltammetric or amperometric transducer the system must contain at least one species that is electroactive, i.e. that undergoes electron transfer with the electrode. In many electroanalytical measurements this species is often the target analyte. The most common alternative to this simple scheme is to select a system in which an electroinactive analyte modifies the behaviour of an electroactive species either through an heterogeneous reaction involving the electrode surface, or through an homogeneous chemical reaction. The electrode surface itself can be modified also to facilitate electron transfer with the analyte species. By and large, voltammetry is a more effective technique for examining the electrochemical behaviour of a chemical system. A typical research strategy would be to design a chemical system that would (it is hoped) supply a convenient interface between the electrode and the analyte, and then to probe the system using conventional voltammetric techniques. Once the chemistry had been appropriately examined and modified, if necessary, it could be transferred to a sensor. In spite of the diagnostic ability of voltammetry, a transducer based on this technique represents a more cumbersome choice for sensing, mainly because of the electronic circuitry needed to scan the applied potential. This is especially so when account is taken of the more complex waveforms that are needed to impart to the voltammetric device detection limit capabilities comparable with amperometric sensors. Accordingly, the majority of sensing applications that have used electroanalytical transducers have employed the cell in an amperometric mode. However, both techniques are essential to the design and implementation of chemical sensors, and both will be discussed in some detail in this chapter. A brief review of electroanalysis and electro-chemical sensors has appeared recently (1). Amperometric biosensors (enzyme electrodes) have formed the entire or partial subject matter of several reviews (2–4). Readers interested in pursuing voltammetric and amperometric transducers in greater depth would do well to consult the specialist texts (5–7).

8.2 Theory of voltammetry

When a slowly changing potential is applied to an electrode that is immersed in an electrolyte solution containing a redox species, a current will be observed to flow as the applied potential reaches a certain value. The current arises from a heterogeneous electron transfer between the electrode and the redox species, resulting in either an oxidation or a reduction of the electroactive species. At sufficiently oxidizing or reducing potentials, the current can become a function of the mass transfer of the redox species to the electrode. Redox couples that give rise to such currents are frequently referred to as reversible couples, or the electrode reaction is referred to as a reversible electrode reaction. Reversibility, like beauty, lies in the eye of the beholder. If the rate of electron transfer between the redox species and the electrode is rapid when compared with the rate of mass transfer, then the electrode reaction is reversible. Under these circumstances the ratio of the concentrations of the oxidized and reduced forms of the couple at the electrode surface is described by the Nernst equation.

If the rate of electron transfer between the redox species and electrode is very slow, relative to the mass transport of solution species to the electrode, then, irrespective of the electrode potential, the observed current will not be a function of mass transport. In this case, the low rate of electron transfer results in a concentration ratio of the two forms of the redox couple at the electrode surface that does not conform with the Nernst equation. Current/voltage curves for these so-called irreversible processes are of limited analytical utility. In extreme cases the curve does not reach a peak or limiting value and so cannot be used for quantitative estimations of analyte concentrations. A reversible reaction may be rendered irreversible, simply by changing the rate of the mass transport to the electrode, and *vice versa*. An intermediate situation arises when the electron transfer and mass transport rates are comparable. These quasi-reversible electrode reactions are quite common, and, clearly, their analytical utility depends to a large extent on the careful control of the mass transfer rate of the electroanalytical method used for their study.

An electrode reaction is a complex affair. The dissolved redox form must move from the bulk solution to the electrode surface region. This step is dominated by the kinetics of mass transport. Once in the electrode surface region, the redox form may be subject to any of a wide range of homogeneous or heterogeneous chemical reactions that may precede, follow, or run parallel to the electron transfer. Typical homogeneous reactions include protonation, dissociation of complexes or conformational changes such as occur in the reduction of aldoses. Heterogeneous chemical reactions include catalytic decompositions on the electrode surface, or the simpler cases of adsorption and desorption. The rates of the steps that take place in the electrode surface region will be determined by the kinetics of the chemical processes. Finally, after electron transfer, the electrogenerated form of the redox couple, or the product of its participation in a chemical reaction must then diffuse back again into the bulk solution. Once again mass transport will determine the rate of this step. The magnitude of the current at any point in the current/voltage curve will be a function of the kinetics of the slowest step in the overall electrode process. In the case of an electrochemical reduction, at potentials well positive of the redox couple's standard electrode potential E^0 (for oxidations this occurs at potentials more negative than E^0), the rate-limiting step is the rate of electron transfer. Around the E^0 value, for reversible reactions, the electron transfer is becoming more rapid, and the rate-limiting step may be either mass transport, or the kinetics of a homogeneous or heterogeneous chemical reaction. At potentials well negative of the E^0 for reductions (positive for oxidations), reversible reactions are entirely rate determined by mass transport, or by the chemical kinetics (electrode reactions of this latter type go under the generic term 'kinetic electrode reactions'). In irreversible electrode reactions, electron transfer is rate-limiting throughout the applied voltage range.

One of the most important mass transfer mechanisms is diffusion: in many measurements it is the only mechanism for bringing fresh electroactive species to the electrode. Diffusion occurs when there is a concentration gradient in a solution; in electrochemical measurements this gradient occurs at the electrode surface. When the potential of an electrode is pulsed from a quiescent region to one in which the electroactive species undergoes a reversible electrode reaction, the surface concentration of the electroactive form of the redox species drops to zero, whereas in the bulk solution it is finite. The layer of solution between the electrode surface and the point in solution at which the redox species concentration is unchanged, is known as the diffusion layer. At first this diffusion layer is thin, the concentration gradient is steep, and the current observed reaches a high value. As the time interval from the imposition of the pulse increases, the diffusion layer moves further out into the bulk solution, the concentration gradient decreases, and the current decays. Accordingly, time is an important function in an electroanalytical measurement, both from the point of view of the reversibility of an electrode reaction, and with respect to the value of the observed current.

Unfortunately, the current arising from heterogeneous electron transfer (the faradaic current) is not the only current to flow at a voltammetric electrode. Changing the potential of an electrode involves changing the charge density at its surface. The layer of solution immediately adjacent to the electrode behaves as a capacitor because, in response to the electrode's surface charge, charged species in this solution zone are orientated in an organized fashion. The orientation of charge in this region (known as the double layer) follows any variation in the electrode's charge density. Consequently, when the potential is scanned, the double layer responds, and this movement of charges in solution gives rise to a current in the electrochemical cell. The current decays with time, as the double layer becomes fully reorientated (charged). This capacitative current limits the lower range of faradaic currents that can be recorded, and hence the detection limit of a technique. For some solid electrodes, the time taken for the charging current to decay is far longer than is predicted from simple capacitor theory. This type of current reflects the slow kinetics of surface processes that occur on materials such as carbon.

8.3 Instrumentation

The instrumentation of voltammetry and amperometry can be divided into three broad areas; wave-form generation; potential control; and the electrochemical cell. A block diagram of the function and interrelationships of these three primary elements is shown in Figure 8.1. Modern electroanalytical systems employ a three-electrode arrangement for the electrochemical cell. A device called a potentiostat maintains a programmed or fixed potential difference between two of the electrodes which are current-carrying (the working electrode and the auxiliary electrode) by reference to a third electrode (the reference electrode) the function of which is to provide a fixed potential reference in the cell. In the last fifteen years or so, potentiostats and waveform generators have benefited substantially from the introduction of operational amplifiers, and latterly from the availability of desk-top computers. The elements of potentiostat design are well covered in the literature (5,7–12), as are the benefits of incorporating micro- or minicomputers into the instrumentation (13–15). Electrochemical cell design is well established both for static systems (5–7), and for detectors in flowing reagent streams (16). In spite of the fact that solid electrodes have been in use for electroanalytical purposes for some time, and are well documented (17), the selection of electrode material and its preparation before use remains an area of high interest and great controversy amongst electroanalytical chemists. A recent

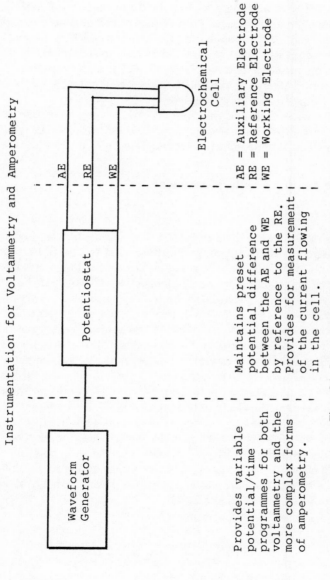

Figure 8.1 Instrumentation for voltammetry and amperometry.

review (18) highlights the complexity of the most frequently used material, glassy carbon, and attempts have been made to formalize and validate the surface pretreatment of this material (19). Solid electrodes for analytical use should be prepared in a rigorous and reproducible manner. This generally involves physically polishing the electrode using successively finer grades of carborundum or diamond paste. The polished electrode should be thoroughly rinsed to remove, as far as possible, all traces of the polishing materials. Many workers subsequently employ electrochemical pretreatment, in which the electrode is immersed in an electrolyte and held at a predetermined potential for a fixed period of time, or cycled between selected potential limits at a given frequency and for a fixed duration.

Although a wide range of applied potential waveforms have been employed, only a few are of mainstream significance; these, along with typical voltammograms and some comments are presented in Table 8.1.

8.4 Linear sweep and cyclic voltammetry

8.4.1 *Introduction*

These two techniques will be discussed in some detail, for they represent powerful methods of investigating electrochemical reactions. Further details may be found in (5–7); a useful introduction to cylic voltammetry is available (20). The seminal paper on these techniques (21) is well worth reading. In linear sweep voltammetry the potential of the working electrode is swept from one selected limit to another at a rate of between $1 \, mV \, s^{-1}$ to $1000 \, V \, s^{-1}$ (these represent extremes; a rate of say, $50 \, mV \, s^{-1}$ would be more typical) and the current is recorded. In cyclic voltammetry two linear sweeps are done as shown in Table 8.1. The essential feature of cyclic voltammetry is that electrogenerated species formed in the forward sweep are subject to the reverse electrochemical reaction in the return sweep. In both methods, as the potential is swept to the electroactive region of the redox couple, the current response rises to a peak value before decaying. This decay is caused by depletion of the electroactive species in the zone close to the electrode surface: i.e. the diffusion zone spreads out further into the bulk solution. A more detailed diagram of a cyclic voltammogram is given in Figure 8.2. Features of the voltammogram that are important in the elucidation of the electrode reaction are given on the diagram.

8.4.2 *Diagnostic features*

Variation of the potential sweep rate of a linear sweep or cylic voltammetric scan causes characteristic changes in three important parameters of the generated voltammogram. These are: the ratio of the magnitudes of the anodic to the cathodic peak currents; the rate of the shift of half peak potential (this is

Table 8.1 Voltammetric techniques

Technique	Applied potential	Measured signal	Comments
Linear sweep voltammetry			Mainly used for diagnostic purposes. Current measured throughout the scan.
Cyclic voltammetry			Reversal of the potential at the end of the forward scan, makes this a powerful diagnostic tool. The current is measured throughout the scan.
Differential pulse voltammetry			The displayed signal is the difference between the current just before, and at the end of the applied pulse. Slow scan rates, but good limits of detection.
Square wave voltammetry			The displayed signal is the difference between the currents measured on the forward and reverse pulses. Rapid scans and good detection limits

Key: → = point at which current is measured
   τ = period of the staircase on which pulses of width $\tau/2$ are superimposed
   E_{SW} = square wave amplitude

Notes: Only a small fraction of a complete applied potential waveform scan is shown for differential pulse and square wave voltammetry. The entire measured signal for a single analyte is shown.

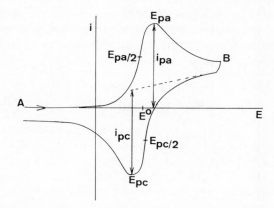

Figure 8.2 Diagnostic features of a cyclic voltammogram. i_{pa}, peak anodic current; i_{pc}, peak cathodic current; E_{pa}, anodic peak potential; E_{pc}, cathodic peak potential; E°, formal reduction potential for the electroactive species; $E_{pa/2}$, anodic half peak potential (potential at which the current is one-half of the peak value); $E_{pc/2}$, cathodic half peak potential; A, start of potential scan; B, point at which the potential scan reverses.

Table 8.2 Electrode reactions

Mechanism number	Equation	Description
1	$Ox + ne \rightleftharpoons Red$	Reversible reduction
2	$Ox + ne \rightarrow Red$	Irreversible reduction
3	$Ox + ne \rightleftharpoons Red$	Catalytic reaction: reversible reduction followed by regeneration of the electroactive species by an
	$Red + X \rightarrow Ox + Y$	irreversible homogeneous chemical reaction.
4	$Ox + ne \overset{k_f}{\rightarrow} Red$	Catalytic reaction: irreversible reduction followed by regeneration of the electroactive species by an
	$Red + X \rightarrow Ox + Y$	irreversible homogeneous chemical reaction.
5	$Ox + ne \overset{k_f}{\rightleftharpoons} Red$	Catalytic reaction: reversible reduction followed by regeneration of the electroactive species by a re-
	$Red + X \underset{k_b}{\overset{k_f}{\rightleftharpoons}} Ox + Y$	versible homogeneous chemical reaction.

given by a plot of $\Delta E_{p/2}/\Delta \log v$ versus $\log v$); and the variation of the current function. The current function is obtained by multiplying the peak current i_p, with the following expression:

$$(n^{3/2} FAD^{1/2} Cv^{1/2})^{-1}$$

where n is the number of electrons involved in the electrochemical step, F is the Faraday constant, D is the diffusion coefficient of the electroactive species, C is the bulk concentration of the electroactive species, A is the area of the electrode, and v is the sweep rate.

Table 8.2 lists some characteristic electrode reactions. The linear sweep and cyclic voltammetric diagnostic features of these are described below. They have been selected as examples of the chemical systems most likely to be incorporated in chemical sensors. Further discussion of a wider range of mechanisms can be found in (21). Cyclic voltammetry of surface-confined species is discussed in Chapter 5.

Mechanism 1. The peak current i_p for a reaction of this type is given by

$$i_p = 0.446nFA(Da)^{1/2}C$$

where $a = nFv/RT$. The peak current function, peak potential, rate of shift of the half peak potential and ratio of anodic to cathodic peak currents are all independent of $v^{1/2}$. The peak occurs at a potential of $28.5/n$ mV more anodic of E^0 for an oxidation at 298 °K, and $28.5/n$ mV more cathodic of E^0 for a reduction. Accordingly, the peak-to-peak separation is $57/n$ mV.

Mechanism 2. The peak current for a reaction of this type is given by

$$i_p = 0.280 nFA (\pi Db)^{1/2}C$$

where $b = \alpha nFv/RT$, and $\alpha =$ the transfer coefficient. A totally irreversible process will not yield a peak on the reverse scan, hence peak separation considerations are meaningless. The current function is independent of $v^{1/2}$, but, unlike the reversible case, the peak potential is a function of the scan rate, even though the rate of the shift of the half peak potential is not.

An interesting case arises when the electrode reaction under study is reversible at low scan rates, but becomes irreversible as the scan rate increases. A characteristic of this type of reaction is the distinct change in the plot of i_p versus $v^{1/2}$, as, first the equation for mechanism 1 holds good, and then the equation for mechanism 2 takes over. In the interval between these two mechanisms, the reaction is called quasi-reversible.

Mechanism 3.   In many biosensors, an enzyme is used to provide a stoichiometric relationship between an electroinactive analyte and an electroactive mediator. The mediator concentration is measured by the working electrode of an amperometric transducer. Such reactions are catalytic in nature, and an important area of voltammetric analysis, therefore, involves the investigation of these reactions, with a view to constructing amperometric biosensors. The effect of a catalytic reaction is to regenerate the electroactive species at the electrode surface *via* an homogeneous chemical reaction. Consequently, for (say) an oxidation, provided that the rate constant for the chemical regeneration of the reduced form is sufficiently fast, the anodic peak height will be increased in size, whereas the cathodic peak will be decreased to the point of extinction. Cyclic voltammograms of catalytic processes are not peak-shaped; instead the current rises to a plateau.

Two distinct cases can be distinguished for catalytic processes. In one, the ratio $k_f C_Z/a$ is low, where k_f is the pseudo-first-order rate constant for the homogeneous irreversible chemical reaction and C_Z is the concentration of the species (Z) that regenerates the electroactive form of the redox couple (a has already been defined). Under these conditions (i.e. slow kinetics, low concentrations of Z and rapid scan rates) the catalytic reaction has little effect on the electrode reaction, and the voltammograms resemble those of a reversible couple and have similar diagnostic features. For large values of $k_f C_Z/a$, the magnitude of the current in the plateau region of the voltammogram is given by

$$i_k = nFAC(Dk_f C_Z)^{1/2}$$

In other words, the plateau current i_k is independent of the potential scan rate, but linearly related to the square root of the concentration of Z. As a result of these two extreme cases, the plot of current function versus scan rate for an irreversible catalytic step following a reversible electrode reaction takes the form shown in Figure 8.3a. Clearly, at low scan rates the kinetics of the catalytic reaction predominate, and values for k_f can be extracted from the data. At high scan rates the kinetic effects are of much less significance. Figure 8.3a also depicts the form of the curve generated by plotting the rate of shift of the half peak potential against the log of the scan

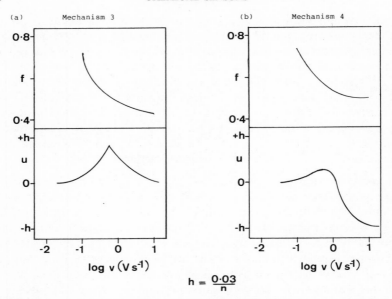

Figure 8.3 Variation of current function (f) and rate of shift of the half peak potential (u) with potential scan rate (v). Left, mechanism 3; right, mechanism 4.

rate. Curves such as those shown in Figure 8·3a, and the i_k dependence on $(C_Z)^{1/2}$, are indications of catalytic effects.

Intermediate values of $k_f C_Z/a$ will give rise to a series of voltammograms in which catalytic activity is apparent at slow scan rates, but the current never reaches a plateau value, always maintaining a peaked appearance, even at slow scan rates. The values of the peak current and appropriate equations for these voltammograms can be found in (6).

Mechanism 4. In many instances the electroactive mediator in a biosensor is oxygen or hydrogen peroxide, in which case, the electrode reaction preceding the catalytic step is irreversible. Once again, two extreme cases can be defined in a manner identical to that used for mechanism 3. For cases where the magnitude of $k_f C_Z/b$ is low, the voltammogram is described by the equations for irreversible electrode reactions. Large values of $k_f C_Z/b$ reflect the increasing influence of the homogeneous chemical reaction on the voltammogram. In this case, the current in the plateau region of the voltammogram is given by an identical expression to the one used for mechanism 3. Again, the effect of these two extremes is to produce a series of voltammograms, for which a plot of current function and rate of shift of the half peak potential versus the log of scan rate take the form shown in Figure 8.3b. At slow scan rates, the homogeneous kinetics exert a major influence on the shape of the recorded voltammogram; at higher scan rates, the kinetics are of diminished significance.

Voltammograms for which intermediate values of $k_f C_Z/b$ occur are not adequately described by the theory outlined in the previous paragraph. A complex function is needed to represent the current voltage curves: once again, further details can be found in (6).

Mechanism 5. Regeneration of the electroactive form of a redox couple at a working electrode can involve a reversible homogeneous chemical reaction. For a reversible electrode reaction, at rapid scan rates, the equations describing the voltammogram are essentially those for mechanism 1. At slow scan rates, the plateau current is given by

$$i_k = nFAC\,[D(k_f C_Z + k_b)]^{1/2},$$

where k_b is the rate constant for the reverse reaction.

8.4.3 *Limitations on use*

The diagnostic ability of these forms of voltammetery are excellent, but their detection limits are poor, being limited to the 10^{-3} to 10^{-4} M level, thus rendering them inappropriate for incorporation into chemical sensors. This rather poor limit of detection arises from the relative contributions to the total cell current of the faradaic and the capacitive currents. The capacitive current contribution is a linear function of sweep rate, whereas the faradaic current is a function of the square root of the sweep rate (for a reversible process). Thus, increasing the sweep rate in an attempt to produce larger faradaic currents inevitably causes a deterioration in the signal to noise ratio. An approximate expression can be used to evaluate the relative contribution of the capacitive (i_C) and the peak faradaic currents (i_P). Assuming typical values for the diffusion coefficient ($10^{-5}\,\mathrm{cm^2\,s^{-1}}$), and the double layer capacitance ($20\,\mu\mathrm{F\,cm^{-2}}$), then

$$|i_C|/i_P = \{(2.4 \times 10^{-8})v^{1/2}\}/(n^{3/2}C)$$

where v takes the units $\mathrm{Vs^{-1}}$, and C is in $\mathrm{mol\,cm^{-3}}$.

Quantitative measurements of analytical concentrations, or kinetic data, are best obtained from step or pulse techniques. The kinetic control that occurs in many cyclic voltammetric experiments means that thermodynamic data (such as values for E^0) obtained from such experiments should be interpreted with caution.

8.5 Pulse and square wave voltammetry

Pulse and square wave voltammetry are much better candidates than linear sweep or cyclic voltammetry for incorporation into a chemical sensor. The primary reason for this is the ability of the former techniques to discriminate between faradaic and capacitive currents. When a potential pulse is applied to an electrode, the capacitive current that flows is proportional to the

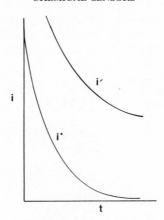

Figure 8.4 Decrease in the faradaic current (i') and the capacitative current ($i*$) with elapsed time (t) from the imposition of a potential pulse.

magnitude of the pulse, and decays exponentially with time. The faradaic current, on the other hand, decays according to the square root of time. The effect of this can be seen from Figure 8.4: judicious selection of the measurement time can radically improve the signal to noise ratio. This feature is used in normal pulse, differential pulse and square wave voltammetry: in the first technique it is the only mechanism for decreasing the effect of the capacitative current. Further rejection of the capacitative current is achieved in differential pulse voltammetry, by limiting the size of the applied pulse, and by subtracting the current just prior to the imposition of the pulse, from the current obtained at the end of the pulse period (see Table 8.1). Better rejection of the charging current means improved detection limits, and for this variant of voltammetry, concentrations as low as 10^{-8} to 10^{-7} M may be readily measured. In order to avoid the problems of depletion of the electroactive species, the delay between application of successive pulses must be approximately half a second; this limits the scan rate of differential pulse voltammetry, and decreases its applicability to sensor work.

The technique of square wave voltammetry (22) (see Table 8.1) has even more to offer as a voltammetric method of probing selective chemistry, because of the speed with which a scan can be carried out. The analytical signal in this technique is the difference between the current for the forward pulse and the current for the reverse pulse. Because of the large amplitude of the square wave, for a reversible reduction, the reduced electroactive species formed at the electrode during the forward pulse is re-oxidized by the reverse pulse. Consequently, the sensitivity of this method is enhanced when compared to differential pulse voltammetry. For identical conditions, an approximately 30% improvement in signal is obtained, but when the higher scan rates that are attainable with square wave are used, this figure increases to around 560%. This element of speed is crucial to square wave voltammetry, because, like all the voltammetric techniques based on pulse waveforms, the measured current is proportional to $t^{-1/2}$. However, unlike the other pulsed voltammetric techniques, square wave voltammetry causes very little depletion of the electroactive species in the zone near the electrode: it is this depletion that gives rise to distortion of the current/voltage waveform. Accordingly, square wave voltammetry is uniquely placed to benefit from rapid scan rates. A typical compromise frequency for the waveform is 200 Hz, and this coupled with a ΔE value of 10 mV, gives rise to a scan rate of 2 Vs^{-1}. A limiting factor to the scan rate is the concomitant increase in the capacitative current. At solid electrodes, square wave is no better than other pulsed voltammetric techniques at rejecting the capacitative current.

The optimum value of the square wave amplitude E_{sw} is $50/n$ mV, irrespective of the values of ΔE or τ. Decreasing this value causes a loss of sensitivity, with no diminution of the peak width; increasing E_{sw} causes peak broadening, with no enhancement in sensitivity. For a reversible couple the current/voltage curve takes on a symmetrical peak shape, the height of the peak being a linear function of the concentration. An early criticism of square wave was that the technique was not particularly good for irreversible electrode reactions. The effect of irreversibility is to shift the peak potential to more negative values (for a reduction), decrease the height of the peak, and broaden it. These effects are common to all voltammetric methods, and are no worse for square wave than for (say) differential pulse voltammetry.

Another significant advantage for square wave is the ability of the technique to reject a wide range of background currents. This ability with respect to capacitative currents is carried out by the subtraction of two currents in a manner akin to differential pulse voltammetry. The slowly varying capacitative currents that arise when surface groups reorganize on certain types of solid electrode are also subtracted, providing that the rate at which they vary is sufficiently slow, compared to the value of τ. Generally, pulsed voltammetric techniques are of diminished value when performed on solid electrodes because of these slowly changing background currents. Once again, square wave has the edge over the other pulse methods. Finally, a significant problem in reductive voltammetry is the presence of a large signal for oxygen. Reductions at more negative potentials than the oxygen wave are often of limited precision because of the variable base line caused by fluctuating oxygen concentrations. The only way to circumvent this is to strip the oxygen out of the system by bubbling nitrogen through the test solution. In square wave voltammetry, the analytical current is a difference current; consequently, even in the presence of substantial concentrations of oxygen, at potentials well negative of the voltammetric peak potential, the square wave voltammetric current is zero, and is hence insensitive to variations in oxygen concentration.

8.6　Theory of amperometry

8.6.1　*Introduction*

Classically amperometry has been concerned with the maintenance of a fixed potential between two electrodes. More recently, pulsed techniques have come to the fore; some details of these are included in Table 8.3. The applied potential at which the current measurement is made is usually selected to correspond to the mass transport limited portion of the corresponding voltammetric scan. Theoretically, in a quiescent solution, this is electrochemical suicide. The current i_l obtained at conventional electrodes under such circumstances gradually decays to zero according to the Cottrell equation (4):

$$i_l = nFAC(D/\pi t)^{1/2}$$

where t is the time elapsed from the application of the potential pulse.

This decrease in current is due to the slow spread of the diffusion layer out into the bulk solution, with a concomitant decrease in the concentration gradient. In practice, this process continues for around 100 s or so, after which time random convection processes in the solution take over, and put an end to the further movement of the diffusion layer. Waiting for nearly two minutes, and relying on chance solution convection for a steady-state current, is not particularly attractive. Accordingly, in amperometric measurements for sensing applications, the spread of this diffusion layer is limited by the use of one or more of the following mechanisms: convective diffusion; the

Table 8.3　Amperometric methods

Method	Applied potential	Measured signal	Comments
Chronoamperometry			The electrode potential is pulsed to a region in which the analyte is electroactive. The decaying current reflects the growth of the diffusion layer.
Pulsed amperometry			The electrode potential is pulsed briefly to a region in which the analyte is electroactive. Between pulses the diffusion layer may be eliminated by forced or natural convection.
Pulsed amperometric detection			Electrode conditioning, analyte sorption and catalytic electro-oxidation are all promoted by the use of this waveform. The current is only measured in the last part of the cycle (point *A*).

imposition of a physical barrier to the limits of the diffusion layer in the form of a membrane; pulsing the electrode from a region of electroinactivity to a region of transport-controlled electroactivity, and then returning the potential to the electroinactive region; microelectrodes.

The advantage of amperometric measurements is that the faradaic currents are observed, at fixed electrode potentials. In these circumstances, capacitative currents no longer contribute to the overall cell current, and much lower detection limits are obtainable compared to linear sweep voltammetry. However, some of the newer variants of amperometry do involve pulsing the electrode potential to the active region: measurements in these cases need to be made carefully to produce optimum signal to noise ratios.

8.6.2 *Hydrodynamic amperometry*

Mass transport in amperometric systems in which the reagent stream is forced to flow along the surface of the electrode may be described in terms of convective diffusion. Effectively this means that at sufficiently high values of P_e, the Peclet number, the liquid above an electrode may be divided into two distinct zones. In one zone, far away from the electrode surface, convection is important, and the concentration profile is substantially flat. In the other zone, adjacent to the electroactive surface, there is a sharp concentration gradient: here diffusion is the predominant mass transport process. The Peclet number is given by $v_m l/D$, where v_m is the main stream fluid velocity, and l is the length of the electrode (measured in the direction of fluid flow). Under these conditions, the mass transport limited current i_L for a reversible electrode couple (i.e. the concentration of the electroactive form is zero at the electrode surface) is given by

$$i_L = nFAD^{2/3} v_k^{-1/6} x^{-1/2} (v_m)^{1/2} C$$

where v_k is the kinematic viscosity and x is the distance along the electrode.

This current is time-dependent too, but in the convective diffusion case, the current rapidly (within a few ms or so) reaches a stationary value. The speed with which this current plateau is reached arises from the establishment of a well-defined steady-state diffusion layer. The thickness d of this diffusion layer is given approximately by

$$d = D^{1/3} v_k^{1/6} (x/v_m)^{1/2}$$

From a practical point of view convective diffusion may be obtained in two ways; moving the solution relative to the electrode or moving the electrode relative to the solution. The equation given above can be used for a stationary electrode in a flowing stream. Of the systems developed to move the electrode, only one merits consideration here, and that is the rotating disc electrode (RDE). The limiting current obtained at a RDE (i_{LRDE}) for a mass transport controlled electrode reaction is given by the Levich equation:

$$i_{\text{LRDE}} = nFAD^{2/3}v_k^{-1/6}\omega^{1/2}C$$

where ω is the angular frequency of rotation of the electrode.

Again, the limiting current obtained after the electrode potential is pulsed from a region of electroinactivity to a region of electroactivity is a function of time, decaying from an initial high value to the steady state within a few hundred ms. Further details on the RDE can be obtained from (5).

From the point of view of sensor design there are limitations to the RDE. These arise from the mechanics of actually rotating the electrode. For the Levich equation to accurately represent the limiting current, the velocity of rotation ω must be maintained between certain limits. At lower values for ω, the hydrodynamic boundary layer becomes large, approaching the magnitude of the electrode radius; at this point many of the assumptions used to derive the Levich equation break down. At higher rotational speeds, turbulent flow sets in. For a perfectly polished disc electrode rotating in a straight, concentric RDE shaft, the value of ω must be less than $2 \times 10^5 (v_k/r)$, where r is the radius of the electrode. For an electrode with an irregular surface, rotating in an eccentric shaft in a vessel of limited volume, the onset of turbulent flow can occur at much lower rates of rotation. On the other hand, exposing the electrode to a flowing stream of reagent is a more straightforward approach, but only for those cases where the species to be sensed already is located in a flowing medium.

8.6.3 *Membrane-covered amperometric devices*

There are four primary advantages to covering the electrodes of an amperometric device with a membrane that is permeable only to the analyte. Poisoning of the electrodes by electroactive or surface-active species is limited. The resolution of the system is enhanced if extraneous electroactive species that would otherwise undergo electron transfer at the electrode can be excluded. The composition of the electrolyte that occupies the space between the membrane and the electrodes remains constant. The membrane forms a physical barrier to the spread of the diffusion layer into the bulk solution. The limiting current obtained at a membrane-covered amperometric device will be a function of time, reaching in due course a steady-state value. The current–time transient cannot be described by a single equation, because different transport mechanisms operate during the time interval. For a typical membrane-covered amperometric oxygen detector of electrolyte thickness 10 μm and membrane thickness 20 μm, an electrode radius of greater than 2 mm and diffusion coefficient of the analyte in the membrane two orders of magnitude lower than in the electrolyte solution (itself around 10^{-5} cm^2 s^{-1}), it can be shown (23) that the current time transient can be described by three equations. For times greater than twenty seconds from the imposition of the potential pulse to the electroactive region, the limiting current is given by

$$i_1 = nFAC(P_m/d)$$

where P_m is the permeability coefficient of the membrane and d is the thickness of the membrane.

An important consequence of this equation is that the limiting current is a function of the permeability of the membrane. Unfortunately, membrane properties such as permeability are a function of time and ambient conditions, particularly temperature. Accordingly, there are advantages to operating membrane-covered detectors away from this steady-state condition. For the typical detector, the limiting current at times less than 100 ms from the imposition of the measurement potential pulse is given by

$$i_1 = nFAC(K_b/K_0)(D_e/\pi t)^{1/2}$$

where t is the time from the imposition of the measuring potential, D_e is the diffusion coefficient of the analyte in the electrolyte, K_b is the distribution coefficient for the analyte at the bulk solution/membrane interface, and K_0 is the distribution coefficient for the analyte at the internal electrolyte/membrane interface.

The ratio K_b/K_0 expresses the salting-out effect of the electrolyte solution. In effect, this equation describes the diffusion-limited current arising within the electrolyte layer. In other words, the diffusion layer has not had time to spread to the membrane and hence, at these short times the limiting current is independent of the membrane. The advantage of this is that alterations in membrane characteristics, and indeed alterations in ambient conditions that affect membrane performance, are no longer contributors to the limiting current. One of the ambient conditions that has no effect on the limiting current is stirring. Under steady-state conditions, stirring of some form is necessary to prevent the diffusion layer from moving out into the bulk solution. The rate of stirring is not critical: just moving the detector up and down in the solution is sufficient. When transient measurements are made, no stirring is required at all. The situation regarding transient measurements with amperometric devices is not quite as simple as it is depicted here; for example, contributions from the capacitative current need to be taken into account. Further details can be found in (23).

8.6.4 Pulsed amperometry

For detectors without membranes, the application of a potential pulse from a region of electroinactivity to a region of electroactivity has some advantages. The duty cycle of the pulse can be arranged such that the electrode spends the majority of the time at the electroinactive potential. Under these circumstances, electrode reactions that give rise to products that poison the electrode are of diminished significance, simply because the deleterious products are produced at a much reduced rate. Secondly, the natural processes of convection can redistribute the electroactive species whilst the electrode is at the rest potential, and thus limit the spread of the diffusion layer into the bulk solution. The drawback to this technique is the reappearance of the capacitative current. Limiting the size of the pulse will decrease the contribution of this current, but a small pulse from a rest potential to the electroactive potential can only be achieved for reversible electrode reactions. The alternative is to allow sufficient time between imposition of the potential pulse and measurement of the current for the capacitative current to have decayed away. As a rule of thumb, the time taken for the capacitative current to decay to 5% of its original value is given by $3R_uC_d$, where R_u is the uncompensated resistance in the cell, and C_d is the capacitance of the double layer. C_d is typically in the region of 10 to $40\,\mu\text{F cm}^{-2}$, and R_d is given approximately by $x_{\text{ref}}/(kA)$, where x_{ref} is the distance between the reference electrode and the working electrode, and k is the solution conductivity.

8.6.5 *Pulsed amperometric detection*

Pulsed amperometric detection (24) is a relatively new technique, and has made possible the direct amperometric determination of many compounds that were at one time considered unsuitable for this type of measurement. These include carbohydrates, amino acids, halides and many sulphur-containing compounds. The waveform for this technique is shown in Table 8.3. This rather complex waveform serves several purposes. On platinum, for example, at the extreme anodic potential, surface-confined oxidation products that tend to poison the electrode are desorbed. This process runs concurrently with the formation of PtO_2 on the surface. At the cathodic potential this surface oxide is reduced and a 'clean' Pt surface is regenerated. In some cases, sorption of the analyte occurs at this cathodic potential. The analytical signal is recorded at the intermediate potential: several different mechanisms can give rise to this signal. Electroactive species that are not absorbed at the electrode may undergo conventional oxidative electrode reactions. Alternatively, electroactive species that are absorbed at the electrode during the cathodic swing of the cleaning cycle may undergo either conventional oxidative desorption, or oxidative desorption that is catalysed by an oxide species on the surface of the electrode. For many compounds this latter reaction is much more favourable from a thermodynamic point of view than oxidation in the absence of a catalyst. This type of reaction occurs most readily when both the analyte and the oxide (in this case it is PtOH) are surface confined. Finally, adsorbed electroinactive analytes may also be detected by virtue of their ability to block the sites on the electrode at which the formation of PtOH takes place. Consequently, the faradaic current arising from the oxidation of platinum is diminished in the presence of adsorbed electroinactive analyte.

8.6.6 *Amperometry at microelectrodes*

A recent review (25) has discussed the electroanalytical application of carbon fibre microelectrodes. These electrodes, constructed from single carbon fibres, present a disc-shaped surface to the solution with a diameter of 8 μm. The rate of electrolysis at these electrodes is approximately the same as the rate of diffusion; consequently, the diffusion zone, once established, does not move out into solution at anywhere near the same speed as the diffusion zone around a conventional electrode. Linear sweep voltammetry gives rise to the S-shaped current–voltage curves seen at rotating disc or dropping mercury electrodes. The time-dependent current (i_t) arising at a microelectrode is given by the equation for the chronoamperometric current at a conventional electrode under *spherical* diffusion conditions, viz.:

$$i_t = nFADC[\{1/(tD\pi)^{1/2}\} + \{1/r\}]$$

where r is the electrode radius.

Indeed, a sphere or a hemisphere is a good model for the diffusion zone that surrounds these electrodes, and because of this enhanced mass transport to the electrode, a steady-state current is rapidly achieved after a potential pulse is applied to the electrode.

A number of other features follow from the relatively large size of the diffusion zone when compared to the electrode radius. One of these is quite significant to biosensors based on enzyme systems. The large diffusion zone radidly dilutes the products of the electrode reaction, thus catalytic mechanisms of the type shown in Table 8.2, mechanisms 3–5, are not perceived at microelectrodes, unless the rate constant for the reaction is fast. It can be shown that for a reversible electrode reaction, the ratio of the catalytic current (i_{kc}) to the faradaic current obtained in the absence of the homogeneous reaction is given by

$$i_{kc}/i = [(k'_c)^{1/2} + D^{1/2}/r]/[(1/(\pi t)^{1/2}) + D^{1/2}/r]$$

where $k'_c = k_f C_Z$ provided that $k'_c \cdot t > 5$.

Although the diffusion zone is large in comparison to the electrode surface, it is small compared to the zone around a conventional electrode. This small size means that the faradaic current obtained at microelectrodes enjoys comparative immunity from the effects of convection in the bulk solution. In flowing streams the current is independent of flow rate.

Decreased effects of capacitance and resistance at microelectrodes, coupled with the high mass transport rates, has enabled electrochemical measurements to be made in cells containing highly resistive solutions. In fact, provided that there is a sufficient quantity of free carrier charge on the surface of the insulator in contact with the microelectrode to charge the double layer, microelectrodes can be used to make direct electrochemical measurements in the gas phase (26).

8.7 Noise considerations

Aside from the considerations relating to the selective chemistry of a sensor, the characteristic that most influences the degree to which a sensor can discriminate between one concentration level and another is the signal to noise ratio. This topic has been discussed for amperometric transducers (27). In the first instance, both amperometric and voltammetric transducers can be considered as high-impedance current sources. Hence, the well-understood techniques of grounding and shielding that can be found in electronic texts (28) are directly applicable. Briefly, the transducers, all connecting leads and any exposed streams of fluid should be shielded from capacitative pick-up of electrical radiation. These shields should be grounded at one end only. Interconnection of signal grounds, shield grounds and power supply grounds should be avoided, each group being kept separate from one another, until they can all be brought together at one point and connected to a good ground system. The quality of the local grounding system can vary immensely; the ground itself can be a significant source of noise. This is often the case in research laboratories! Temperature variations are also a source of noise. The use of a combination of active and passive thermal shielding has been shown to decrease the low-frequency noise power by an order of magnitude (27). This is of considerable importance to a sensor system in which an electroinactive analyte is detected *via* an homogeneous chemical reaction with an electroactive mediator. The effect of temperature on the rate constants of the chemical reaction may introduce significant fluctuations in the final current.

Apart from the external noise sources mentioned above, the effects of which can be decreased by good shielding, there are several noise sources that are intrinsic to the transducer, arising either from the electrochemical cell, or the associated electronics. Three major contributors are: the input voltage noise of the current to voltage convertor; the impedance noise of the electrochemical cell; and the area of the working electrode. The operational amplifier (op-amp) used to carry out the current to voltage conversion, should have as low an input voltage noise as possible. This figure is available as part of the manufacturer's specification of an op-amp, and varies quite considerably depending on the device selected. A value around $2.0-4.0\,\mu V$ peak to peak in the bandwidth 0.1–10 Hz is sufficient. In the low-frequency region used to make most amperometric or voltammetric measurements, impedance noise, to a large extent, is represented by the capacitative noise in an electrochemical cell. Accordingly it is of most significance when making measurements against large background currents. Decreasing the electrode area may help to limit this source of noise. The root mean square current noise is a linear function of the electrode area, again limiting the electrode area will limit this noise contribution. However, the important parameter is the signal to noise ratio, improvements in this value depend not only on the reduction of noise, but also on the relationship between signal size and electrode area. Accordingly, a plot of signal to noise ratio against electrode area will show a distinct peak (27), the position of which is a complex function of various cell parameters.

It would be premature to suggest that optimization of the signal to noise ratio for this type of chemical sensor could be done entirely theoretically. Careful sensor design with respect to electrical and thermal shielding are essential, as is the use of low-noise electronic components in the measuring circuits, and to a large extent these considerations can be taken account of in the design stage of the sensor. However, selecting the optimum area of the working electrode is best done empirically.

References

1. M.L. Hitchman and H.A.O. Hill, *Chem. in Britain* **22** (1986) 1117–1124.
2. L.D. Bowers and P.W. Carr, *Immobilized Enzymes in Analytical and Clinical Chemistry*, John Wiley, New York (1980).
3. G. Davis, *Biosensors* **1** (1985) 161–178.
4. P. Vadgama and G. Davis, *Med. Lab. Sci.* **42** (1985) 333–345.
5. A.J. Bard and L.R. Faulkner, *Electrochemical Methods*, John Wiley, New York (1980).
6. Z. Galus, *Fundamentals of Electrochemical Analysis*, Ellis Horwood, Chichester (1976).
7. D.D. Macdonald, *Transient Techiques in Electrochemistry*, Plenum, New York (1977).
8. R. Kalvoda, *Operational Amplifiers in Chemical Instrumentation*, Ellis Horwood, Chichester (1975).
9. D.T. Sawyer and J.L. Roberts, *Experimental Electrochemistry for Chemists*, John Wiley, New York (1974).
10. D.E. Smith, in *Electroanalytical Chemistry*, ed. A.J. Bard, Vol. 1. Edward Arnold, London (1966) 1–155.
11. W.M. Schwarz and I. Shain, *Anal. Chem.* **35** (1963) 1770–1778.
12. Southampton Electrochemistry Group, *Instrumental Methods in Electrochemistry*, Ellis Horwood, Chichester (1985).
13. L. Kryger, D. Jagner and H.J. Skov, *Anal. Chim. Acta* **78** (1975) 241–249, 251–260.
14. L. Kryger, *Anal. Chim. Acta* **133** (1981) 591–602.
15. A.M. Bond, H.B. Greenhill, I.D. Heritage and J.B. Reust, *Anal. Chim. Acta* **165** (1984) 209–216.
16. T.H. Ryan, *Electrochemical Detectors: Fundamental Aspects and Analytical Applications*, Plenum, New York (1984).
17. R.N. Adams, *Electrochemistry at Solid Electrodes*, Marcel Dekker, New York (1969).
18. W.E. van der Linden and J.W. Dieker, *Anal. Chim. Acta* **119** (1980) 1–24.
19. D.G. Thornton, K.T. Corby, K.A. Spendel and J. Jordon, *Anal. Chem.* **57** (1985) 150–155.
20. P.T. Kissinger and W.R. Heineman, *J. Chem. Educ.* **60** (1983) 702–706.
21. R.S. Nicholson and I. Shain, *Anal. Chem.* **36** (1964) 706–723.
22. J. Osteryoung and J.J. O'Dea, in *Electroanalytical Chemistry*, ed. A.J. Bard, vol. 14, Marcel Dekker, New York (1986) 209–308.
23. M.L. Hitchman, *Measurement of Dissolved Oxygen*, John Wiley, New York (1978).
24. D.S. Austin, J.A. Polta, T.Z. Polta, A.P.C. Tang, T.D. Cabelka and D.C. Johnson, *J. Electroanal. Chem.* **168** (1984) 227–248.
25. T.E. Edmonds, *Anal. Chim. Acta* **175** (1985) 1–22.
26. J. Ghoroghchian, F. Sarfarazi, T. Dibble, J. Cassidy, J.J. Smith, A. Russell, G. Dunmore, M. Fleischmann and S. Pons, *Anal. Chem.* **58** (1986) 2278–2282.
27. D.M. Morgan and S.G. Weber, *Anal. Chem.* **56** (1984) 2560–2567.
28. P. Horowitz and W. Hill, *The Art of Electronics*, Cambridge University Press, Cambridge (1980).

9 Potentiometric transducers

B.J. BIRCH and T.E. EDMONDS

9.1 Introduction

The essential component of a potentiometric measurement is an indicator electrode, the potential of which is a function of the activity of the target analyte. Many types of electrodes exist (see Table 9.1), but those based on membranes are by far the most useful analytical devices. The broader field of potentiometry has been reviewed recently (1). The potential of the indicator electrode cannot be determined in isolation, and another electrode (a reference electrode) is required to complete the electrochemical cell. Undoubtedly the best known of the potentiometric indicator electrodes is the glass pH electrode, the operation and use of which has been adequately discussed (2). Ion-selective electrodes (ISEs) are also commonplace, and have been the subject of several books (3–5): there is even a review journal for ISEs (6). Unfortunately, the simplicity of fabrication and use of ISEs has given rise to the idea that ISEs are chemical sensors. At the best this is a half-truth; certainly, they can behave like chemical sensors under well-controlled laboratory conditions, but in the real world their performance leaves much to be desired. Moreover, from a manufacturing point of view important features of a sensor are that it can be fabricated in relatively large numbers, and that each device is identical to all the others. Although some ISEs can be 'mass-produced', many cannot, and even those that do lend themselves to this form of production invariably require calibration before use. None-theless, in spite of the limitations of ISEs, transducers based on potentiometric membrane electrodes have much to contribute to the field of chemical sensing.

9.2 Membrane electrodes

The electrochemical cell that is formed when a membrane electrode is used in conjunction with a reference electrode is shown below.

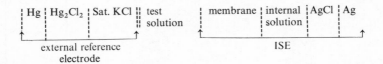

It is clear from this diagram that the function of the membrane is to provide a

Table 9.1 Potentiometric indicator electrodes

Electrode type	Example
Redox	Pt or other 'inert' electrode for the determination of redox potentials
Electrode of the first kind	Metal electrode in reversible equilibrium with a solution of its own ions ($Ag/AgNO_3$)
Electrode of the second kind	Metal electrode in contact or coated with one of its own compounds: the interface is reversible ($Ag/AgCl$)
Electrode of the third kind	Metal electrode in reversible equilibrium with a solution of one of its chelates ($Hg/HgEDTA^{2-}$)
Membrane electrode	Electrode in which a potential difference occurs across a membrane (pH glass electrode)

well-defined zone across which a potential difference develops. Ideally this occurs only when there is a difference in the analyte activities of the solutions on either side of the membrane. At first sight, conventional analysis of a chemical sensor as a device consisting of a more or less non-specific transducer interfaced with a zone of selective chemistry appears quite adequate when applied to potentiometric membrane devices. Thus, the membrane represents the selective chemistry of the sensor. (A classification of membrane types, and an indication of their diversity, is shown in Table 9.2.) The two reference electrodes shown in the cell diagram could be thought of as the non-specific transducer that enables the potential difference across the membrane to be detected by external electronic equipment. This analysis is too facile; even a rudimentary consideration of the physical chemistry of the membrane shows that it too plays an integral part in the transduction process.

Consider the origin of the potential difference across a membrane that separates two zones of solvent containing a solute X of charge Z. Suppose that the membrane exhibits ion-exchange activity towards X, and that the activity of X in solution is fixed on one side of the membrane (side F), and variable on the other (side V). This system may be analysed in a very simple manner along the lines of Nickolskii's work (7). The following equations can be written:

$$\mu_{X(F)} = \mu_X^0 + RT \ln a_{X(F)} \qquad \text{(for side F)}$$

$$\mu_{X(V)} = \mu_X^0 + RT \ln a_{X(V)} \qquad \text{(for side V)}$$

where μ_X = the chemical potential of the solution species X, μ_X^0 = the standard chemical potential of the solution species X, a_X = the activity of species X in the solution and R and T have their usual meaning. The electrochemical potential $\bar{\mu}_X$ of the solutions on each side of the membrane must include the term $zF\phi$, where ϕ is the Galvani potential, and F is the Faraday constant. Hence

$$\bar{\mu}_{X(F)} = \mu_X^0 + RT \ln a_{X(F)} + Fz\phi_{(F)} \qquad (9.1)$$

Table 9.2 Membrane types

Membrane type	Comments
Solid	
Glass	Doped glasses of various types give rise to H^+, Na^+, Li^+, K^+ and NH^+-sensitive electrodes
Crystalline	Single-crystal membranes (e.g. LaF_3), and pressed or sintered mixtures of Ag_2S with AgX or MS (X = halide, M = Pb, Cu or Cd) to give electrodes sensitive to F, S, X or M ions respectively.
Heterogeneous	Polythene, polypropylene, silicone rubber etc. with various halide, sulphate, phosphate, sulphide or oxalate precipitates form membranes for Ba, SO_4, PO_4, X, S etc. PTFE + graphite with ionic compounds for range of ions.
Liquid	
Homogeneous	PVC or (occasionally) other polymer mixed with plasticizer and sensor/solvent; membrane cast from this is sensitive to, K, Ca, M, NO_3 ions etc.
Heterogeneous	Porous material such as sintered glass, glass fibre or polymer (PVC or PTFE) holds liquid sensor by (e.g.) capillarity: sensitive to similar range of analytes as for the homogeneous membranes.
Composite	
Enzyme	A secondary layer containing (e.g.) an enzyme is placed on a (usually) glass membrane. Production or consumption of (e.g.) H^+ or OH^- when analyte reacts with the secondary layer is monitored as a measure analyte concentration.
Gas	A thin layer of solution is trapped between a glass membrane and a gas-permeable membrane. H^+ or OH^- is used to monitor reaction of NH_3, SO_2, NO_x, CO_2, H_2S within the entrapped layer.

and

$$\bar{\mu}_{X(V)} = \mu_X^0 + RT \ln a_{X(V)} + Fz\phi_{(V)} \tag{9.2}$$

In the same manner equations can be written for the electrochemical potentials on either side of the membrane. Thus:

$$\bar{\mu}_{X(F)M} = \mu_{X(M)}^0 + RT \ln a_{XM} + Fz\phi_{(F)M} \tag{9.3}$$

and

$$\bar{\mu}_{X(V)M} = \mu_{X(M)}^0 + RT \ln a_{XM} + Fz\phi_{(F)M} \tag{9.4}$$

where the subscript 'M' refers to the membrane phase. At the interface between the membrane and solution, the electrochemical potential of the adjacent phases must be identical, in which case equations (9.1) and (9.3) can be set equal, as can (9.2) and (9.4). By rearranging these identities, expressions can be obtained for $\theta_{M/F}$ and $\theta_{M/V}$, the potential difference across each phase boundary of the solution/membrane/solution system. Thus:

$$\theta_{M/F} = -\frac{\delta}{Fz} + \frac{RT}{zF}\ln\left[\frac{a_{X(F)}}{a_{X(M)}}\right] \tag{9.5}$$

and

$$\theta_{M/V} = -\frac{\delta}{Fz} + \frac{RT}{zF}\ln\left[\frac{a_{X(V)}}{a_{X(M)}}\right] \tag{9.6}$$

where

$$\delta = \mu^0_{X(M)} - \mu^0_X$$

The potential difference E_{memb} across the membrane is given by (9.6) − (9.5)

$$E_{memb} = \frac{RT}{zF}\ln\left[\frac{a_{X(V)}}{a_{X(F)}}\right] \tag{9.7}$$

If the potential difference across a membrane varies with activity in the manner depicted by equation (9.7), then the potentiometric device is said to behave in a Nernstian manner. For many membranes, this sort of behaviour, or an approximation to it, is observed.

The problem now arises as to how this potential difference may be sensed, without introducing further analyte-sensitive phases or electrodes into the cell. As has been mentioned previously, when an ISE is used a reference electrode is introduced into the bulk solution (i.e. the variable analyte activity side of the membrane), while an electrode of the second kind is incorporated into the solution on the fixed activity side of the membrane; typically Ag/AgCl is used. Equation (9.7) must be modified to include the finite potential of this internal electrode:

$$E_{ISE} = E_{int} + \frac{RT}{zF}\ln\left[\frac{a_{X(V)}}{a_{X(F)}}\right] \tag{9.8}$$

where E_{int} is the potential of the internal electrode.

Finally, both the potential of the external reference electrode (E_{ref}), and the junction potential (E_{ij}) associated with it are included in the measured potential E_{meas}:

$$E_{meas} = E_{ref} + E_{ij} + E_{ISE} \tag{9.9}$$

The simple theory developed above furnishes the basis for an analysis of membrane potentiometric devices as chemical sensors. Later, the effect of interferents will be introduced, first in an extended form of equation (9.8), and then in a more detailed form. The key parameters in membrane potentiometric sensor fabrication are the selective membrane; internal potentials; and external reference electrodes. Each of these will be discussed in turn.

9.2.1 Selective membranes

The basic role of the selective membrane has been discussed above. The methods of fulfilling this role are legion; even a simple classification of membrane types (Table 9.2) reveals the diversity. Generally, though, the membrane may either be capable of ion sensing in its own right, by virtue of its structure, e.g. doped lanthanum fluoride for F^-, or modifiers may be added to the membrane to give selectivity. An extensive discussion of the addition of organic ionophores to membranes may be found in Chapter 3. The use of (usually) plasticized polymer matrices containing relevant ionophores is an attractive route to selective membranes, particularly so for chemical sensors, for organic solvent solubility would seem to give opportunities of forming the membranes directly on to solid-state electronic structures. A range of other materials has been added to these membranes, including enzymes, bacteria and tissues (5). Undoubtedly the best ion-selective material is pH glass, and no other ion-selective membrane comes even close to its performance. Of the rest, doped lanthanum fluoride is the best. Generally, for ion-selective membranes, other than pH glass, the detection limit in pure solution is around 10^{-5}M. In the presence of interferents this situation is much worse. Thus, the potentiometric determination of free potassium ions in blood is just possible over the normal physiological range (8). There is a great and pressing need for better ion-selective membranes, i.e. improved ionophores, particularly to allow potentiometric sensors to be better used in biomedical applications. Until then, direct potentiometric determination of clinically important material will be limited to hydrogen, potassium, bicarbonate and sodium ions.

Ion-exchange membranes as a class are 'permselective', that is, there is a finite current-carrying capacity in the membrane associated with ionic charge transport. In a fairly simple model of the operation of these membranes, the ion exchange activity can be considered as localized in the surface of the membrane that is directly in contact with the solution. The inner portion of the membrane is impervious to either the solvent or the analyte ions. This model precludes the presence of diffusion potentials, but some form of ion transport within the membrane must take place in order for measurement to be made of the potential difference across the membrane. The first chapter of reference (4) is a good starting point for those wishing to get to grips with this area.

At a more detailed level, it is still not certain how even a relatively simple selective membrane operates (e.g. PVC containing plasticizer and valino-mycin, selective to K^+). The idea that PVC plus plasticizer acts as an inert structureless membrane in which the sensor species (the neutral carrier valinomycin in this case) dissolves, has been abandoned. A more heterogeneous structure is proposed, in which free dissociated counter-ions contribute to the membrane's conductivity. The counter-ions are presumed to arise from the dissociation of impurities associated with the method of manufacture of the PVC. Both fixed and mobile charged sites have been identified, and may represent, via the process of dissociation, an important source of counter-ion charge carriers (9–11). It is known that water penetrates the PVC film, but the exact role of this water is a matter still open to debate. Even the method of entrance of the water molecules is not clear; do they enter in a uniform manner, or are there hydrophilic chains arising from impurities in the PVC, along which the water travels? Recent impedance studies of membranes (12) have confirmed that the presence of fixed and mobile sites with weakly dissociable charge is associated with impurities in the PVC and the valinomycin, but not the plasticizer. Similarly, hydrophilic heterogeneities were identified in some membranes. It has been suggested that three distinct regions can be identified in the bulk of a plasticized PVC membrane: plasticized polymer, plasticizer, and water. Exudation of both plasticizer and valinomycin could give rise to the formation of a site-free high-resistance surface film (12).

Uncertainty in the understanding of the way in which membranes function does not detract from their usefulness. Even a cursory glance at Chapter 3 will reveal the extent and validity of the role of membranes in chemical sensing. However, as the authors point out, current knowledge permits only the use of a few simple ground rules in membrane design, and much still depends on an intuitive rather than an exact approach.

9.2.2 *Internal potentials*

In conventional ion-selective electrodes, the internal potential is kept constant by interfacing an aqueous solution (internal filling solution) between the membrane internal face and an internal reference element (Ag/AgCl). Addition of constant activities of Cl^- and the ion to which the membrane is selective ensures constant potentials at the relevant phase boundaries. This type of approach is also applicable to chemical sensors, although the added requirement for robustness precludes the use of a simple aqueous system. The appropriate ions are localized in gels or immobilized in a hydrophilic support. Although satisfactory laboratory devices can be made in this way, large-scale production appears to be difficult (13), as does the production of devices with repeatable standard potentials (14). Circumvention of this problem by direct placement of a selective membrane on to an 'inert' substrate, although simple in principle, turns out to be difficult in practice, because of the formation of a 'blocked' interface between the membrane (an ionic conductor) and the substrate (an electronic conductor). Coated wire electrodes (CWE) are an example of this type of device. There is no defined internal potential in a CWE, because of the blocked interface, and this and related devices are susceptible to stray capacitance effects (15). Any potential-forming process at the interface probably arises from inevitable impurities present in the membrane materials. Consequently, large variations in standard potentials occur with these devices, and again, this is not appropriate for medium- or large-scale manufacture. A more fruitful approach seems to be to place suitable ion-conducting layers between the sensing membrane and final conductor wire. This is reported (16) to be successful for the system $LaF_3/AgF/AgCl/Ag$, and could in principle be applied to polymer membranes.

9.2.3 *External reference electrode*

A complete potentiometric transducer includes an external reference electrode. This point is often ignored in the ISE literature, for it is assumed that some suitable device can be incorporated into the cell, and that calibration will overcome any problems associated with variability in this cell component. This approach is inadmissible for a chemical sensor, because the reference electrode must be included as part of the package. Too often, publication of an elegant sensor is marred by a test set-up which includes a large, conventional reference electrode.

Extensive discussion on the subject of reference electrodes can be found in the literature (17–19). The role of the reference electrode in an ISE/reference electrode pair is to maintain an invariant potential, independent of changes in the test solution, against which potential change at the selective membrane/test solution interface can be measured. This is usually achieved by placing an electrode of the second kind (e.g. Ag/AgCl) in a strong solution of KCl, then contacting this reference solution via a liquid junction (frit cone capillary, etc.). The constant Cl^- activity of the reference solution (assuming no chemical change or dilution with time) ensures a constant reference potential, whilst the high ionic strength bridge solution minimizes the generally unknown liquid junction potential. To mimic such an arrangement in chemical sensor terms, other than by direct miniaturization (see, for example, work by Armstrong-James (20), but note too the problems (21) associated with this sort of approach) is clearly very difficult. Attempts have included the use of gels (22), but with no great success. A miniature liquid-junction reference electrode that can be integrated with a solid-state chemical sensor has been described (23). The device can be batch-produced using planar integrated circuit technology. A less conventional reference device has been made using an ISFET with a Parylene gate as a site-free ion-blocked membrane (24). An alternative approach involves rejecting the concept of a universal reference element, opting instead for a system in which the operating environment of the sensor provides a reference feature. For example, suppose that the near-constant level of Cl^- in blood were to be exploited by using potentiometric sensing devices based on differential principles. The typical range of Cl^- concentrations in blood lies between 96 and $105 \, mML^{-1}$; this corresponds to a potential range of around $5 \, mV$ for a chloride-sensitive reference element. Alternatively it might be possible to add trace materials (e.g. I^-, F^-) in industrial monitoring situations.

9.3 Limitations to the use of potentiometric devices

9.3.1 *Incomplete selectivity*

In the derivation of equation (9.8), no attempt was made to include the effect of an interfering ion: it is the nature of many of the ion-exchangers used in membrane electrodes that they show exchange activity with a range of species. If the following exchange is considered:

$$X_{(M)}^z + Y_{(S)}^{z_1} \rightleftharpoons X_{(S)}^z + Y_{(M)}^{z_1}$$

where the subscripts 'M' and 'S' refer to the membrane-bound and free solution forms respectively of the ionic species, then the exchange constant is given by

$$K_{X/Y} = \frac{a_{X(S)} \cdot a_{Y(M)}}{a_{X(M)} \cdot a_{Y(S)}}$$

Incorporation of this constant into equation (9.8) yields

$$E_{ISE} = E^0 + \frac{RT}{zF} \ln [a_{X(V)} + K(a_{Y(V)})^{z/z_1}] \qquad (9.9)$$

where

$$E^0 = E_{int} - \frac{RT}{zF} \ln a_{X(F)}$$

This equation has been developed by Nikolskii (7). A more complete statement of the effect of interferents on the ISE potential can be seen from equation (9.10).

$$E_{ISE} = E^0 + \frac{RT}{zF} \ln [a_{X(V)} + \sum_{J=1}^{J=n} K(a_{J(V)})^{z/z_J}] \qquad (9.10)$$

A more general solution has been proposed, a summary of which can be found on page 19 of reference (25).

$$E_{memb} = \frac{RT}{zF} \ln \frac{a_{X(V)} + k_{XY}^{pot}(a_{Y(V)})^{z/z_1}}{a_{X(F)} + k_{XY}^{pot}(a_{Y(F)})^{z/z_1}} \qquad (9.11)$$

where

$$k_{XY}^{pot} = \frac{U_Y}{U_X} K_{X/Y} \quad (k_{XY}^{pot} \text{ is the 'selectivity coefficient'})$$

U_X is the mobility of X in the membrane, and U_Y is the mobility of Y in the membrane. Usually $a_{Y(F)} = 0$, and equation (9.11) becomes

$$E_{ISE} = E^0 + \frac{RT}{zF} \ln [a_{X(V)} + k_{XY}^{pot}(a_{Y(V)})^{z/z_1}]$$

Clearly the effect of an interferent depends on the magnitude of the selectivity coefficient. At the very least, the detection limit of the electrode is degraded because of the uncertainty in measurements at low analyte concentrations. In the worst case, analysis may not be possible at all.

9.3.2 Potential drift

In potentiometry with ISEs, there are two significant causes of potential drift (other than variation in the analyte activity): variations in the liquid junction potential and temperature. There are, of course, many other causes, such as photoeffects, pressure variation, high cell resistance etc., but these are fairly uncommon. A rather more common cause of drift for ion-exchange type membranes is gradual leaching of the sensor material from the membrane. Improvements in covalent bonding of the ion exchanger to the polymer matrix (see for example refs. 73, 75–77 of Chapter 3) should go some way to eliminating this source of potential drift. The situation should be much the same for potentiometric transducers, with temperature effects and variation in the liquid junction potential accounting for most of the drift. Differential measurements employing another membrane electrode sidestep the issue of liquid junction variation, but, of course, this approach may not always be feasible. A detailed discussion of liquid junction problems can be found on page 323 *et seq.* of reference (4).

Conventional potentiometric measurements rely on the use of these electrodes under isothermal conditions; this may be an unattainable luxury in a real sensing situation. The temperature effect on a potentiometric measurement is manifest from equation (9.9), where temperature appears explicitly in the numerator of the activity-dependent term. However, the crucial issue is the difference between the temperature coefficients of the external reference electrode, and the indicator electrode. The temperature-dependent variation in the potential of external reference electrodes is well documented (26), and can be measured to include the temperature effects on the liquid junction potential. The temperature coefficient of the indicator electrode depends on the nature of the internal reference electrode as well as on the membrane. In the case where, even for the optimum design, there is a finite temperature coefficient for the potentiometric transducer, and where the close control of temperature is out of the question, then it might be possible to include a temperature sensor alongside the chemical sensor. Data from the temperature sensor could be used to correct the potentiometric response, in a manner analogous to a system recently described for piezoelectric crystal transducers (27). Such a correction may help to decrease the temperature effect, although the presence of thermal gradients may prove more difficult to cope with, particularly when the hysteresis of the different components of the potentiometric transducer is considered. The use of passive thermal shielding in a manner advocated for amperometric sensors (see Chapter 8) should limit the effects of thermal transients.

9.4 Nature of the response

Membrane potentiometric devices respond in a logarithmic fashion to analyte activities. The majority of chemists are more used to working with concentrations than with activities, and unless the ionic strength of a solution is known, the concentration cannot be calculated. Addition of a total ionic-strength adjusting buffer can overcome this problem, but is not always possible. The problem will be compounded where the ionic strength of the measurand solution is subject to variation, as happens, for example, in estuarine waters. In this case the concentration of the analyte may be constant, but its activity will vary. A recent report has described a fluorescence sensor for monitoring ionic-strength values (28), and it is possible that the use of this probe in conjunction with suitable data processing may help to alleviate the ionic-strength problem. Below about 10^{-4} M, the difference between activity and concentration is negligible when compared with other errors involved in the measurement. The logarithmic relationship between activity and potential has implications for the effect of error on the measurement. Thus, for a monovalent ion, a 0.1 mV error results in 0.4% error in the estimated activity, and a 1 mV error causes a 4.0% error. The magnitude of the errors is doubled for a divalent ion.

9.5 Instrumentation

An ISE may be viewed as a high-impedance voltage source, and measurement of the generated potential requires the use of high-input-impedance volt-meters. As a rule of thumb, the impedance of the measuring device should exceed that of the source by three orders of magnitude. Hence, for classical ISEs, with impedances in the range 10^6 to $10^7 \Omega$, the input impedance of the voltmeter should be at least $10^{10} \Omega$. For micro-ISEs, the input impedance needs to be two orders of magnitude greater. Essentially, the problem under discussion here is the measurement of a potential difference across a network of resistances. Thus, the following should be borne in mind also: the impedance between the reference electrode and the ISE must be substantially less than the input impedance of the voltmeter; the impedance of the insulation between the test solution and ground should be at least two orders of magnitude greater than that between the reference solution and ground. Potentiometric trans-ducers may be constructed with substantially lower impedances (see, for example, the thick in Chapter 11, p. 236 film devices described), and this lessens the associated problems. High-impedance devices are also subject to electrostatic pick-up, and careful attention must be paid to shielding the potentiometric device. As a result of these various effects, the performance of potentiometric devices can depend on the ambient humidity. At low humidity, electrostatic disturbances predominate; at high humidities electrical leakage can be a source of error.

9.6 Conclusions

The thrust of chemical sensor development is to produce devices which are small, inexpensive, disposable and readily manufactured. These need not be advantageous for all (e.g. industrial) measurement scenarios. In this context, the need to calibrate each device individually before use is clearly a major disadvantage. Potentiometric sensors will never be fully exploited until manufacturing processes are able to produce many devices with identical (and known) calibration parameters. This can best be done by addressing the fundamental issues relating to the key sensor components, coupled with innovative development of manufacturing methods. Even then, major problems remain, such as biocompatibility, lifetime under arduous conditions and consumer acceptance.

References

1. E.A.M.F. Dahmen, *Electroanalysis: Theory and Applications in Aqueous and Non-Aqueous Media and in Automated Chemical Control*, Elsevier, Amsterdam (1986) Chapter 2.
2. P.W. Linder, R.G. Torrington and D.R. Williams, *Analysis Using Glass Electrodes*, Open University Press, London (1984).
3. A.H. Covington, *Crit. Rev. Anal. Chem.* **3** (1974) 355.
4. H. Freiser (ed.), *Ion-Selective Electrodes in Analytical Chemistry*, Vol. 1, Plenum, New York (1978).
5. R.L. Solsky, *Crit. Rev. Anal. Chem.* **14** (1982) 1.
6. *Ion-Selective Electrode Reviews*, ed. J.D.R. Thomas, Pergamon Press, New York.
7. B.P. Nikolskii, *Zh. Fiz. Khim.* **10** (1937) 495.
8. D. Amman, W.E. Morf, P. Anker, P.C. Meier, E. Pretsch and W. Simon, *Ion-Selective Electrode Reviews* **5** (1983) 3.
9. C.A. Kumins and A. London, *J. Polym. Sci.* **46** (1960) 395.
10. M. Perry, E. Lobel and R. Bloch, *J. Membr. Sci.* **1** (1976) 223.
11. D.G. Rance and E.L. Zichy, *Polymer* **20** (1979) 266.
12. G. Horvai, E. Graf, K. Toth, E. Pungor and R.P. Buck, *Anal. Chem.* **58** (1986) 2735, 2741.
13. C.J. Battaglia and J.C. Chang, US Patent 4,214,968.
14. R.P. Buck, personal communication.
15. R.G. Kelly and A.E. Owen, *IEEE Proc.* **132** (1985) 227.
16. T.A. Fjeldy and K. Nagy, *J. Electrochem. Soc.* **127** (1980) 1299.
17. D.J.G. Ives and G.J. Janz (eds.), *Reference Electrodes*, Academic Press, New York (1971).
18. J.N. Butler, *Adv. Electrochem. Electrochem. Engg*, **7** (1970) 77.
19. D.T. Sawyer and J.L. Roberts, *Experimental Electrochemistry for Chemists*, John Wiley, New York (1974).
20. M. Armstrong-James, K. Fox, Z.L. Kruk and J. Millar, *J. Neurosci. Meth.* **4** (1981) 385.
21. M. Lavalee, D.F. Schanne and N.C. Hebert (eds.), *Glass Microelectrodes*, John Wiley, New York (1969).
22. S.J. Pace, US Patent 4,454,007.
23. R.L. Smith and S. Collins, US Patent 4,592,824.
24. T. Matsuo and H. Nakajima, *Sens. Act.* **5** (1984) 293.
25. J. Vesely, D. Weiss and K. Stulik, *Analysis with Ion-selective Electrodes*, Ellis Horwood, Chichester (1978).
26. R.G. Bates and V.E. Bower, *J. Res. Natl. Bur. Std.* **53** (1954) 283.
27. S.M. Fraser, T.E. Edmonds and T.S. West, *Analyst* **111** (1986) 1183.
28. O.S. Wolfbeiss and H. Offenbacher, Ger. Offen., DE 3,430,935.and *Sens. Act.* (in press) (1986).

10 MOSFET devices

I. ROBINS

10.1 Introduction

The first reported gas sensor based on a metal oxide semiconductor (MOS) device was the palladium gate MOSFET for the sensing of hydrogen described by Lundstrom (1) in 1975. Work since then has been based mainly on MOS devices having various transition-metal gate electrodes, primarily for the sensing of hydrogen-containing gases, although some work has been reported on the sensing of carbon monoxide with these devices. These sensors are based on standard microelectronic fabrication techniques and should therefore be easily mass produced at a relatively low unit cost.

10.2 Background physics

The metal oxide silicon field effect transistor (MOSFET) has the basic structure as shown schematically in Figure 10.1 (an n-channel device is shown and will be discussed, although p-channel devices work equally well). The device consists of two n-type diffusions (called the source and the drain) in a p-type substrate separated by a p-type area. This area is covered by a thin (50 nm) insulator layer (generally silicon dioxide or silicon dioxide overlaid with silicon nitride) which is called the gate insulator. On top of the gate insulator there is a metallic contact called the gate electrode, which in conventional devices is made from aluminium. The source and drain diffusions are

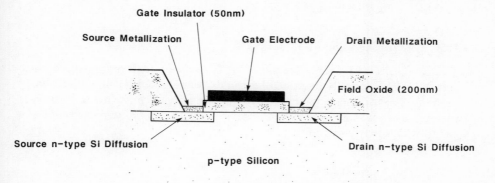

Figure 10.1 Schematic diagram of an n-channel MOSFET.

contacted by metallizations and the remainder of the device is covered by a thick (200 nm) layer of insulator.

The current that flows between the source and the drain in the device upon applying a voltage across the two diffusions is controlled by the voltage that is applied between the gate electrode and the substrate. The field produced across the insulator layer by the gate voltage is able to influence the properties of a thin layer of silicon close to the insulator and can cause inversion to occur in this region, which alters the p-type silicon to n-type silicon. This inverted area of silicon between the source and the drain diffusions is called the channel of the device.

Prior to inversion, the device acts as two back-to-back diodes, hence no drain current flows; after inversion, a totally n-type conduction path exists between the source and the drain, and current can readily flow—for a more detailed discussion see Sze (2). For an FET, operating in the saturation region, the drain current I_d flowing through the channel is given by

$$I_d = (WC\mu/2L)(V_g - V_T)^2$$

where W and L are respectively the width and the length of the channel, C is the capacitance of the insulator layer, μ is the mobility of the charge carriers within the channel, V_g is the gate voltage and V_T is the threshold voltage.

The threshold voltage of the device is the smallest voltage that is required to be applied to the gate electrode of the device (with respect to the substrate) in order for a current to be able to flow through the channel of the device (i.e. for inversion to occur in the channel), and it has been found that when the gate electrode is made from certain catalytically active metals, the threshold voltage of the device is dependent on the concentration of certain hydrogen-containing gases in the atmosphere above the device.

In many cases MOS capacitors have been used in place of MOSFETs, as these are easier to fabricate, although they are less useful as a production device. These capacitors sense gases in the same manner as MOSFETs and are therefore a useful experimental device.

10.3 Gas sensing using MOS devices

The gas-sensitive FET structure most closely studied to date is the palladium-gate MOSFET initially described by Lundstrom in 1975 (1), and which has been the subject of many papers since (3–7). This device can be made sensitive to a range of gases, and therefore the devices will be discussed in terms of the gas to which they respond.

10.3.1 *Hydrogen sensors*

For the gate electrode of a MOSFET device to be a hydrogen sensor, the surface of the electrode has to be able to catalytically decompose the hydrogen gas to hydrogen atoms, and these atoms must be able to diffuse into the bulk of the electrode and through to the electrode/insulator interface. The obvious choice of metals for hydrogen sensors are therefore palladium, platinum and certain other of the so-called 'noble metals'. The first and most widely used MOSFET hydrogen sensor is the palladium gate device. Palladium is a highly catalytically active metal and can solvate the largest amount of hydrogen of any of the elements. The palladium gate devices are operated at an elevated temperature (generally 150 °C, although they operate at both higher and lower temperatures), and respond readily to hydrogen in the concentration range 1–5000 ppm in air, inert atmospheres and in vacuum.

A mechanism of operation of these devices was proposed by Lundstrom (1, 3) and is outlined as follows. Hydrogen molecules are catalytically dissociated to atomic hydrogen on the palladium surface. The hydrogen atoms then diffuse into the palladium bulk and then diffuse through the metal where they are adsorbed at the palladium/silicon dioxide interface (Figure 10.2). At this interface, the hydrogen atoms become polarized and generate a voltage, which is seen as an effective change in the threshold voltage of the device. If there are only a certain number of adsorption sites at this interface, then there is a maximum voltage change that can occur corresponding to complete filling of these sites. Therefore, if ΔV_{max} is the maximum voltage change, then

$$\Delta V = \Delta V_{max} \theta$$

where ΔV is the change in the voltage of the device (threshold voltage for an

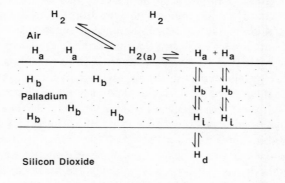

H_a = Absorbed Hydrogen Atoms

H_b = Absorbed Hydrogen Atoms

H_i = Hydrogen Atoms at Metal/Insulator Interface

H_d = Hydrogen Atoms Drifting in SiO_2

Figure 10.2 Section of gate electrode/insulator.

FET and flatband voltage for an MOS capacitor) and θ is the fractional coverage of the sites at the interface.

The filling of these sites is assumed to be governed by Langmuir kinetics and as such is dependent only on the number of unfilled sites, and on the concentration of hydrogen atoms near the interface. The hydrogen atoms at the interface are assumed to be in equilibrium with the dissociated hydrogen atoms at the outer surface of the metal and the concentration of these is dependent on the concentration of hydrogen gas above the metal, and the number of unfilled adsorption sites on the palladium surface (desorption at the outer surface is also governed by Langmuir kinetics). This yields an equation relating the change in voltage of the device to the square root of the hydrogen gas concentration as

$$1/\Delta V - 1/\Delta V_{max} = K/(\Delta V_{max}[H_2]^{1/2})$$

where $[H_2]$ is the concentration of hydrogen gas and K is a term that depends on the dissociation constants of hydrogen on palladium, and on other kinetic rate constants (4). The Langmuir adsorption kinetics gives the equation its general form

$$1/A - 1/A_{max} = C/(A_{max}P)$$

where A is volume adsorbed, C is a constant and P is the pressure, while the square root of the hydrogen concentration arises purely because of the dissociation of hydrogen gas into two hydrogen atoms. Lundstrom presented results to support this theory both in vacuum and in an inert gas. The theory was modified slightly to allow for the action of oxygen on the system (1, 3, 4). The oxygen is considered to remove hydrogen atoms from the surface of the palladium by a water formation reaction and therefore decreases the voltage shift observed for a set hydrogen concentration. A typical response of a palladium gate MOSFET to hydrogen in air is shown in Figure 10.3.

This initial theory has been modified several times by Lundstrom and his co-workers. The first modification was to explain the long response and recovery times of the devices (8). The change in threshold voltage of the devices upon exposure to hydrogen was shown to consist of an initial rapid change in voltage followed by a much slower response. The same was observed for the recovery of the devices after exposure to hydrogen. This effect was explained by proposing that the fast response was caused by the hydrogen dipoles at the palladium/silicon dioxide interface whilst the slow part of the response was due to the slow diffusion of hydrogen atoms into the bulk of the silicon dioxide. The signal due to the slow response was shown to be responsible for about 60% of the total signal. Dobos et al. (9) produced devices where the gate insulator was less permeable to hydrogen atoms (e.g. silicon dioxide overlaid with either silicon nitride or aluminium oxide), thereby reducing the slow part of the signal and producing devices with a more rapid response and recovery to hydrogen.

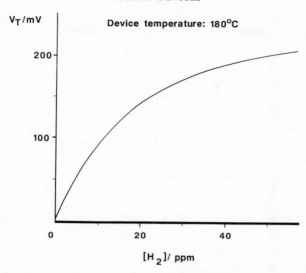

Figure 10.3 Graph showing response of a palladium gate MOSFET to hydrogen gas in air.

The second modification was by Petersson *et al.* (10). Palladium gate MOS capacitors were exposed to hydrogen in an ultra-high-vacuum system. They showed that the response of the devices was dependent on the logarithm of the hydrogen pressure, not the square root as predicted by the initial theory. The Lundstrom theory is based on adsorption at the palladium/silicon dioxide interface being Langmuir-like (i.e. the adsorption enthalpy of a site being independent of the number of filled sites). Petersson modified the theory to allow for a variation in the adsorption enthalpy upon filling. The Tempkin isotherm (11) was used and yielded a theory that predicted that the change in threshold voltage would be dependent on the logarithm of the hydrogen pressure (as found). Robins *et al.* (12) also found the response of palladium gate devices to be dependent on the logarithm of the hydrogen gas concentration with a gradient of about 26 mV per decade. This work was carried out, however, at atmospheric pressures in air, and with silicon nitride-insulated FETs. A different theory was proposed to explain this result. The metal/insulator interface was considered to behave as an electrochemical half cell with hydrogen atoms (being produced by catalytic dissociation on the surface of the palladium) being ionized at the metal/insulator interface and the resulting protons (or possibly hydroxyl groups) entering the insulator. By application of the Nernst equation to this situation, a logarithmic dependence of the change in threshold voltage with hydrogen gas concentration was predicted and, provided that the activity of the protons (hydroxyls) in the insulator did not vary, a gradient of about 30 mV per decade was predicted.

Only a small amount of work has been published on platinum-gate FET-based hydrogen sensors. Lundstrom and co-workers (13) showed that platinum-gate devices respond in a similar manner to palladium-gate devices, although they showed a smaller response to similar hydrogen concentrations.

There has been very little reported work on the use of other metals as gate electrodes for FET-based hydrogen sensors, although Winquist (14) presented results on a number of different metal electrode devices, comparing their responses to both hydrogen and ammonia.

Hydrogen-sensitive devices have been used in several specialized applications apart from their use as general hydrogen detectors. The palladium-gate devices have been used in ultra-high vacuum systems for studying the fundamental reactions occuring between hydrogen and oxygen (15) by using the device both as a reaction site (the palladium surface) and as a detector for the adsorbed hydrogen at the site of reaction. These devices have also been used for the detection of hydrogen that has been generated by corrosion occuring within an oil pipeline and has subsequently diffused through the pipe wall (16). The hydrogen emitted during the early stages of a fire has also been monitored by these devices (17), and they therefore have possible applications in fire detection systems.

10.3.2 *Ammonia sensors*

The earliest report of an ammonia-sensitive MOSFET device was by Lundstrom (4) in 1975, investigating the recovery of a palladium-gate MOSFET in nitrogen. Theory predicted that the device should be ammonia-sensitive, and data were presented to confirm this. However, other reports by Lundstrom (3) show that only a few palladium-gate devices are ammonia-sensitive. Ross *et al.* [18] presented results showing that MOSFET devices having platinum-gate electrodes deposited by a thermal evaporation process were ammonia-sensitive, while those having the gate electrode deposited by a sputtering technique were insensitive to ammonia. X-ray photoelectron spectroscopy (XPS) results showed that there was no chemical difference between the two types of gate electrodes. Secondary electron microphotographs revealed that the evaporated platinum electrodes (those that were ammonia-sensitive) were highly porous, while the sputtered platinum electrodes were coherent. The authors postulated that the ammonia sensitivity was due to interaction between the platinum metal and the silicon dioxide insulator, causing a modification of the catalytic activity of the platinum metal and enabling it to be able to partially decompose ammonia gas, yielding hydrogen atoms. The hydrogen atoms produced are then sensed by the device in a similar way to palladium- or platinum-gate hydrogen sensors.

Spetz *et al.* (19) presented results on a new type of MOS capacitor, but arrived at similar conclusions. The devices used were MOS capacitors having a thick contact electrode which had a second thin electrode (about 3 nm thick) covering the contact electrode and also spreading on to the insulator layer (Figure 10.4). The thick contact electrode was made from palladium; the thin

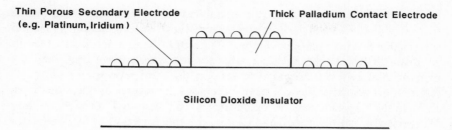

Thin Porous Secondary Electrode
(e.g. Platinum, Iridium)

Thick Palladium Contact Electrode

Silicon Dioxide Insulator

Figure 10.4   Detail of electrode structure.

secondary electrode was made from one of a range of metals. It was found that
the devices were ammonia-sensitive when the thin electrode was made from
platinum or iridium. The authors proposed that the interaction between the
thin porous secondary electrode and the silicon dioxide insulator caused the
catalytic activity of the metal to be enhanced. Ammonia gas could therefore be
decomposed, generating hydrogen atoms, which then dissolved in the thick
palladium electrode and produced a response in the normal manner.

Further work by Winquist et al. (20) using iridium/palladium dual-gate
capacitors developed this theory, and produced an equation relating the
change in voltage of the device to the partial pressure of ammonia gas in a
similar manner to that of the response characteristics of hydrogen-sensitive
devices. The equation was, however, only approximate, due to uncertainties
over the exact mechanism of decomposition of the ammonia molecule (i.e. the
number of hydrogen atoms split from the ammonia, and the nature of the
interaction of oxygen). By limiting the equation to the steady-state response of
the device and by considering only the response at low surface coverages (up to
100 ppm ammonia concentration), the equation could be simplified to

$$\Delta V = K[NH_3]^{1/m}$$

where K and m are constants for a particular system.

Further work by Spetz et al. (21) showed that the thick contact electrode did
not have to be made from palladium. Their initial theory of operation was
therefore modified to postulate that the hydrogen atoms, generated from the
catalytic decomposition of the ammonia, caused an interfacial dipole layer to
be formed at the platinum (or iridium)/silicon dioxide interface, not at the
palladium/silicon dioxide interface.

Ross et al. (22) presented results on hydrogen and ammonia sensitivity of
thin platinum-gate MOSFET devices. They concluded that the mechanism of
response of these devices was as previously suggested, and cited literature on
the dissociation of ammonia on silicon dioxide supported platinum catalysts
as additional evidence. Lundstrom et al. (23) presented further results on the
two-metal capacitors. They outlined three possible modes of response of these
devices, the first the same as the hydrogen sensitivity on palladium-gate
devices (which they immediately ruled out), the second as already described,

and the third due to capacitative coupling between the insulator and charge formed on the surface of the porous platinum due to the production of reaction intermediates on the platinum surface. The authors stated that it was difficult to distinguish between the last two possible mechanisms but that the experimental evidence appeared to support the third process.

Ammonia-sensitive devices have been used in several applications even though their mechanism of operation is still disputed. Danielsson and Winquist (24–26) have used the devices for the detection of ammonia gas generated from enzymic reactions and have related this to the concentration of the enzyme substrate in the test solution. This system can therefore be used for the detection of any enzyme/substrate pair that generates ammonia. Ross *et al.* (27) have used an ammonia-sensitive MOSFET in conjunction with a sacrificial coating as an overheat detection system. The sacrificial coating contains a material that decomposes at a set temperature, generating ammonia gas. This can then be detected by the sensor. Such a system provides volume protection of components with only one sensor and produces a warning signal on overheating prior to the onset of a fire.

10.3.3   *Sensors for other hydrogen-containing gases*

Apart from hydrogen and ammonia, MOS-based devices have also been shown to respond to a number of other hydrogen-containing gases.

The palladium-gate devices have been shown to respond to hydrogen sulphide (28). The devices were shown to follow a half-power law similar to the response to hydrogen. The response was considered to be due to the formation of atomic hydrogen from the catalytic dissociation of the hydrogen sulphide (again similar to the response of hydrogen).

Arsine has also been detected by palladium-gate MOS devices (29). These devices were different from the hydrogen-sensitive devices (although they also respond to hydrogen) in that the gate electrode was fabricated with a series of holes with a diameter of $2\,\mu m$ along its length. The sensors were shown to be able to detect at least 1 ppm of arsine in air. No response mechanism was proposed for these devices.

Palladium-gate MOS capacitors can be used to detect ethanol vapour when they are operated at temperatures above $150\,°C$ (30). The mechanism is believed to be the catalytic dissociation of ethanol yielding hydrogen atoms, which are then sensed in the same manner as that of the hydrogen-sensitive devices. The high temperature of operation of the device is required to enable the ethanol decomposition to occur. Sensitivity to various organic vapours similar to that of the thin porous dual-metal capacitors was observed when operated at temperatures above $190\,°C$ (31). Again the proposed mechanism was decomposition of the parent molecule to yield hydrogen atoms.

10.3.4 *Carbon monoxide sensors*

The sensors discussed up to this point have all detected the presence of hydrogen-containing gases, and are probably all detecting hydrogen atoms generated from catalytic reactions on the surface of the gate metal (the case of ammonia-sensitive devices being in doubt). Carbon monoxide, however, is different, as the molecule does not contain any hydrogen atoms.

The first report of a carbon monoxide-sensitive MOSFET was by Dobos *et al.* in 1979 (32), using a palladium-gate MOSFET which was manufactured with holes in the gate electrode (similar to the arsine-sensitive device described above). The initial theory for the sensitivity of the device was that the carbon monoxide molecules were absorbed on to the surface of the palladium metal (including down the holes). The carbon monoxide molecules near the bare gate insulator then affected the field experienced by the devices because of the inherent dipole moment on carbon monoxide.

Further reports by Dobos *et al.* (33, 34) confirmed the initial results, but also showed that the response of the device was variable, depending on its previous history. The authors predicted a response characteristics of the device based on a Langmuir type of adsorption of the carbon monoxide on to the palladium surface:

$$1/\Delta V = k/(\Delta V_{max}[CO]) + 1/\Delta V_{max}$$

where k is a constant and $[CO]$ is the concentration of carbon monoxide in the atmosphere. This equation was obeyed by only some of the devices, depending on their previous thermal history. The authors also showed that the cross-sensitivity of these devices to hydrogen could be decreased by covering the surface (but not the hole walls or the bare insulator) of the palladium gate electrode with aluminium. The theory of operation of these devices was modified in 1985 (35). The response was now considered to be due to the change of the field seen by the bare gate insulator areas (i.e. the areas in the holes) caused by the reduction of any palladium oxide coating on the palladium gate electrode. This oxide coating was simply produced by a thermal oxidation or by a sputtering technique.

A different carbon monoxide-sensitive MOSFET in which the gate contact was made from tin dioxide was described by Dobos *et al.* (34). This device sensed the gas by a change in the Fermi energy of the tin dioxide on adsorption of carbon monoxide. The device was shown to respond well to carbon monoxide, but had poor baseline stability, due to the poorly conducting nature of the tin dioxide required for high gas sensitivity.

10.4 Conclusions

Several types of gas-sensitive MOS devices have been developed since 1975. Already sensors for hydrogen, ammonia, hydrogen sulphide, ethanol vapour, arsine and carbon monoxide have been developed. The devices have the advantages over other types of gas sensors of being small, rugged and easily interfaced to control equipment. However, the most important advantage of the MOS-based sensors is that they should be easily mass-produced using conventional silicon microelectronic techniques and should, therefore, have a relatively low cost.

References

1. I. Lundstrom, S. Shivaraman, C. Svensson and L. Lundkvist, *Appl. Phys. Lett.* **26** (1975) 55–57.
2. S.M. Sze, *Physics of Semiconductor Devices*, 2nd edn., John Wiley, New York (1981).
3. I. Lundstrom, *Sensors and Actuators* **1** (1981) 403–426.
4. K.I. Lundstrom, M.S. Shivaraman and C.M. Svensson, *J. Appl. Phys.* **46** (1975) 3876–3881.
5. I. Lundstrom and D. Soderberg, *Sensors and Actuators* **5** (1981–2) 105–138.
6. M. Armgarth, D. Soderberg and I. Lundstrom, *Appl. Phys. Lett.* **41** (1982) 654–655.
7. T.L. Poteat, B. Lalevic, B. Kuliyev, M. Yousef and M. Chen, *J. Electron. Mater.* **12** (1983) 181–214.
8. C. Nylander, M. Armgarth and C. Svensson, *J. Appl. Phys.* **58** (1984) 1177–1188.
9. K. Dobos, M. Armgarth, G. Zimmer and I. Lundstrom, *IEEE Trans. on Elect. Dev.* **ED31**(4) (1984) 508–510.
10. L.-G. Petersson, H.M. Dannetun, J. Fogelberg and I. Lundstrom, *J. Appl. Phys.* **58** (1985) 404–413.
11. P.W. Atkins, *Physical Chemistry*, Oxford University Press, Oxford (1978) 946.
12. I. Robins, J.F. Ross and J.E.A. Shaw, *J. Appl. Phys.* **60** (1986) 843–845.
13. I. Lundstrom and T. Distefano, *Solid State Comm.* **19** (1976) 871–875.
14. F. Winquist, A. Spetz, M. Armgarth, C. Nylander and I. Lundstrom, *Appl. Phys. Lett.* **43** (1983) 839–841.
15. I. Lundstrom, M.S. Shivaraman and C. Svensson, *Surf. Sci.* **64** (1977) 497–519.
16. J. van Dijk, P.M. van der Velden, J.F. Ross, I. Robins and B.C. Webb, *6th Eur. Symp. on Corrosion Inhibitors*, Ferrara (Sept. 1985).
17. I. Lundstrom, M.S. Shivaraman, L. Stiblert and C. Svensson, *Rev. Sci. Instrum.* **47** (1976) 738–740.
18. J.F. Ross, I. Robins and B.C. Webb, *1st Conf. on Sensors and their Applications*, Manchester (1983).
19. A. Spetz, F. Winquist, C. Nylander and I. Lundstrom, *ACS Symp. Ser.* **17**: *Proc. Int. Meeting Chem. Sens.*, Fukuoka, Japan, Elsevier, Amsterdam (1983) 479–487.
20. F. Winquist, A. Spetz, I. Lundstrom and B. Danielsson, *Anal. Chim. Acta* **164** (1984) 127–138.
21. A. Spetz, M. Armgarth and I. Lundstrom, *VTT Symp.* **45**, *11th Nord. Semicond. meeting* (1984) 233–236.
22. J.F. Ross, I. Robins and B.C. Webb, *Sensors and Actuators* **11** (1987) 73–90.
23. I. Lundstrom, M. Armgarth, A. Spetz and F. Winquist, *Proc. Int. Meeting on Chem. Sensors*, Bordeaux (1986) 387–390.
24. B. Danielsson, F. Winquist, K. Mosbach and I. Lundstrom, *Biotech '83: Int. Conf. Appl. Biotech.* (1983) 679–687.
25. F. Winquist, A. Spetz, M. Armgarth, I. Lundstrom and B. Danielsson, *Sensors and Actuators* **8** (1985) 91–100.
26. F. Winquist, A. Spetz, I. Lundstrom and B. Danielsson, *Anal. Chim. Acta* **163** (1984) 143–149.
27. J.F. Ross, C.I. Terry and B.C. Webb, *J. Phys. E: Rev. Sci. Instrum.* **19** (1986) 143–149.
28. M.S. Shivaraman, *J. Appl. Phys.* **47** (1976) 3592–3593.
29. W. Mokwa, G. Zimmer, K. Dobos, D. Kohl and G. Heiland, *2nd Conf. on Sensors and their Applications*, Southampton (1985).
30. U. Ackelid, M. Armgarth, A. Spetz and I. Lundstrom, *IEEE Electron. Dev. Lett.* **EDL-7** (6) (1986) 353–355.
31. U. Ackelid, F. Winquist and I. Lundstrom, *Proc. 2nd Int. Meeting on Chem. Sensors*, Bordeaux, (1986) 395–398.

32. K. Dobos, B. Hofflinger and G. Zimmer, *Vide Couches Minches* **201** (1980) 743–745.
33. D. Krey, K. Dobos and G. Zimmer, *Sensors and Actuators* **3** (1982/3) 169–177.
34. K. Dobos, D. Krey and G. Zimmer, *Proc. Int. Meeting on Chem. Sensors*, Fukuoka, Japan, Elsevier, Amsterdam, (1983) 464–467.
35. K. Dobos and G. Zimmer, *IEEE Trans. on Electron. Dev.* **ED-32** (7) (1985) 1165–1169.

11 Thick film devices

R.E. BELFORD, R.G. KELLY and A.E. OWEN

11.1 Introduction

The realization of a standard chemical or ion-selective electrode in a thick film (T-F) or 'hybrid' form is not merely an engineering exercise. Many problems, both chemical and physical in nature, have to be solved. A general guide for the development of hybrid devices is contained in four basic questions:

(i) How do T-F methods of deposition affect the sensing materials, i.e. will there be any change in morphology or chemical composition, and if so, how can this be remedied?
(ii) Can the sensing material, normally an ionic conductor, be interfaced successfully with an electronic conductor? Most ion-selective materials (ISMs) are conventionally used as membranes and electrical connection is via a solution phase. A solid back contact poses questions of reversibility and hence stability.
(iii) Will the active surface of the sensor withstand the relatively high temperatures used in the T-F deposition process?
(iv) Has the ISM the physical requirements necessary for planar structures, e.g. a thermal expansion coefficient compatible with that of the substrate over the required temperature range? Is the chemical polarity of the material compatible with that of standard encapsulants?

All of the above questions can be investigated experimentally but the requirements are different for each ISM. The following section gives a brief account of what exactly a thick film or hybrid structure is.

11.2 Background

Glass pH electrodes and most other practical ion-selective electrodes (ISEs) are *membrane* devices. The term 'electrode' is, strictly, a misnomer because the active ion-selective membrane has on one side an analyte (the test solution) and on the other an internal (reference) solution containing an internal reference electrode; the internal solution and electrode together provide the electrical contact to the membrane. A solid-state ion-selective electrode is one in which the internal components are replaced by a direct 'back contact' to a metal or semiconductor, i.e., an electronic conductor. This is not a trivial matter, because most ion-selective materials are ionic conductors, and interfacing them to electronic conductors has practical and mechanistic consequences.

There are two basic types of solid-state chemical sensor: (i) potentiometric devices, and (ii) field effect devices, e.g. ion-selective field effect transistors (ISFETs or CHEMFETs). Electrodes of the potentiometric type usually have a metal as the back contact and they also have a high output impedance. Field-effect devices are a variant of the metal oxide field-effect transistor (MOSFET) familiar in electronics, and they have a low output impedance. Hybrid devices attempt to combine the advantages of both.

11.2.1 *Potentiometric devices*

A potentiometric sensor is one in which the potential difference developed across the ISM and its interfaces is measured with respect to a reference electrode, and this potential represents directly the concentration of the ion of interest. Most ISMs have a high impedance, and the voltmeter used to measure the potential must be at least of two orders of magnitude higher in impedance.

The so-called coated wire electrodes (1) are typical examples. A recent review (2) and a book by Freiser (3) give very detailed accounts of the fabrication and range of coated wire sensors studied. Coated wire electrodes are most commonly formed by dipping a copper wire into a polymer solution which, once dry, leaves the metal covered with plastic. The composition of the plastic is often tailored to reproduce compositions which have been used in membrane electrode configurations. A well-documented example is poly-vinylchloride (PVC) impregnated with valinomycin (1), which is selective to K^+ ions in solution. An extensive collection of ISE literature titles by Moody and Thomas (4) can be used to locate, *inter alia*, most new devices of this type.

Similar processes of manufacture have been applied to glass pH electrodes, i.e. 'dipped electrodes' have been fabricated (5) with apparently good results (6). An example of a dipped glass electrode (7) is illustrated in Figure 11.1.

This structure is of particular interest as it employs a Pt metal wire and it can be easily adapted to the thick film methods; Pt metal is added to thick film conductor pastes to aid solderability, thus a Pt-based conductor is very desirable. Most of the electrodes of the kind described above are generally reported to perform 'adequately', but little information is available on their longevity and operating mechanisms.

11.2.2 *Field effect devices*

The ion-selective field-effect transistor (ISFET) is derived from the well-established MOSFET device familiar in silicon integrated circuits. Field effect devices have three major advantages to offer (see Chapter 10 of this volume).

(i) *Size*. The active area or gate can be as small as 2.5 μm × 2.5 μm in area.

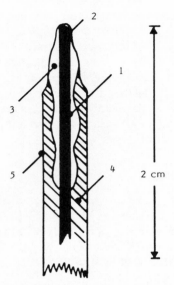

Figure 11.1 Dipped glass electrode. 1, a wire conductor, Au, Pt or Tl; 2, etched wire section on top of which is an oxide layer containing small quantities of alkali metal ion $5\text{Å} \rightarrow 2\,\mu\text{m}$ thick, making a hemispherical tip; 3, ion-selective glass; 4, glass insulating jacket; 5, plastic coating. (Redrawn from ref. 7.)

The exceptionally small size lends itself to biomedical applications, e.g. for intravenous measurements.

(ii) *Selectivity.* Experimentally-determined selectivity parameters could be programmed into comprehensive 'back-up' circuitry to compensate for individual sensor error in multi-sensor packages.

(iii) *Output impedance.* This is the most important advantage, as in analogous potentiometric sensors a high impedance voltmeter is required to measure the potential across the ISM. By contrast, the output from an ISFET device is not only of low impedance but is also completely compatible with standard integrated circuit operation.

In the early literature on ISFETs there was much controversy concerning the need or not for an external reference electrode (8, 9). In practice, current measurements, which are the most obvious, are not made, as a better arrangement makes use of a feedback loop. Most workers are now agreed that a reference is needed. Janata and most others use the ISFET in the feedback mode (11, 12) and obtain a low-impedance potential measurement via the reference electrode.

Encapsulation of ISFETs is required to expose only the active ISM area. This is an extremely difficult problem due to the difference in polarizability of the ISM and the encapsulating materials. Electrical leakage renders the device inoperable and it is probably for this reason that, at present, the use of ISFETs is restricted to situations where a disposable sensor is acceptable. Longevity, or rather the lack of it, is still a major problem.

11.2.3 Hybrid devices

Hybrid techniques are a combination of thick film, thin film and integrated circuit technology. The basis of hybrid circuits is thick film processing to produce complex patterns of insulating, resistive, conducting and capacitive layers, all on an electrically insulating substrate, usually alumina (19). Complete circuits with active components can be fabricated, but the active elements (transistors, operational amplifiers, etc.) are bonded separately into appropriate places in the thick film circuit of passive elements. Hybrid circuits are relatively inexpensive by comparison to silicon integrated circuits (ICs) in terms of initial capital costs, and they are the basis of the fabrication techniques for the research in the authors' laboratory. Hybrid technology has the following advantages.

Production scale. Although it can be a mass production technique, it is also adaptable to small batch production. The thick film process is therefore extremely useful in research where varying parameters may be tested on a small number of samples. Also, the eventual commercial market for such devices is probably not sufficient to warrant silicon IC fabrication on a massive scale.

Size of device. Although the extremely small scale of the ISFET is not achievable, a hybrid sensor could be relatively small and still retain the other advantages of the ISFET. A possible practical realization of a hybrid thick film electrode is illustrated in Figure 11.2. This would include a bonded MOSFET

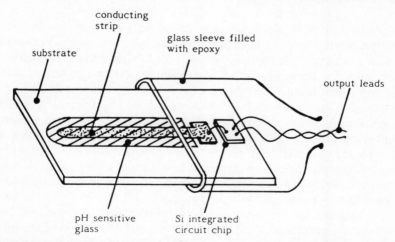

Figure 11.2 Hybrid (thick film) electrodes. (Redrawn from M.A. Afromowitz and S.S. Yee, *J. Bioeng.* **1** (1977) 55, Pergamon Journals Ltd.).

connected to the conducting strip so that the device as a whole has a low output impedance.

The lateral dimension of approximately 20 mm × 6 mm makes laboratory assembly and handling much easier than the ISFET analogue.

Encapsulation. The larger scale of the device makes encapsulation for laboratory testing much simpler, and the processes involved are more easily carried out than the corresponding procedure required to encapsulate ISFET devices.

The first attempt to fabricate a thick film pH electrode was reported by Afromowitz (13) and it is his device which is illustrated in Figure 11.2. The Si-integrated circuit chip in this case was simply an *n*-channel MOSFET. The overall size of the device need be no more than a few centimetres. Subsequent work by Leppavuori *et al.* (14) reported good responses for similar devices, but although the electrodes gave a good pH response initially, the devices functioned only for 24 hours, at best (15). A weakness in the design shown in Figure 11.2 is that it does not take into account the possibility of the substrate being porous to aqueous solution. Unless carefully encapsulated, a metal–solution connection results and renders the device insensitive; this problem is dealt with in more detail in the following sections.

It is important to emphasize that the thick film method of production is extremely versatile and adaptable to many experimental variations. Thin films may be interposed between thick film layers to experiment with 'back contact' materials, and geometries can be altered as desired. Attempts to rationalize the various possible mechanisms and to construct thermodynamically reversible 'back contacts' is made possible via this versatile technique.

There is also a wide variety of known and well characterized ISMs ranging from ionically-conducting organic polymers (3) to electronically conducting semiconductors (2) (electronically conducting ISMs are much simpler in their electronic interfacing). For example, a fluoride electrode (16) and other ISEs have involved thick film techniques where the ISM, in pressed pellet form, is connected to a thick film conductor layer via silver conducting epoxy. Alternative deposition techniques are also available, such as sputtering. In sputtering, the composition of the sputtered film may be different from that of the initial target but this may be compensated for by adjusting the initial composition. Thin film sensors fabricated by sputtering include a pH-sensitive iridium oxide device (17).

In this chapter, however, attention will be focused entirely on T-F hybrid electrodes based on inorganic ISMs which are ionic conductors, as these are by far the most common and general case. Tailoring the T-F process to the requirements of each individual sensor involves many considerations. Those appropriate to producing the T-F version of a pH glass electrode based on Corning 015 are discussed (18) as an example, but much of the following will be appropriate to many ionically-conducting sensing materials.

11.3 Practical considerations

The term 'thick film' (T-F) technology is accepted to mean that field of microelectronics in which specially-formed pastes are applied and fired on to a ceramic substrate in a defined pattern and sequence to produce a set of individual components such as resistors and capacitors, or a complete functional circuit (19). Figure 11.3 shows a flow diagram of a standard thick film process.

11.3.1 *Instrumentation and techniques*

Pastes are applied to the substrate using silk-screen printing methods. High-

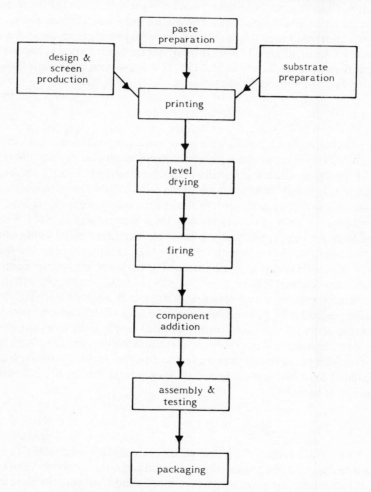

Figure 11.3 General T-F production flow diagram.

temperature firing drives off the binding agents present in the pastes and bonds the resulting film to the ceramic. The typical thickness of films ranges from 10–25 μm, as compared with so-called 'thin films' which are usually of submicron thickness. Thick film techniques have been greatly refined in the past ten years, and although size and resolution is limited by the screen mesh dimensions, laser and other trimming methods can yield extremely well-defined dimensions. Fine resolution was not the prime consideration in the present work, as the qualitative physical parameters were much more important. The finest dimension limits achieved were those of the inter-conductor pathways which were 0.2 mm. Surface contours of the printed films were monitored with a Sloan Dektak II surface profile system, to a precision of 1000 Å. The grain size of the ground glass pastes was determined using a Vickers image shearing eyepiece fitted to a Leitz microscope. The masks used for defining screens were high-resolution negatives taken with a Canon 50 mm lens. To deposit a thick film paste or ink using the DEK 65 printer, a squeegee is pulled across a screen in which the required pattern is defined. This deposits the ink on a substrate placed directly under the screen pattern.

11.3.2 Screen preparation

Screens for T-F printing are available in various mesh sizes and emulsion thicknesses. The screens used for glass printing in the present work had relatively thick emulsions (typically 23 μm), permitting a fairly thick print. Mesh sizes as low as 65 counts per cm were used to allow deposition of as much material as possible. Conductor prints, by contrast, were deposited using metal screens typically of emulsion 18μm thick with a mesh size of 128 counts per cm, to give more precision in the patterns. Negatives of the required mask were photographed full frame on to 35 μm high resolution film and the film negative was used directly as the mask. Examples of masks for conductor patterns are shown in Figure 11.4.

The particular designs in Figures 11.4(a) and (b) are for those T-F devices which incorporated an isolated junction gate FET (JFET) die; the gate, drain and source of the JFET were Au-ball-bonded to the pads as indicated. The masks were clamped to the screens and the emulsion developed by exposure to UV light which polymerized the non-blacked out areas. The printing pattern is then defined by washing with warm (30 °C) water to remove the undeveloped polymer coating.

11.3.3 Material compatibility

(i) *Choice of substrate.* The low thermal expansion coefficient (TEC) of alumina, along with the need to limit alumina contamination of the pH glass, made the standard thick film substrates impractical. Corning 015 glass in bulk form has a measured TEC of $11.0 \times 10^{-6}\,°C^{-1}$. Empirical calculation of TECs,

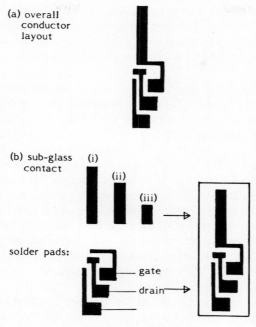

Figure 11.4 Conductor masks.

carried out by multiplying the relevant compositional fractions by their individual TECs, give a fairly good approximation (22) to the experimental value: more theoretical treatments are also available (20). *Thick films* of 015 Corning glass have TECs between 11 and 13 ($\times 10^{-6}\,°C^{-1}$)(15). A compatible substrate must have a similar expansion coefficient to the 015 glass print, be chemically and physically stable up to 900 °C and not form a less stable phase by reaction with the glass.

Not surprisingly, there are few materials which satisfy these requirements. Previous workers have used a ceramic known as forsterite (23) which is fairly compatible with Corning 015 glass. A range of ceramic and glass ceramic substrates was evaluated, but forsterite and an enstatite material (PF2, supplied by Steatite and Porcelain Products Ltd) gave the best results. One major disadvantage of these substrates is that they are porous, and extreme care is therefore needed for the encapsulation of the completed devices. Even with closely matched substrates, however, printed films of glass often developed lateral and vertical fissures. Sometimes the fissures were not observed immediately but became noticeable after some days. One of the main parameters governing the quality of glass prints is the firing cycle and in particular the cooling processes. Well-annealed samples have a greater tendency to crack than those which are air-quenched. It should be noted that all of the substrates used have a TEC comparable with or lower than that of printed 015 glass. If a substrate with a slightly higher TEC could have been found, the glass film would be under compression rather than under tension on cooling. This would significantly reduce the incidence of cracking, and further investigations using tailored substrates are promising.

(ii) *Choice of conductor metal.* Ideally, pure metallic conductors would be desirable for 'chemical' simplicity of the resulting device, but this is not feasible. Various types of conductor materials were tried. Evaporated metal films dissolved easily into the glass thick film and/or substrate during the subsequent firing processes. Only T-F conductors were substantial enough to resist this dissolution. It is quite probable that additional components (mainly glass) in commercial pastes aid the cohesive nature of these contacts and limit diffusion. Many different Ag-based pastes were examined because of the good quality of both the Ag/glass contact and resultant glass print. Silver, however, diffuses rapidly through the glass even when mixed Ag/Pt and Ag/Pd pastes are used. It was found essential to limit the sub-glass conductor print thickness to less than $10\mu m$; conductor films of $2\text{-}3\,\mu m$ thickness were generally satisfactory. Conductor films made with standard inks are usually at least $10\,\mu m$ thick, but subsequent glass coverage of these prints was unsatisfactory. Pinholes resulted, and SEMs showed that they were caused by bubbles in the glass layer. Possible reasons for the evolution of bubbles are sodium metal (gas) in glass as a product of reaction with the metal contacts, or insufficient degassing of organic components in the conductor ink during firing. Making the inks less viscous alleviated this problem. Care was taken to ensure that a low resistivity of typically $10^{-3}\text{-}10^{-4}$ ohm cm was maintained. Thinning the conductor ink also allowed penetration of the ink into the porous substrate which, to an extent, may act as a large surface for degassing of the ink during firing. Glass prints over silver contacts were notably devoid of fissures and pinholes but unfortunately the Ag diffuses through the glass films during the firing process. Temperature and/or time variations in printing were also tried to limit this diffusion but were unsuccessful. To obtain glass/metal interfaces of known composition, evaporated metal films of Cu or Fe were deposited on top of the T-F conductors prior to printing the glass layer. A remarkable observation made when using Au metal contacts was that the glass would not adhere at all. In fact, during firing the glass print would separate, leaving the gold surface exposed and a peripheral annulus of thick glass. Movement of the glass was over distances of up to 5 mm; the high surface tension of glass at these temperatures is a major factor (22). Large contact angles (24) are also obtained in sessile drops of molten glass on Au substrates.

A range of metal contacts, thick film, thin film and combinations of these were used along with oxidized contacts. There was a marked improvement in the quality of subsequent glass layers when oxidized metal contacts were used, presumably due to the increased strength of the glass-to-metal interface. According to Pask (26), in order to realize continuity of atomic and electronic structure a transition zone is required which is compatible and in equilibrium with both metal and glass at the interfaces. In general, it must contain at least a monomolecular layer of the metal oxide and both metal and glass should be saturated with it. The processes involved in T-F processing are in the main conducive to the formation of such a layer forming on all but the more inert metals, such as gold.

(iii) *Glass prints and paste preparation.* Corning 015 glass was obtained in fritted form and hand-ground using an onyx mortar and pestle, until an average particle size of 5 μm was achieved. Ball milling was used to reduce the particle size further to submicron dimensions, but sodium leaching into the mill solvent, and surface hydration, were problems (26). Dry hand milling to a finer particle size was just as effective, however, and ball milling was therefore abandoned. Once the glass was ground to a fine state, pastes were made by mixing the glass powder with pine oil and thinners to form a thixotropic paste. Several layers of glass were sometimes required, and to produce successive glass prints screens were prepared with slightly smaller print areas. After firing, the glass layers, whether single- or multi-layered, often developed fissures and these were responsible for device failure.

(iv) *Thermal history.* The firing cycle of the glass print is presumably the most crucial factor affecting both the physical strength and chemical sensitivity of the planar structure. The major reason for device failure is cracking in the glass film, and this is almost certainly a result of thermally induced stress frozen into the glass on cooling from its glass transition temperature (T_g). Altering the thermal cycle can prevent fissure formation but unfortunately air quenching from 900 °C, which is optimum for good fissure-free glass, is not good for the pH response.

This was a difficult problem to isolate. Initially, failure of pH devices in early samples was attributed to unsatisfactory back contacts and thus much effort was fruitlessly directed to that problem until a lower firing temperature of 700 °C was tried. This produced a pH-sensitive glass surface, presumably because the active surface groups are not removed (21). One disadvantage of the lower firing temperature, however, was that the glass required a longer time (30 minutes) to flow properly. In most cases sintering was incomplete, and many printed layers were required to obtain good coverage of the underlying contact. In some cases glass pastes were spread with a palette knife to obtain full coverage. The lower-temperature prints (700 °C) were slightly opalescent, indicating some phase separation. Alkali metal evaporation from the surface (27) is probably insignificant at this temperature.

The thermal history is particularly critical when planar devices are involved and when the heating is from the substrate side. In this case, it is possible that there is a greater sodium loss compared with that which occurs during the fabrication of pH-sensitive glass bulbs when they are flame annealed. The stresses of planar glass layers in contact with other planar ceramics would not allow flame working, and glass films shattered when this was attempted. Firing at high temperatures (900 °C) results in the surface becoming silica-rich (26), with, according to Kanazawa *et al.* (21), loss of active -OH sites. Prolonged (> 20 minutes) high temperature treatment has two main effects:

(i) pH sensitivity is lost
(ii) Depletion of sodium from the surface lowers the TEC markedly (22) and so the surface cools under compression from the underlying soda-rich glass. This results in a much better glaze.

Therefore, samples fired at high temperature and air-quenched are usually the best and mechanically strongest, but they are also the least sensitive to pH. An acceptable compromise to these opposing constraints was found (7). The firing temperature was raised very quickly to 900 °C, while ensuring that devices were held for no more than 7 minutes above 850 °C, followed by air quenching to 550 °C and annealing at that temperature, before cooling to ambient. Firing in this way retained the glass pH sensitivity. Surface analysis of printed glass films fired at different temperatures was attempted by x-ray photoelectron spectroscopy (XPS) but there was no evidence of differences in the surface concentration of sodium. It is significant that similar analyses of float glass surfaces (27) were also insensitive to sodium, due to its migration from the charged surface—similar charging of the surface would be present under the conditions used for the XPS analysis. Another feature which must be considered is that of fissures developing after firing. Accurate measurement of tensile strength in glass is difficult due to its brittle nature. Marked reductions in strength are caused by the presence of surface flaws, which may be too small to be visible, but are readily propagated by atmospheric moisture, CO_2 and acidic vapours (28, 29). A marked increased in mechanical strength may be achieved by air-blast quenching from temperatures above the strain point.

The present investigation was confined to the Corning 015 composition, as the more recently developed lithia-based pH-sensitive glass requires temperatures over 950 °C for flow to occur, and this was not possible using standard thick-film equipment. The success rate of glass film fabrication, even under the best parameters investigated, was not good, but could be improved by using a substrate with a slightly higher thermal expansion.

11.3.4 *Encapsulation*

Encapsulation of pH glass sensors is basically a problem of material compatibility. A more polar surface will ahere to a surface of similar polarity. Glass surfaces are hydrophilic and have a stronger affinity for water than for any organic encapsulant, even the more polar epoxys which, once applied, are eventually undercut by water erosion. During this project a great many different approaches to encapsulation were attempted; different encapsulants and designs, e.g. encapsulating the metal contact through the porous PF2 substrate with polymer solutions, etc. All proved inadequate.

An alternative approach to encapsulating glass is to use another glass or glass ceramic, but one possible complication is that those glasses used in conventional pH electrodes as stem glasses (i.e. lead borosilicates), which would be the obvious choice, may themselves have a pH response, as yet undiscovered because of their extremely high impedances. (The system described in Figure 11.7 may detect these responses, but this was not a factor in the present case.) The only suitable glass (with regards to its firing temperature) is a lead borosilicate glass paste (EMCA 92)*. This was used as a T-F printed encapsulant. Although satisfactory in terms of its adhesion and temperature of application (650 °C), it dissolved in acid solution. Non-porous glass ceramics were also tried by fusing them to the glass. In all cases where a glass ceramic was used, however, stresses in the glass films resulted in cracks and flaws developing.

11.4 Fabrication of T-F pH electrodes

11.4.1 *The basic process*

The following is an account of a method developed in the authors' laboratory for fabricating T-F pH electrodes. The flow diagram for the process is set out in Figure 11.5. Substrates of PF2 (25 mm × 6 mm × 1 mm) were thoroughly cleaned, pre-baked at 900 °C, and a metallic contact of Pt/Au (EMCA 180) was first printed. After printing, the ink was allowed to settle on the flat substrate surface for 10 minutes prior to drying at 150 °C for 10 minutes. This film was fired to its maximum temperature for 10 minutes and allowed to cool slowly to room temperature. The glass paste was then printed on top and the drying and firing processes repeated. The maximum temperature of 900 °C for glass prints was not sustained. Immediate air quenching to 550 °C was followed by slow cooling to ambient temperatures. Some prints were also fired at 700 °C and cooled slowly. The masks used gave a resulting glass film of 5 mm × 10 mm, totally covering the sub-glass contact depicted in Figure 11.4(*b*). Where more than one glass print was required, screens of fractionally smaller areas were used for each print in the sequence. Although this was the basic design, many variations were tried; some successful examples are described in the next section.

11.4.2 *T-F sandwich structures*

Sandwich structures are those which have some intermediate layer between

---

*EMCA is a proprietary brand of thick film materials.

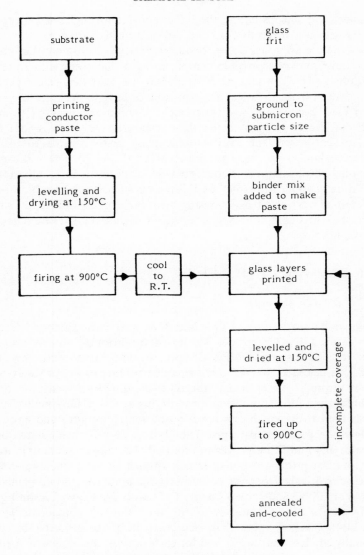

Figure 11.5 T-F electrode fabrication process.

the printed T-F conductor and the printed glass film, as shown in Figure 11.6. They were made to explore the possible reversible characteristics of different back contacts. Three main variations to the conductor used in the basic process are listed below.

(i) *Pt oxidized contacts.* In order to obtain oxidized Pt contacts (7), EMCA

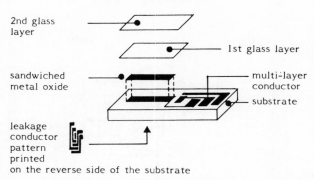

Figure 11.6 Sandwich structures.

180 Pt/Au conductor was printed as shown in Figure 11.4(*b*). The sub-glass contact (i) was printed and fired first followed by (ii) and (iii). In this way the 'single-print' conductor 'tip' was completely oxidized, having been fired three times, but electronic contact was still present through (iii). The subsequent glass prints were generally of better quality than prints on non-oxidized surfaces.

(ii) *Evaporated intermediate layers.* Iron or copper was evaporated over printed Au/Pt contacts to give intermediate layers of mixed conduction. These were either printed over immediately with glass or oxidized first, but this made little difference to the resultant device. The structures described above in general gave better glazes than the basic structure. Several different approaches were tried to produce intermediate layers, e.g. mixing finely divided metal with glass paste and T-F printing the mix, but the methods described above proved to be the most successful and produced both physically stronger structures and some of the most stable electrodes made in the study.

11.4.3 *Low output impedance system*

To incorporate the benefits of the ISFET system, the sub-glass electronic conductor was ball-bonded to the gate of a MOS- or J-FET. Preliminary tests were made using an *n*-channel MOSFET (No. 2N 3819). Low-leakage discrete JFETs were also used successfully. They were cemented to the T-F device substrate, and the gate, source and drain were ball-bonded to relevant pads. Figure 11.7 is a diagram of the circuitry used for pH measurements. The difference in potential can be monitored via the reference electrode at a low impedance level. The length of sub-glass conductor must be kept short (~ 10 mm) for stable readings.

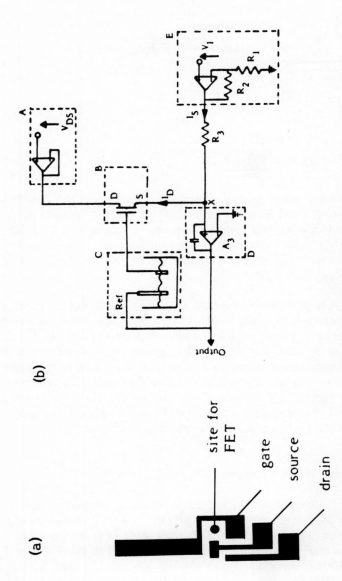

Figure 11.7 Low output impedance system. *A*, Buffered voltage supply to provide constant drain voltage; *B*, FET, e.g. 2N3819; note the gate connection in reality is short (*a*); *C*, cell; *D*, feedback amplifier to give low measurement impedance; *E*, constant-current source; *X*, virtual earth. A_3 is in a feedback loop with the reference electrode and cell. This provides appropriate values of V_G to give V_{DS} and I_D, i.e. the operating point is independent of the transistor used. V_{DS} and I_D are kept constant, this being the case I_G and V_{GS} are also constant under steady state conditions. Change in potential is monitored through the calomel reference electrode. A similar feedback system was used by Janata *et al.* (11).

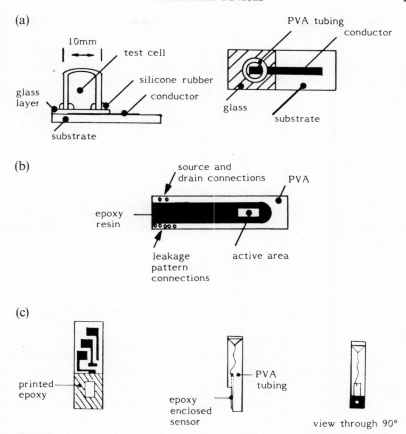

Figure 11.8 T-F electrode structures. (*a*) Plan view. (*b*) A solid finger of PVA was milled out and the T-F cemented in place using epoxy resin. An active area is left clear. (*c*) An alternative probe structure.

11.4.4 *Encapsulation*

(i) *Structures for laboratory experiments.* T-F structures required assessment of their pH response. Each needed therefore to have an active surface exposed to the test solution, and it was necessary to ensure that no test solution leaked to the electrical connections. The most reliable way of achieving such a structure is illustrated in Figure 11.8(*a*). The sealing medium is silicone rubber. One of the main advantages of its use is that electrodes can be dismantled, dried and reassembled, if leaks develop. This is ideal for laboratory situations where the quality of device is under test, but not appropriate for general use when a robust structure is required.

(ii) *Encapsulation for general use.* To evaluate the effectiveness of different encapsulants and designs, the interdigitated conductor pattern shown in Figure 11.6 was printed on the reverse side of the substrate and aligned to be opposite the active area. These devices were housed in the structure illustrated in Figure 11.8(*b*). The electrode spacing in the interdigitated pattern was 0.2 mm, which on a dry substrate gave a resistance greater than 10^{12} ohms. When leakage occurred, the resistance fell by six orders of magnitude. The form required for general use is that of a probe structure. Figure 11.8 shows the designs of some probe devices constructed, and a range of different encapsulants were tried. Encapsulation of planar glass devices for general handling was not satisfactory and remains a major problem.

11.4.5 *pH response*

It is generally accepted that as an optimum operational objective a Nernstian response is desired, i.e. 59 mV per pH unit.

$$E = E_0 + 59\,\mathrm{pH} \quad \text{at room temperature}$$

where E = emf of the cell (in millivolts).

E_0 = standard potential of the electrode (millivolts)

and $\mathrm{pH} = -\log_{10} a_{\mathrm{H}^+}$

The successful devices made by the methods described here generally exhibited responses between 45–50 mV/pH. Figure 11.9 shows the typical development of linear response. All glass pH electrodes require an inital period for hydration of the surface (gel layer formation) before full sensitivity is achieved, and this is illustrated in Figure 11.9. By contrast, prolonged firing at 900 °C results in insensitive devices, and a typical response is illustrated in Figure 11.10. A more detailed account (18) of the authors experimental results and their interpretation will be published elsewhere.

11.5 Conclusions and future work

The problems involved in producing solid-state sensors are by no means solved in this study. Nevertheless, many feasible routes toward that objective have been extensively investigated, and promising directions have been revealed. The fabrication of thick film pH electrodes, which function in a consistently stable fashion, has been accomplished, but the exact procedure and materials required for robust devices have still to be developed. The major reason for the low yield was cracking of the T-F glass layer; hence if cracking can be prevented the yield should be much improved. The cause of cracking is almost certainly thermally-induced tensile stresses frozen into the glass film on cooling from T_g. Simple calculations show that even with an apparently well-matched combination of Corning 015 glass (TEC between 11 and $13 \times 10^{-6}\,°C$),

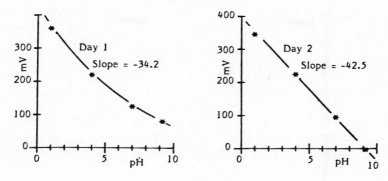

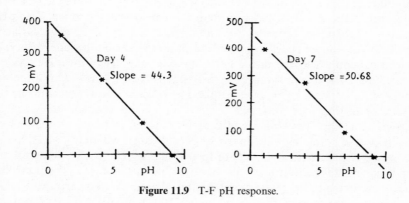

Figure 11.9 T-F pH response.

and PF2 (TEC $= 10.5 \times 10^{-6}\,^{\circ}C^{-1}$), the *tensile* stress could be as high as $100\,MNm^{-2}$, which is not very different from the fracture strength of typical brittle solids. The solution would be to find a suitable substrate material with a TEC slightly higher than that of the glass, so that the glass film is under compression and thus able to withstand much greater stress. Suitable substrates are unavailable commercially at present, but could be developed. The pH response of the electrodes fabricated with materials presently available was slightly lower than ideal. Robust fissure-free structures provided with adequate encapsulation would doubtless improve these results, given the firing cycles reported.

The crucial factor governing the response of T-F pH electrodes has been experimentally established in this case to be the glass surface condition (18).

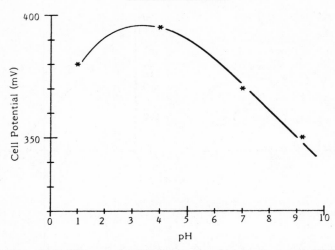

Figure 11.10   pH response of devices made at high temperature (mV/pH).

The ability of the surface to hydrate and form a gel-layer was found to be more important to the electrode function than the type of 'back contact'. The pH sensitivity is therefore dependent on the thermal history of the glass surface. Physically robust electrodes with stable responses were obtained for each type of electrical contact investigated. The long-term stability of these electrodes revealed no particular preference for the type of contact. The only noticeable feature was that glass layers over oxidized metal contacts were less prone to fracture, but longevity was not a function of the type of glass–metal interface. The fact that all the different types of electrodes constructed provided some examples which were long-lived, albeit with a low yield, has a bearing on the theoretical aspects of their function. The back contact of all long-lived and stable electrodes must be reversible with respect to the electrode process occurring at the glass–metal interfacial zone or plane, irrespective of the electrode mechanism. The other major factor involved in the eventual commercial success of planar hybrid pH electrodes rests with their encapsulation. This is not trivial, especially when the close proximity of the FET limits further heat treatments (like the fusing of other glasses).

Acknowledgements

The authors are grateful to ELF (UK) Ltd for financial support of the work reported in this paper and particularly to M. Bernard Michaux of ELF for many helpful discussions.

References

1. R.W. Cattrall and H. Frieser, *Anal. Chem.* **43** (1971) 1905.
2. R.W. Cattrall and I.C. Hamilton, *Ion-Selective Electrode Rev.* **6** (1984) 125–172.
3. H. Frieser (ed.), *Ion-Selective Electrodes in Analytical Chemistry*, vol. 2, Plenum Press, New York (1980).

4. G.J. Moody and J.D.R. Thomas, *Ibid.*, vol. 6 (1984) 209–263.
5. Patent, 1,018,024 Maatschappij, The Netherlands (1966).
6. US Patent 1,260,065 (Beckman Instruments) (1972).
7. US Patent 4,312,734 (Nicols) (1982).
8. P. Bergveld, *IEEE Trans. Biomed. Engg.* **BME 19** (1972) 342.
9. R.G. Kelly, *Electrochim. Acta* **22** (1977) 1–8.
10. P. Bergveld, N.F. DeRooij and J.N. Zemel, *Nature* **273** (1978) 438.
11. J. Janata *et al.*, *Sensors and Actuators* **5** (1984) 127–136.
12. J. Janata *et al.*, *Ion-Selective Electrode Rev.* **1** (1979) 31–79; **1** (55) (1977) 55–60.
13. M.A. Afromowitz and S.S. Yee, *J. Bioeng.* **1** (1977) 55.
14. S.I. Leppavuori and P.S. Romppainen, *3rd Eurs. Hybrid Microelectronic Conf.*, Avignon (1981).
15. (a) S.I. Leppavuori and P.S. Romppainen, *Electrocomponent Sci. Technol.* **10**(2–3) (1983) 129–133. (b) P.S. Romppainen, private communication.
16. T.A. Fjeldy, K. Nagy and J.S. Johannesen, *J. Electrochem. Soc.* **26**(5) (1979) 793–795.
17. I. Lauks, J.N. Zemel *et al.*, *Sensors and Actuators* **2** (1982) 399–410.
18. R.E. Belford, Ph.D. Thesis, Edinburgh University (1985).
19. C.A. Harper (ed.), *Handbook of Thick-Film Hybrid Microelectronics*, McGraw-Hill, New York (1974).
20. A. Makishima and J.D. Mackenzie, *J. Non-Crystalline Solids* **22** (1976) 305–313.
21. T. Kanazawa *et al.*, *J. Ceram. Soc. of Japan* **92**(11) (1984) 654–549.
22. F.V. Tooley (ed.), *Handbook of Glass Manufacture I*, Ogden Pub. Co., New York (1953).
23. R.A. Chappell and C.T.H. Stoddart, *Phys. and Chem. of Glasses* **15**(5) (1974) 130–135.
24. V.K. Nagesh, A.P. Tomsta and J.A. Pask, *J. Mater. Sci.* **18** (1983) 2173–2180.
25. J.A. Pask, *Prof. A.I. Andrews memorial lecture*, Ohio, USA, October 1971.
26. J.O. Isard, in *Glass Electrodes for Hydrogen and Other Cations: Principles and Practice*, G. Eisenman, (ed.), Marcel Dekker, New York (1967).
27. A. Wikby and B. Karlberg, *Electrochim. Acta.* **19** (1974) 323–328.
28. W.B. Hillig, in *Modern Aspects of the Vitreous State*, Vol. 2, J.D. Mackenzie (ed.), Butterworth (1962), Chapter 4.
29. R.H. Doremus, *Treatise on Materials Science and Technology*, Vol. 22, Glass III, Academic Press, New York (1982).
30. J. Janata *et al.*, *Sensors and Actuators* **5** (1984) 127–136.

5

NON-ELECTROCHEMICAL TRANSDUCTION

12 Catalytic devices

S.J. GENTRY

12.1 Introduction

Catalytic devices are widely used throughout industry as a convenient means of estimating the concentration of flammable gases in air. The range of hazards encountered and the number of instruments available from many different manufacturers are increasing annually. However, the vast majority of these instruments use the same principles of detection and hence have many features in common. The purpose of this chapter is to describe these features and the advantages and disadvantages which result from them.

All gas detection instruments contain four basic components, shown schematically in Figure 12.1. These are:

(i) The sensing unit which converts the gas concentration into an electrical signal
(ii) A processing unit which controls the power supplied to the sensor and processes the signal for display
(iii) A display unit which presents the signal in an appropriate form and/or takes appropriate action
(iv) A power supply.

The processing unit can range in complexity from a simple bridge circuit in a portable instrument to complex circuitry with constant voltage or current supplies to the sensing unit (either continuous or intermittent), alarm circuits, etc. The display unit may be a simple meter or digital display, but can also include automatic cut-off of dangerous machinery, remote audible and visual

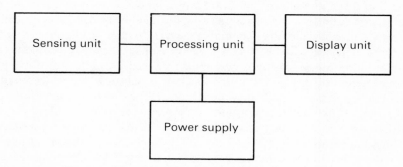

Figure 12.1 Instrument components.

alarms, etc. The power supply is generally a battery for portable instruments or a mains supply in continuously operating fixed instruments, where a standby battery is included in case of mains failure. These components are common to detection systems other than catalytic devices, and detailed discussion of their circuitry is beyond the scope of this chapter.

12.2 Principle of operation

In the vast majority of catalytic devices, the concentration of the gas of interest is measured as the heat liberated in a controlled chemical reaction. Thus, for example, in the oxidation of one mole of methane

$$CH_4 + 2O_2 \rightarrow CO_2 + 2H_2O$$

800 kJ of heat are liberated. To sustain such a reaction at reasonable temperatures, a solid-phase catalyst is used. Since reaction then occurs at the catalyst surface, it is convenient to supply heat to the catalyst rather than the reacting gases, and to measure the heat of reaction as a temperature rise of the catalyst itself. Thus, catalytic gas transducers consist essentially of (i) a catalyst surface; (ii) a temperature sensor and, where necessary, (iii) a heater to maintain the catalyst at operating temperature. Any chemical reaction accompanied by the liberation of heat may be monitored by such a device. In practice catalytic sensors are generally used to monitor flammable gases in air, where the combination of high heats of oxidation and a ready oxygen supply are ideally suited for this type of instrument.

12.3 Types of sensor

The most convenient heat-sensitive device is a platinum resistance thermometer. This system allows a single platinum coil to act as heater and temperature sensor. Since platinum is also an active catalyst for hydrocarbon oxidation, the simplest form of sensor is a single platinum coil (1).

A typical example would be a fifteen-turn coil of 75 μm diameter wire with a length of 5 mm and a diameter of 0.5 mm. The coil is used as one arm of a Wheatstone's bridge. The current passing through the bridge heats the coil to a temperature at which the flammable gas will oxidize on the surface of the platinum. The heat of oxidation causes a further rise in temperature of the coil. The resulting resistance rise is determined by measuring the out-of-balance voltage across the bridge. Although this type of sensor is still widely used in spot reading instruments, it does not lend itself readily to use in continuous operation. Bulk metallic platinum is a relatively poor catalyst for the oxidation of hydrocarbons, especially methane, hence the sensor must operate at high temperatures (c. 1000 °C). At this temperature platinum evaporates from the coil at a significant rate, especially in the presence of combustible gas. This results in a reduction of the cross-sectional area of the wire, and hence an increase in resistance, causing a 'zero drift' and an early 'burn-out' of the coil.

Most improvements in sensor design have arisen from the development of more active catalysts. In each case the platinum wire coil is retained as the temperature-sensing device. However, the catalyst is separated from it. This technique permits improvement in sensor activity in two ways. Firstly, catalysts with greater specific activity than platinum may be employed. Secondly, the catalyst may be prepared in such a way as to expose a far greater surface area per unit mass, and hence greater activity, than a metallic coil.

One example is the pellistor-type sensor first described by Baker (2) which is shown in Figure 12.2. This sensor uses palladium supported on thoria as the catalyst. This is deposited on the surface of a refractory bead of c. 1 mm diameter encapsulating the platinum coil. Palladium is more active than platinum for hydrocarbon oxidation and readily oxidizes methane at temperatures of about 500 °C. Encapsulation of the coil within a spherical bead in this way produces a device which is insensitive to orientation and also resistant to shock. This type of sensor is widely used in all types of flammable gas detection instruments.

The major limitation to the operation of catalytic gas-sensing elements of this type is their loss of sensitivity on exposure to atmospheres containing certain gases and vapours. Such a loss of sensitivity is particularly important in catalytic gas sensors since the device will fail to indicate danger; that is, a false low reading may be obtained in a flammable atmosphere.

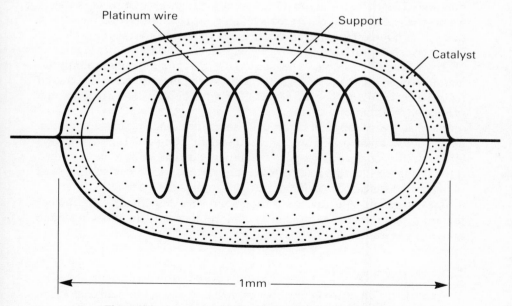

Figure 12.2 A typical catalytic gas-sensing element: the pellistor.

To overcome the problems of gas-sensor failure due to poisoning of the elements, several manufacturers have used specific filter devices in front of the elements. For example, filters to remove alkyl lead compounds from petrol vapour have included paradichlorobenzene and fibrous silica (3). However, the most commonly used material is based on active charcoal. Such filters (4) effectively absorb many common poisons and inhibitors including silicones, alkyl lead compounds and halogen-containing species. However, along with the absorption of potentially harmful vapours, carbon filters will also absorb hydrocarbons. This effect in practice limits the use of such filters to detectors for C_3 and lower hydrocarbons. They cannot be employed in general flammable gas detectors.

The rapid growth in demand for catalytic flammable gas sensors to operate in hostile atmospheres (leaded petrol atmospheres, gas compressor stations, etc.) and areas where routine calibration and maintenance are difficult and costly (North Sea gas and oil platforms) has stimulated the search for sensing elements which are resistant to poisons. These devices generally take the form of a platinum wire coil embedded in a porous bead comprising active catalyst either dispersed throughout a porous substrate (5–7) or as discrete layers within the substrate (8).

The improved poison resistance of these devices is obtained in three ways. First, the catalyst is chosen to have a high intrinsic resistance to poisoning. This can be achieved by consideration of the interaction of the poison with the catalytic sites through adsorption effects, poison-induced surface reconstruction and compound formation (9). Secondly, the dispersion of the catalyst increases the effective surface area available for reaction. When the rate of reaction is limited by diffusion of reactants to the catalyst surface there is effectively 'spare' surface available. Increasing this surface results in a lower observed rate of poisoning of the sensor. Thirdly, and most importantly for high-molecular-weight poisons, the porosity of the support can be selected to restrict access of the poison while allowing relatively free access of the reactants to the catalyst (10).

The effects of a range of catalyst poisons, including sulphur-containing, halogen-containing and high-molecular-weight compounds, on the performance of porous and non-porous catalytic gas sensors have been studied (11–13). Figure 12.3, taken from reference (11), shows the degree of poison resistance which can be achieved.

Many forms of catalytic gas sensor have been described in the patent literature; however, the vast majority conform to one of the types described above.

12.4 The gas supply system

There are many gas-sensing head designs in use. However, they all conform to one of the basic designs shown in Figure 12.4. Type (a) represents the most common system. Gas reaches the sensing element by diffusion through a sinter exposed to the atmosphere under investigation. Type (*b*) represents the system common on aspirated detectors. Here air is drawn along the tube and reaches the elements by diffusion along the 'chimney'. The third system (type (*c*)) is not commonly used in field instruments, since the output of the device may be flow- and orientation-sensitive. However, this system can be adapted (14) to give devices with fast response ($\sim$ 1s).

Figure 12.5 shows the characteristic signal temperature plots obtained with these systems. It can be seen that a plateau is reached for types (*a*) and (*b*). At this point the output of the device is independent of temperature, indicating that the reaction is diffusion-controlled at a point remote from the sensor, that is, the sinter in type (*a*) and the 'chimney' entry in type (*b*). No such plateau exists for the type (*c*) system. Operation of the detector under diffusion-controlled conditions limits the response time to several seconds, but offers several advantages:

(i) Stability: minor voltage (and hence temperature) variations do not induce changes in the observed signal
(ii) Linearity: the rate of diffusion is linearly dependent on gas concentration, thus the output is a linear function of concentration
(iii) Direct measure of flammability (see below).

12.5 Mode of operation

12.5.1 *Circuitry*

Almost all catalytic gas transducers currently used employ resistance thermometry as the temperature sensor. This section will deal only with this type of transducer, although the general principles are applicable to all types of transducer.

Two methods (15) of using these transducers can be identified. The first method, the most widely used in commercial gas detectors, is a non-isothermal method. In this case the temperature of the sensing element is allowed to rise as a result of chemical reaction at the catalyst surface and the rate of reaction, and hence the concentration of flammable gas is derived from the increase in temperature. In the second method the temperature of the sensing element is maintained constant during the reaction, and the reaction rate is obtained in terms of a difference in electrical power dissipation of the element under reacting and non-reacting conditions.

In the non-isothermal method the catalytically active sensing element and a similar but catalytically inert element form two arms of a Wheatstone's bridge. Power is supplied to the circuit to heat the elements to their operating temperature, and the values of the fixed resistors selected such that the bridge balances in air. When flammable gas is passed over the elements, reaction takes place on the sensing element, imparting power to the element and thus

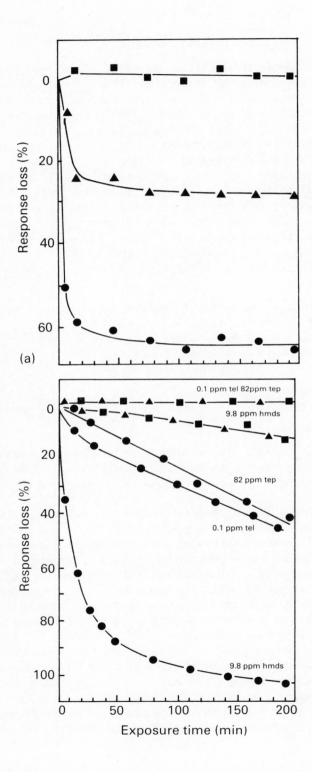

(a)

0.1 ppm tel 82ppm tep

9.8 ppm hmds

82 ppm tep

0.1 ppm tel

9.8 ppm hmds

Exposure time (min)

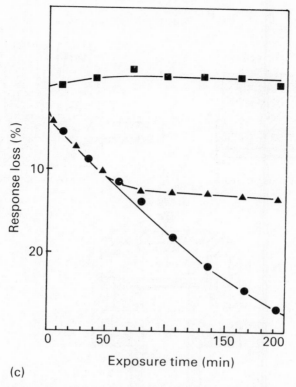

(c)

Figure 12.3 Effect of (*a*) 105 ppm chlorine, (*b*) high molecular weight poisons and (*c*) 107 ppm sulphur dioxide on the response to 1% methane in air: temperature 550° C. Rhodium porous element ▲, platinum porous element ■ and palladium pellistor ●.

raising the temperature and hence resistance of the element, until a new thermal balance is achieved.

This resistance change produces an out-of-balance signal (*V*) across the bridge which, for small changes, can be shown (16) to be proprotional to the rate of reaction and heat of combustion. Thus:

$$V = Kr\Delta H \tag{12.1}$$

where ΔH is the heat of combustion, r the reaction rate and K a constant defined by the components of the system. The out-of-balance voltage of the bridge is therefore proportional to the concentration of the flammable gas.

There may be superimposed effects on the temperature of the element (and hence *V*) arising from changes in the heat loss characteristics in different gas mixtures. These effects are particularly marked for flammable gases with thermal conductivities vastly different from that of air. The incorporation of the non-active element compensates for such effects in that this element

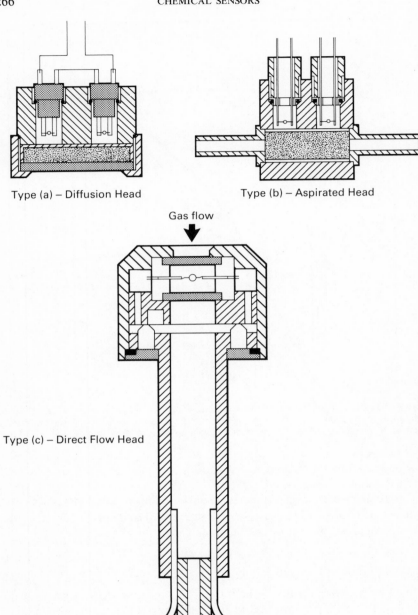

Type (a) – Diffusion Head

Type (b) – Aspirated Head

Gas flow

Type (c) – Direct Flow Head

To pump

Figure 12.4 Sensor head designs.

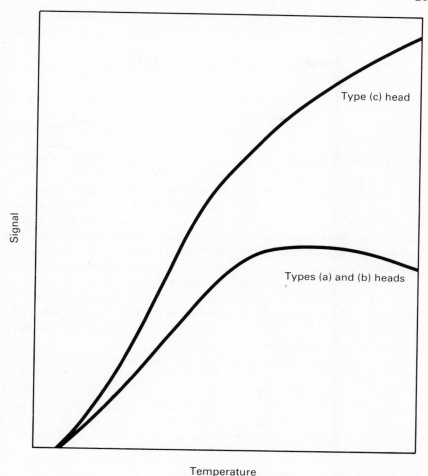

Signal

Type (c) head

Types (a) and (b) heads

Temperature

Figure 12.5 Signal $v.$ temperature characteristics for different sensor head designs.

responds in the same manner as the active element to thermal conductivity changes. Since the out-of-balance voltage corresponds to the difference in resistance of the two elements, only changes in V due to catalytic reaction are recorded.

The isothermal method employs a similar Wheatstone's bridge circuit to the earlier method. However, the compensator is replaced by a fixed resistor, and the voltmeter is replaced by a feedback circuit (17) which governs the electrical power supplied to the catalytic element. In this way the heat liberated during reaction is sensed by the feedback circuit and the electrical power is reduced to maintain the element at a constant temperature. Thus, if P is the electrical power required to maintain the element at a temperature T in the absence of reaction, and P' the electrical power required to maintain the same temperature in the presence of a flammable gas, then

$$r = (P - P')/\Delta H \tag{12.2}$$

$$\Delta P = r \, \Delta H \tag{12.3}$$

12.5.2 *Comparison of the methods*

There are essentially two limitations to the use of the non-isothermal method. Both follow inevitably from the fact that the temperature is allowed to rise. The constant K cannot be assumed to remain constant for large changes in temperature, which arise from high concentrations of flammable gas or very exothermic reactions. Secondly, a finite time is required to re-establish thermal balance at a new temperature following a change in the reaction rate. This 'thermal lag' is a function of the sensing elements, and for a typical 'pellistor'-type element is approximately three seconds for the temperature rise associated with an incremental change in methane concentration from 0 to 1% in air (about 20°). The isothermal method overcomes both these limitations. The change in electrical power is a direct measure of the rate of heat generated by the reaction, and the time constant for the circuitry used to maintain constant temperature is of the order of milliseconds.

The non-isothermal method does, however, have advantages which have ensured that it has been the most widely applicable method in the past. These are simplicity of circuitry, direct voltmeter display and compensation for variations in ambient conditions. The isothermal method has been used successfully for several fundamental studies (18, 19), and is now beginning to be exploited in practical devices where rapid response is required (14).

12.6 The detection of flammable atmospheres

As stated in section 12.4, it is usual to mount the sensing elements in a diffusion head of the type shown in Figure 12.4a. In such an assembly the characteristic plateau in the signal as a function of temperature (Figure 12.5a) is reached when the rate of reaction becomes controlled by the rate of diffusion of gas through the sinter. Under these conditions, the signal is directly dependent on gas concentration and independent of both element temperature and oxygen concentration.

In a binary mixture the rate of diffusion of gas 1 into gas 2 through an area A is given as

$$R_D = D_{1,2} A \frac{dG}{dx} \tag{12.4}$$

where $D_{1,2}$ is the diffusion coefficient and dG/dx is the concentration gradient.

Under normal operating conditions the sensing element consumes all the fuel which reaches it, thus dG/dx is directly related to $[G]$.
Now

$$V = Kr\,\Delta H \tag{12.1}$$

and

$$\Delta P = r\Delta H \tag{12.3}$$

where r is the reaction rate. Thus, under diffusion controlled conditions

$$V = K'D_{1,2}\,\Delta H[G] \tag{12.5}$$

and

$$\Delta P = K''D_{1,2}\,\Delta H[G] \tag{12.6}$$

Thus for any gas at its lower explosive limit (LEL) concentration, the signal generated is given by

$$V_{LEL} = K'\,\Delta H D_{1,2}[LEL] \tag{12.7}$$

It has been shown (20) for many fuels that there is an inverse linear relationship between $[LEL]$ and heat of combustion, i.e. $[LEL] \times \Delta H \simeq$ constant. This may be appreciated qualitatively in that the LEL of a gas/air mixture is a measure of the total energy of the system. Thus a gas with a large heat of combustion will have a correspondingly small LEL. This relationship means that for many fuels, equation (12.7) may be simplified to

$$V_{LEL} \simeq K_2 D_{1,2} \tag{12.8}$$

Furthermore, since the diffusion coefficients for hydrocarbons vary by only a factor of 4 over the range C_1–C_{10} and by a factor of 2 over the range C_3–C_{10}, equation (12.8) may be further reduced to

$$V_{LEL} \simeq \text{constant} \tag{12.9}$$

The three equations (12.7)–(12.9) summarize the most important characteristic of catalytic gas detectors. An instrument calibrated to read, say, 0–100% LEL in a standard fuel will give an immediate estimate of the explosiveness of any unknown vapour or mixture of vapours based on equation (12.9). If the composition of the vapour is known, simple correction factors based on the constants in equations (12.7) and (12.8) may be applied to give more accurate readings.

Table 12.1 shows typical results obtained for a series of fuels at their respective LEL concentration. Methane is used as the standard and the concentrations of the other flammable gases estimated using equations (12.7)–(12.9).

As shown in Table 12.1, only equation (12.9) underestimates the true concentration. For general flammable gas detection, it is desirable to calibrate the instrument in a fuel more representative of the potential hazards than methane; butane and n-pentane are commonly used.

Table 12.1 Comparison of the validity of equations (12.7)–(12.9) in predicting the explosiveness of fuel air mixtures

100% LEL of gas	Signal (mV)	Concentration expressed as % LEL derived from		
		equation (12.9)	equation (12.8)	equation (12.7)
Methane	168.8	100	100	100
Ethane	129	76.4	114	113
Propane	106.8	63.3	121	115
n-butane	98.5	58.4	118	100
n-heptane	73.1	43.3	143	112
Methanol	154.6	91.7	124	107

Firth, Jones and Jones (20) have published a table containing the parameters required to calculate correction factors for 93 flammable species.

12.7 Effects of high concentrations of gas

Since oxidation of the combustible gas is used to obtain a signal from the instruments, it is apparent that if the concentration of oxygen in the gas mixture is too low for complete oxidation of the combustible gas to occur, then the signal obtained will be low. Hence, as the concentration of a flammable gas is increased in air from zero, the signal from the transducer increases to a maximum at the stoichiometric ratio for complete oxidation and then decreases to zero at a concentration of 100% combustible gas (2). Therefore there is a range of high concentration of gas in air which will give readings which indicate that the gas is below the LEL. For methane, this concentration range is approximately 40–100%.

Various methods of overcoming this problem are employed. The simplest method involves careful observation of the instrument reading as the gas enters the sampling chamber. As the concentration inside the cavity containing the transducer rises, the signal will rapidly rise above 100% LEL before returning on scale. As the instrument is withdrawn from the gas to air, the signal will again rise above 100% LEL before returning on scale.

It is possible to incorporate a mixing valve which allows air to be added to the gas being sampled before it reaches the transducer. The true concentration of gas can be estimated by multiplying the instrument reading by the mixing ratio. Other instruments employ a separate transducer measuring the thermal conductivity change produced by the combustible gas. The signal from this transducer can be used independently to measure high gas concentrations, or added to the signal from the catalytic transducer so that when the concentration is above the stoichiometric ratio, the combined signal is always above that at the LEL of the gas (21, 22). When a transducer is operated in high concentrations of hydrocarbons the incomplete oxidation which occurs results in the deposition of carbon on the catalyst. If the catalyst is porous, then disruption of the catalyst can occur and permanent damage of the transducer can result. Such damage only occurs during prolonged exposures, but was a problem with early forms of the catalytic element if exposure exceeded 45 minutes (2).

This hazard has been eliminated by changing the mechanical base to a less porous base having greater mechanical strength (23), and by the dispersion of the palladium catalyst in thoria (24).

12.8 Future developments

In the past, gas-sensing elements have been operated at a single temperature and calibrated to a single acceptable standard gas, such as pentane. This method provides catalytic gas sensors with their most important characteristic, direct measurement of explosiveness. However, in consequence of this, catalytic gas sensors are of course non-specific. The published work on the kinetics of oxidation reactions on these devices, for example (11), (12), (13), (18), (19), (25), suggests that is may be possible to distinguish between some flammable gases through interrogation of the sensor over a range of temperatures. The recent strides in microprocessor technology bring this potential within reach of a gas detection instrument. For example, Figure 12.6 shows some results obtained for the response of a catalytic sensor to a mixture of hydrogen, butane and methane in air as a function of sensor temperature. The signal at high temperature is a measure of the total flammability of the mixture, while between 300–350 °C the signal is dominated by the concentrations of hydrogen and butane and below 200 °C is a measure of hydrogen only. In practice the 'light-off' temperature for most hydrocarbons is similar to that of butane; thus it may be possible to discriminate only between hydrogen, general hydrocarbons and methane. Furthermore there is evidence (27) that certain flammable gases may modify the behaviour of others. A further application of this technique may permit an early indication of poisoning. This is illustrated in Figure 12.7, taken from reference (13), where the low-temperature activity of the sensor for propene oxidation is progressively removed by exposure to HMDS before there is any observable loss of activity at high temperature. Thus by referring the signal at 600 °C to those at 200 °C and 300 °C, it is possible to obtain an early warning of poisoning before the high-temperature signal has become unreliable.

12.9 Conclusions

The catalytic gas sensor has proven to be a most reliable and rugged means for the quantitative measurement of flammable gases in air, giving a measure of explosiveness irrespective of the composition of the gas mixture. The recent development of poison-resistant devices has overcome the problem of

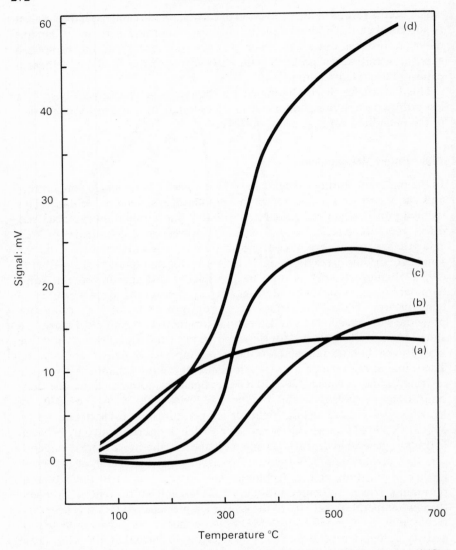

Figure 12.6 Response of a catalytic sensor to (a) hydrogen, (b) butane, (c) methane and (d) a mixture of all three.

operation in many previously hostile environments and thus ensures that catalytic sensors will continue to make a major contribution to industrial safety for many years to come.

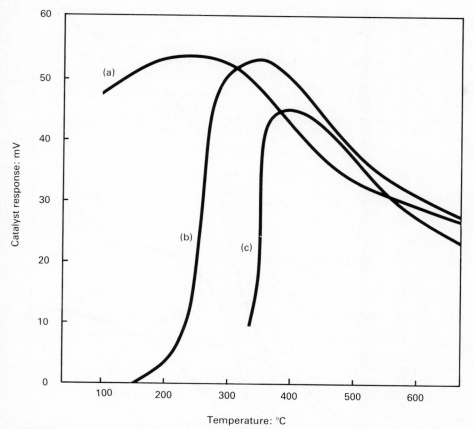

Figure 12.7 Influence of exposure to 40 ppm hexamethyldisiloxane on the activity of a catalytic sensor for the oxidation of propene.

References

1. A.R. Baker, *Mining Engineer* **128** (1969) 643–653.
2. A.R. Baker and J.G. Firth, *Mining Engineer* **128** (1969) 237–244.
3. J.G. Firth, A. Jones and T.A. Jones, *Proc. Symp. Chemical Process Hazards V*, Inst. Chem. Eng. Symp. Ser. 39(1), UMIST (1974).
4. S.J. Gentry and S.R. Howarth, *Sensors and Actuators* **5** (1984) 265–273.
5. D.W. Dabill, S.J. Gentry, N.W. Hurst, A. Jones and P.T. Walsh, UK Patent 2,083,630 (1982).
6. J.M. Sonley, Euro Patent 0,004,184 (1979).
7. G.S. Wilkinson-Tough, UK Patent 2,066,963 (1981).
8. B.R. Edgington and E. Jones, UK Patent 2,121,180 (1983).
9. L.L. Hegedus and R.W. McCabe, *Cat. Rev. Sci. Eng.* **23** (1981) 377–476.
10. A. Wheeler, *Adv. Catal.* **3** (1951) 249–327.
11. S.J. Gentry and P.T. Walsh, *Sensors and Actuators* **5** (1984) 239–251.
12. S.J. Gentry and P.T. Walsh, in *Preparation of Catalysts III*, G. Poncelot, P. Grange and P.A. Jacobs (eds.), Elsevier, Amsterdam (1983) 203–212.
13. S.J. Gentry and A. Jones, *J. Appl. Chem. Biotechnol.* **28** (1978) 727–732.
14. D.W. Dabill, S.J. Gentry and P.T. Walsh, *Sensors and Actuators* **11** (1985) 135–143.
15. A. Jones, J.G. Firth and T.A. Jones, *J. Phys. E. Sci. Instr.* **8** (1975) 37–40.

16. J.G. Firth, *Trans. Faraday Soc.* **62** (1966) 2566–2576.
17. R. Wilson, UK Patent 1,451,231 (1976).
18. S.J. Gentry, J.G. Firth and A. Jones, *J. Chem. Soc. Faraday I* **70** (1974) 600–604.
19. S.J. Gentry and P.T. Walsh, *J. Chem. Soc. Faraday I* **76** (1980) 2084–2095.
20. J.G. Firth, A. Jones and T.A. Jones, *Combustion and Flame* **21** (1973) 303–311.
21. T.M. Sporton, UK Patent 1,447,488 (1976).
22. F. Klauer, *Gluckauf* **103** (1967) 379.
23. J.G. Firth and H.B. Holland, *J. Appl. Chem. Biotechnol*, **21** (1971) 139–140.
24. S.J. Gentry and P.T. Walsh, *Sensors and Actuators* **5** (1984) 229–238.
25. J.G. Firth and H.B. Holland, *Trans. Faraday Soc.* **65** (1969) 1121–1127.
26. D.W. Dabill, S.J. Gentry, H.B. Holland and A. Jones, *J. Catal.* **53** (1978) 164–167.

13 Spectroscopic and fibre-optic transducers

A.L. HARMER and R. NARAYANASWAMY

13.1 Introduction

Chemical sensing with optical fibres is one of the most interesting of the emerging sensor technologies. Optical fibres, which act as light carriers, impart several attractive features to the chemical sensor. The optical fibre allows the transmission of optical signals through great distances with low power loss and minimum or no corruption due to ambient electrical noise or electromagnetic interference. Thus optical fibres and hence optical-fibre transducers are ideally suited for remote monitoring. The optical fibre also bestows total electrical isolation at the transducing end, thus making the device safe for use in explosive environments and in *in-vivo* medical applications. The small physical dimensions lend the fibre to miniaturization and allow geometric flexibility for the transducer. It is possible to carry out real-time, on-line analysis with these devices. The robust nature of the optical fibre allows the transducer to be operated in hostile environments and the relatively low cost of the fibres introduces an economic advantage to the sensing device. Like many other chemical sensors, these transducers can be multiplexed to enhance their versatility.

The great potential of optical fibre chemical sensors has been recognized, and a variety of sensors for the measurement of a number of different parameters have already been studied. Several reviews on optical chemical sensors have also been published (1–7).

13.2 Optical fibres

The optical fibre in a chemical sensor acts as a waveguide, where the incident light is trapped at one end of the fibre and is transmitted within the structure along to the other end with minimal loss of optical energy. The propagation of light in an optical fibre can be explained rigorously using electromagnetic theory or, with less accuracy, in physical terms utilizing geometric optics. Total internal reflection is the basic mechanism involved in optical fibres. The simple scheme in Figure 13.1 illustrates the principle of light guidance in an optical fibre with a core of refractive index N_1, and a cladding of refractive index N_2. When $N_1 > N_2$, total internal reflection is possible. An optical ray launched into the fibre at an angle θ_i is reflected at the critical angle θ_c from the core/cladding interface. Any ray which is incident on the end face at an angle

K

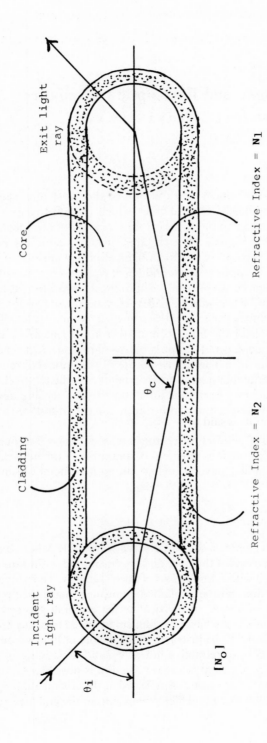

Figure 13.1 Light through optical fibre.

greater than θ_i will escape from the core. Therefore it is important to establish the value of θ_i.

The critical angle is given by

$$\sin \theta_c = \frac{N_2}{N_1} \tag{13.1}$$

and it can be shown that

$$N_o \sin \theta_i = (N_1^2 - N_2^2)^{1/2} \tag{13.2}$$

where N_o is the refractive index of the outside medium. The expression $(N_1^2 - N_2^2)^{1/2}$ is known as the 'numerical aperture' (NA) of the optical fibre. If the outside medium is air ($N_o = 1$), then NA $= \sin \theta_i$. Putting this into physical terms, NA is the measure of the light-gathering capacity of the fibre end, and is equal to the sine of the half-angle of the maximum core of the incident light which can be guided by the fibre. It is dependent only on the refractive indices of the core and cladding.

Two types of optical fibre can be employed in the construction of optical-fibre transducers. One type is a single optical fibre of a prescribed diameter and is made of optically transparent material such as quartz, glass or plastic. The second type is a fibre-optic bundle composed of numerous strands of light-transmitting fibres. Optical fibres can be bent easily within practical limits, while maintaining their light-transmitting capabilities. Various factors affect the actual transmission of light through optical fibres, including absorption of light by the fibre material, reflection losses and the condition of the ends of the fibres.

An optical fibre bundle is said to be *non-coherent* if the fibres are not aligned within the bundle, and is therefore capable of transmitting only light. A bundle is said to be *coherent* if the individual fibres are aligned within the bundle and at the ends; such a bundle is capable of transmitting both light and images. A single fibre can transmit light but not images.

13.3 Spectroscopic techniques for optical sensors

Optical transducers are devices that can be used for the detection and determination of physical or chemical parameters through measurements of changes in some optical property. The optical property measured can be absorbance, reflectance, scattering or luminescence. Changes in parameters such as refractive index can also be employed with these sensors. In chemical sensing the phenomenon employed gives information concerning the chemical composition of the sample.

13.3.1 *Absorbance measurements*

The change in light intensity in an absorbance measurement is determined by the number of absorbing species in the optical path and is related to the concentration C of the absorbing species via the Beer–Lambert relationship

$$A = \log \frac{I_0}{I} = \varepsilon l C \qquad (13.3)$$

where A is the absorbance, l is the pathlength of the light, ε is the molar absorptivity and I_0 and I are the incident and the transmitted light respectively. When employing this technique with optical fibre sensors, the medium used to support the selective chemistry, and indeed the selective chemistry itself, should be optically transparent.

13.3.2 Reflectance measurements

When the medium and/or the selective chemistry used in an optical sensor is opaque or when it transmits light only weakly, then measurement of the intensity of the reflected light may be used. Reflection takes place when light infringes on a boundary surface, and two distinct types of reflection are possible. The first is a 'mirror' type or specular reflection, which occurs at the interface of a medium with no transmission through it. The second type is diffuse reflection, where the light penetrates the medium and subsequently reappears at the surface after partial absorption and multiple scattering within the medium. Specular reflection can be minimized or eliminated through proper sample preparation or optical engineering.

The optical characteristics of diffuse reflectance have been recognized to be dependent on the composition of the system. Several models for diffuse reflectance have been proposed based on the radiative transfer theory, and all these models consider that the incident light is scattered by particles within the medium. The most widely used is the Kubelka–Munk theory, in which it is assumed that the scattering layer is infinitely thick. The reflectance R is related to the absorption coefficient K and the scattering coefficient S by the equation

$$f(R) = \frac{(1-R)^2}{2R} = \frac{K}{S} \qquad (13.4)$$

where $f(R)$ is known as the Kubelka–Munk function. The absorption coefficient K can be expressed in terms of the molar absorptivity ε and the concentration C of the absorbing species, as $K = \varepsilon C$. Equation (13.4) then becomes

$$f(R) = \frac{\varepsilon C}{S} = kC \qquad (13.5)$$

where S is assumed to be independent of concentration. Equation (13.5) is analogous to the Beer–Lambert equation and holds true within a range of concentrations for solid solutions in which the absorber is incorporated with the scattering particles and for systems in which the absorber is adsorbed on the surface of a scattering particle. Absolute reflectance values are evaluated using standard reflectance samples such as $BaSO_4$ (9, 10).

Alternative models have been formulated and found to be valid for other types of systems. For example, the Pitts–Giovanelli theory is shown to be valid for scattering particles suspended in an absorbing aqueous medium (11), while the Rozenberg model is applicable to mixtures of absorbing and non-absorbing powders (12).

13.3.3 *Luminescence measurements*

Measurement of luminescence, more commonly called fluorescence, is an extremely sensitive technique capable of measuring low analyte concentrations. It is particularly suited for optical sensing and is compatible with a single optic measurement because the wavelength of luminescence radiation detected is different to that of the incident radiation. At very low analyte concentrations, the response of an optical fluorescence sensor is linear according to the equation

$$I_F = kI_0\phi_F\varepsilon l\cdot C \tag{13.6}$$

where I_F is the intensity of measured fluorescence, I_0 is the intensity of incident light, C is the concentration of reagent, ϕ_F is the quantum yield of fluorescence, ε is the molar absorptivity, l is the optical pathlength in the sample and k is a constant related to instrument and sensor configuration.

13.3.4 *Scattering*

Unlike the processes of absorption and luminescence, scattering of light need not involve a transition between quantized energy levels in atoms or molecules. Instead, a randomization in the direction of light radiation occurs. Particles with sizes that are small compared to the wavelength of radiation give rise to Rayleigh scattering. This type of scattering is exhibited by all atoms and molecules. If the particles are large compared to the wavelength of radiation, the scattering of light is known as Mie scattering. In this case, the intensity of the scattered radiation can be related to the concentration of the scattering particles.

In both Rayleigh and Mie scattering, the polarization of the particle remains constant. However, with molecules, the incident radiation can promote vibrational changes which can alter the polarization of the molecule. The frequency of the light scattered by these molecules will be different from that of the incident light, and with much weaker intensity. Such a scattering phenomenon is known as Raman scattering and can be observed if intense light sources such as lasers are used.

13.4 Instrumentation

13.4.1 *Spectrometers*

Special spectrometers need to be developed for fibre optics, with appropriate optics to match the fibre numerical aperture and the small core size. It may also be necessary to miniaturize the spectrometer for mounting and integration with the processing electronics, and to allow multiple connection of separate sensing heads (fibres) with a centralized instrument. Conventional spectrometers have been modified to allow coupling with several fibres, or a star coupler can be used to divide the light into multiple channels for monitoring several locations (13), or to monitor a single location from different positions (14). Mechanical switching techniques are being developed to connect a single fibre from the spectrometer with an array of fibres connected to different probes (15).

A number of fibre-optic spectrometers or wavelength selection devices have been developed for wavelength multiplexing of communication data channels on a single fibre. One such spectrometer is a small block of BK 7 glass ($20 \times 20 \times 98$ mm) with a spherical mirror at one end and a grating at the other (16). The fibres are attached to a window in the middle of the grating. This is designed for a NA of 0.3. Losses of 1.8 dB to 2.2 dB are reported for a 9-channel multiplexer with a graded-index 50/125 fibre as input and 9-step-index 80/125 fibres as output (17). The cross-talk between neighbouring channels (40 nm separation) is around -30 dB, but decreases to -50 dB with fewer channels.

A miniaturized spectrometer with CCD array for mounting on a printed circuit card inside the electronics has been developed for a film thickness monitor (18, 19). Light from the input fibre is focused by a 40 mm achromatic lens on to a blazed reflection grating with 300 lines/mm. A 50-element CCD array gives a spectral resolution of 10 nm. The spectrometer is 10 mm thick and 80 mm long. Spectrometers made by integrated optics have also been suggested (20).

13.4.2 *Optical sensors*

The basic instrumentation associated with optical fibre transducers is simple. Along with the optical fibre, there is a light source, a photodetector, monochromator or filters and optical couplers (Figure 13.2). The source should be capable of providing an intense and stable output. Several types of sources have been used in optical chemical sensors, including incandescent lamps, light-emitting diodes (LED) and lasers. The photodetection systems used in conjunction with optical chemical sensors include photomultiplier tubes, PIN photodiodes and avalanche photodiodes. Optical couplers are employed to focus the source light rays on to the optical fibre. Since laser beams are coherent and of small cross-section, they can be coupled very efficiently to optical fibres. The divergent rays from LEDs and incandescent

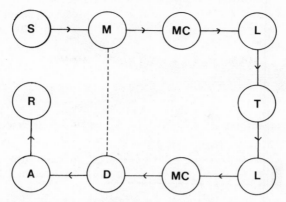

Figure 13.2 Schematic of instrumentation for use with optical fibre transducers. *S*, source; *M*, modulator; *MC*, monochromator; *L*, lens; *T*, optical chemical transducer; *D*, detector; *A*, amplifier; *R*, readout. (*MC* and *L* are optional).

lamps, however, require the use of lenses to focus their optical output on to the fibre. Finally, spectral resolution can be achieved by the use of filters or monochromators.

13.5 Optical transducer configurations

Initially fibre optic chemical sensors were simply devices for modifying the light path of a conventional spectrometer. The optical fibres were just used to conduct light to and from the sample. Sometimes conventional colorimetric reactions were carried out on the sample, and specialized probes could be used in conjunction with the optical fibres to make absorption measurements (for example). Such devices can be described as photometric or spectroscopic transducers.

Recent developments of optical-fibre chemical sensors are based on the use of immobilized chemical reagents (reagent phase) interfaced with the optical fibres (3, 7). The reagent phase provides the selective chemistry by which chemical information pertaining to the analyte is converted into spectroscopic information. The fibre optic transducer converts this spectroscopic information into an electrical signal.

Optical fibre chemical transducers may involve the use of either optical fibre bundles or single optical fibres. Fibre bundles can be bifurcated so that separate fibres transmit incident and detected radiation (Figure 13.3*a*). In optical transducers employing fibre bundles, the optical fibres observe only the part of the reagent phase that fall within both the cone of incident fibres and the cone of detection fibres. The chemical transducer can be interfaced at the common end of the fibre bundle. A similar configuration is adapted when using two single optical fibres in the construction of a sensor, where one fibre transmits the incident radiation and the second fibre the detected radiation.

When employing a single optical fibre, the chemical transducing element can

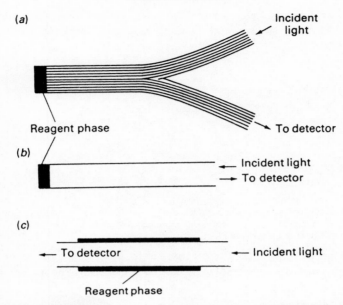

Figure 13.3 Optical fibre transducer, based: (*a*) on bifurcated optical fibre bundle with reagent phase at the end of the fibre; (*b*) on a single optical fibre with reagent phase at the end of the fibre; (*c*) with reagent phase coated on the outside of the single optical fibre. (Reproduced with permission from reference 7).

be interfaced at the end of the fibre or applied as a coating on the outside of the optical fibre as a cladding material (Figure 13.3*b, c*).

When the reagent phase is coated on the optical fibre, it undergoes an optical change due to its chemical reaction with the analyte, and the transmission characteristics of the fibre are modified due to alteration of the refractive index of the coating. Such a device employs the technique of evanescent wave detection. In transducers employing single optical fibres, it is necessary to distinguish between the incident and detected radiation either temporally or by wavelength differences.

Unlike potentiometric devices, optical fibre transducers do not require an 'external reference'; simple integral referencing is carried out optically by the use of the ratiometric method. In this method, a part of the conducted light that is not affected by the measurement variable is used to correct for other optical variations. A high degree of stability can be achieved in this manner.

13.6 Applications

13.6.1 *Spectroscopic sensors*

These are straightforward applications of optical fibres as light carriers in spectrophotometry. The sample is placed remotely from the instrumentation system, often without any pretreatment or conditioning. Several sensors of this type are based on the absorption characteristics of molecules in the IR or UV/visible region of the electromagnetic spectrum.

An important development has been the use of fibre-optic transducers for gas detection, particularly of explosive gases. A significant limitation in this area is the fibre transmission spectrum, which for silica fibres extends only to about 1.8 μm, whereas most of the standard 'fingerprint' absorptions used to identify simple molecules (CO, CH_4, C_2H_6, N_2O, etc.) occur in the mid-IR region (2 μm to 6 μm). Thus overtone spectroscopy is used, in which the higher harmonics of these fundamental absorptions (21–23) are measured between 1.2 and 1.7 μm. Absorption coefficients are low: for the CH_4 harmonic the absorption coefficient is 9.3×10^{-6}/ppm.m. The lower explosive limit for CH_4 is 5% and the higher explosive limit 15%. Detections of 0.8% (or 400 mm) of the lower explosive limit has been achieved with fibre-optic cavities (24).

To increase the sensitivity a long cavity length is used, sometimes with multiple passes, achieved by reflection from a corner cube or concave mirrors (25). Mechanical design of a stable cavity is critical for multiple reflections and to refocus the light back into the small core of a fibre. The optical measurement system generally includes an IR LED, interference filter and dual photodetectors (25) and a differential absorption technique for signal and reference channels (24). Further developments need to be made to provide a stable high-resolution optical detection system at low cost.

Remote absorption measurements have been made at 5 km distance (26) with a resolution of 5.3% of the lower explosive limit of CH_4. Field trials have been conducted over a six-month period on a North Sea oil rig with excellent results (27).

One of the early uses of a photometric sensor was in oximetry (28) (i.e. blood oxygen saturation), where the measurements are based on the fact that fully oxygenated haemoglobin (oxyhaemoglobin) and fully reduced haemoglobin have different absorption and reflectance spectra. Measurements at a wavelength of about 650 nm, with a reference wavelength at about 850 nm (isosbestic point), is related to blood oxygen saturation. The concentration of oxyhaemoglobin (HbO_2) is given by

$$\% HbO_2 = A - B(R_1/R_2) \tag{13.7}$$

where A and B are constants dependent on the geometry of the probe, and R_1 and R_2 are the reflectance at the measurement and reference wavelengths respectively. Such devices have been combined with the fibre-optic pressure sensors for intravascular monitoring, and for determination of cardiac output by measuring dye dilution of injected indocyanine green dye (which absorbs strongly at 805 nm near the isosbestic wavelength) into the bloodstream (29).

Fluorescence techniques with fibre optics have been developed for remote monitoring of different chemical species such as UO_2^{2+}, actinides, sulphates, inorganic chloride, H_2S, etc. (30–33). Applications are in nuclear power stations, pollution monitoring, and detection of trace impurities in ground and surface water. Fibre optic probes offer advantages of localized measurement in remote locations (the optical instrumentation can be 1 km away), with high sensitivity and high coupling efficiency of the exciting and fluorescent light.

Fluorescence is excited by laser (e.g. argon), and the back-scattered fluorescent light picked up by the fibre and analysed by spectrometry and photon-counting techniques. Where a single fibre is used to carry both the exciting and fluorescent light, the natural fluorescence in the fibre can limit the lower detection limit (3×10^{-6}) for Rhodamine B dye detected by a 600 μm polymer-clad silica fibre (33). This can be overcome using two separate fibres.

Different probe geometries have been used which include a simple fibre end (34), a ball-shaped lens (32) and a semi-permeable membrane (32). For uranium ions, detection sensitivities of better than 10^{-14} molar can be achieved with enhancement by phosphate coprecipitation (30).

Other applications of fibre-optic fluorimetry include the *in-vivo* determination of certain drugs in biological fluids (34, 35), and measurement of protoporphyrin in blood (36, 37).

Raman spectroscopy has also been performed using optical fibres in gas flames, for example, measuring N_2 in a CH_4/air flame at $2100°$ K (38). A 60 μm diameter, 20 m long fibre was used. Raman spectroscopy has also been performed with fibre-optic bundles (39). Light from an argon laser was used, and conducted by a glass bundle to a cavity of two spherical mirrors where multiple passing enhances the excitation. The Raman light was picked up at right angles through a horn-shaped fibre bundle with a rectangular aperture 0.8 by 9.7 mm and analysed by a double spectrometer. Basic problems in improving sensitivity include (i) the efficient coupling of light in and out of fibres, (ii) minimization of unwanted fluorescence and Raman light generated in the transmitting fibre, and (iii) improvement of the excitation intensity in the sample volume where scattered light is collected.

Light scattering is an important measurement in many applications, for the determination of impurities in liquids and for measurement of particle size. One sensor which has been commercialized is an oil-in-water monitor for measuring the oil pollution in effluent water. For marine use the specifications for ballast dumping are 1000 ppm $\pm$ 10 ppm. For other areas, such as electricity generating stations, oil content is measured at 0–2 ppm $\pm$ 0.1 ppm (40). The marine oil-in-water monitor consists of measurement of scattered light from a laser diode at 850 nm connected to a fibre-optic lead placed in the water flow (41). Straight-through and sideways scattered light is collected and the angular dependence allows distinction between large and small particles. The device shows some temperature sensitivity. For light crude oils the change is less than $0.2\%\,°C^{-1}$, for heavy crudes there is a larger change. The assembly includes an automatic window-cleaning system to counter dirt accumulation.

High-pressure cells, up to 250 bars pressure, for turbidity measurement, have been built with optical pathlengths of up to 50 cm (42). The construction requires accurate alignment of the optics and high-pressure seals. These have been used for differential measurement of particle contamination in cooling water. Alternatively, dual fibre-optic probes measure back-scattered light through 180° (43).

Measurement of surface scattering from opaque surfaces (44) has been made with a fibre-optic reflectance probe, for determination of demineralization of tooth enamel, milk-fat content and the whiteness of the eyeball. A reference fibre optic probe on a white $BaSO_4$ calibrated surface provides a differential measurement.

13.6.2 Chemical reagent-based fibre-optic sensors

Many optical fibre chemical sensors employ optical fibre transducers that are interfaced with immobilized chemical reagent(s). At the chemical transducer end, active chemical/biochemical reagents are held on supporting polymeric matrices and covered by membrane materials in certain applications. The following sections review some developments in this field.

(i) pH sensors. A number of pH sensors have been developed based on the use of a hydrogen-ion permeable membrane which encapsulates an immobilized colorimetric or fluorimetric pH indicator. These devices utilize the measurement of light intensity at a reagent wavelength together with a similar measurement at a reference wavelength insensitive to pH changes.

The pH sensor based on absorbance measurements of phenol red (45) indicator dyestuff, immobilized by covalent bonding on to polyacrylamide microspheres (5–10 μm in diameter), is a reversible sensor. The probe design is similar to that in Figure 13.3a, which consists of the dye bound on to polyacrylamide microspheres together with polystyrene microspheres ($\sim$ 1 μm size) to scatter light. The particles are packed into an ion-permeable cellulose dialysis tube (0.3 mm i.d.), and fabricated into a probe configuration by inserting two plastic optical fibres (0.15 mm diameter). Phenol red shows maximum absorption at 560 nm. In this sensor, light of wavelengths 560 nm and 600 nm (reference) is passed to the sensing end of the transducer, and the ratio R of absorbed light to non-absorbed light, is measured. R is related to pH by the equation

$$R = k \times 10[-D/(10^{-\Delta} + 1)] \tag{13.8}$$

where k is an optical constant of the system, D is the optical density of the sensor ($\lambda = 560$ nm) when the dye is totally in the base form, and Δ is equal to (pH $-$ pK_{In}) where K_{In} is the indicator constant. The shape of the curve of R v pH is sigmoidal with an approximately linear region near pH = pK_{In}. The system is capable of measuring pH in the physiological range (7.0–7.4) with a precision of ± 0.01 pH unit and a temperature coefficient of 0.017 pH unit °C^{-1}.

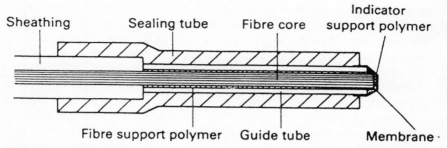

Figure 13.4 Cross-section schematic diagram of the reflectance probe design. (Reproduced with permission from reference 46).

When solid support matrices are employed for the immobilization of reagents in optical transducers, it is difficult to measure the transmitted light. In these cases, the intensity of light reflected may be used as a measure of the change in colour of the reagent phase, with an intensity measurement at a wavelength where the reagent does not absorb, as a reference. The functional relationship between reflectance and concentration will depend on the nature of the reagent phase and the optical arrangement. A reflectance pH sensor has been developed based on immobilized bromothymol blue which is retained at the tip of the polished end of an optical fibre (single or bundle) by a membrane of polytetrafluoroethylene (PTFE) (46). The devices are *c.* 2.0 mm in overall diameter and their design is shown in Figure 13.4.

Changes in the pH in the vicinity of the sensor cause variation in the attenuation of specific reflectance bands. In the pH probe, the wavelength of measured response corresponds to the attenuation of the reflected optical radiation at 593 nm, with a reference wavelength at 800 nm, to provide the optimum dynamic range with respect to pH. The reflectance spectra obtained from this pH sensor are shown in Figure 13.5. The probe shows a stable and reversible response within a few minutes.

Measurement of pH can also be performed by monitoring the fluorescence characteristics of fluoresceinamine immobilized by covalent bonding to cellulose (47). In this sensor, only the base form of the reagent is fluorescent. Therefore, as pH increases, there is an increase in intensity of fluorescence. This pH sensor has a useful working range of pH 3–6 and reaches a steady fluorescent intensity in about 15–30 seconds. However, it is reported that concentration quenching occurs due to the proximity of the immobilized fluorescent molecules.

(ii) *Oxygen sensor.* A number of indicators have been investigated for the development of optical sensors for the measurement of oxygen concentrations, and some of these can be used to measure physiological pO_2. An optical oxygen sensor based on immobilized haemoglobin utilizes reflectance measurements (48), and exploits the shift of absorption band when deoxy-

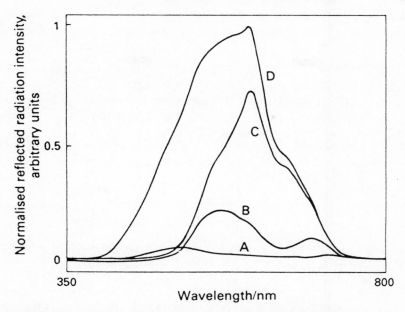

Figure 13.5 Reflectance spectra from a pH probe based on immobilized bromothymol blue. (A) with the indicator predominantly in the base form; (B) with approximately equal concentrations of each conjugate form of the indicator; (C) with the indicator predominantly in the acid form; (D) without indicator present. (Reproduced with permission from reference 46.)

haemoglobin combines with oxygen to form oxyhaemoglobin. The ratio of the haemoglobin reflectance at 405 and 435 nm is used as a measure of pO_2. This sensor involves a true chemical equilibrium requiring a mass transfer of oxygen to form the oxyhaemoglobin, which may lead to some errors in measurement. In addition, the immobilized haemoglobin is degraded with time, which imposes serious limitations on the useful lifetime of the sensor.

Luminescence of a number of molecules is quenched by oxygen via the transfer of energy from the photoexcited molecule to molecular oxygen. The greater the collision frequency of oxygen molecules with the indicator reagent, the fewer excited molecules are left for observation by their luminescence. In general, the luminescence intensity decreases with increasing pO_2. Usually there is no chemical reaction involved in these sensors and, as a result, the response is rapid and reversible, The response of the sensors depends on the relative rates of fluorescence and non-radiative decay of the excited state of the indicator reagent to its ground state *via* interaction with oxygen. The decrease in fluorescence intensity is related to oxygen concentration by the Stern–Volmer equation. The quenching of fluorescence of pyrene butyric acid (49), perylenedibutyrate (50), and 9, 10-diphenylanthracene (51) have been studied in the development of optical fibre oxygen sensors. The sensor described by Peterson *et al.* (50) has a diameter of 0.6 mm and can be used to measure physiological pO_2 in the range 0–150 Torr (0–20 kPa) with a precision of 1 Torr. The geometry of this sensor is suitable for *in-vivo* blood gas measurements.

(iii) *Glucose sensor.* An optical fibre sensor for the measurement of glucose concentrations has been developed (52, 53), based on the concept of affinity binding and direct measurement of fluorescence. This sensor is an implantable optical fibre glucose sensor that offers several advantages over the more conventional electrochemical sensors. The principle of detection in this type of optical sensor involves the competitive binding of glucose and fluorescein-labelled dextran (Fl-dextran) with a carbohydrate receptor, concanavalin A (Con A), as represented below:

$$\text{Con A} + \text{glucose} \rightleftharpoons \text{Con A-glucose}$$

$$\text{Con A} + \text{Fl-dextran} \rightleftharpoons \text{Con A-Fl-dextran}$$

Increasing concentrations of glucose will displace the first reaction to the right and the second reaction to the left, releasing the Fl-dextran, whose fluorescence intensity is measured and related to glucose concentrations.

The glucose transducer consists of a hollow dialysis tube of 0.3 mm internal diameter, sealed at one end and attached to a single optical fibre at the other. Con A is immobilized on the inner surface of the tube and the Fl-dextran is bound to this natural substance. Con A has a high affinity to glucose and as a result of increasing glucose concentration, say from blood, the Fl-dextran will be liberated into the optical path of the transducer. The intensity of fluorescence measured is proportional to glucose concentration. This sensor is reversible and has been shown to be stable for several days. The response is linear in the range of 50 to 400 mg% glucose with a response time of 5–7 minutes at 24°C.

(iv) *Carbon dioxide sensor.* pH sensors are normally used for measurements of pCO_2 because a change in CO_2 concentration causes a change in pH. The difference between a normal pH sensor and a pCO_2 sensor is that a membrane is used in the latter which separates the sample from the reagent and is permeable only to gases. Depending on the reagent used, the fibre-optic pCO_2 sensor can be based on absorbance or fluorescence measurements. For example, the changes in absorbance of phenol red (54) and the changes in fluorescence of trisodium salt of 8-hydroxy-1, 3, 6-pyrenetrisulphonic acid (55) and 4-methyl umbelliferone (56) have been utilized in the development of fibre-optic CO_2 sensors.

(v) *Ammonia sensor.* The first optical sensor for ammonia was based on the use of a ninhydrin-coated quartz rod (57), which changed colour on exposure to NH_3. Though the device was capable of detecting NH_3 vapour at concentrations of below 100 ppb, it had limited practical use because the chemical reaction utilized was irreversible. Subsequently, an optical waveguide sensor for ammonia has been developed which is reversible (58). This sensing device utilizes the indicator dye oxazine perchlorate, which changes colour from blue to red on exposure to ammonia. The principle of reaction in the hydrated film of the dye coated on the waveguide involves an acid–base equilibrium reaction of oxazine perchlorate, as summarized below:

$$NH_3(vapour) + H_2O(film) \rightleftharpoons NH_4{}^+OH^-$$

$$NH_4{}^+OH^- + H^+dye^- \rightleftharpoons NH_4{}^+dye^- + H_2O$$

$$NH_4{}^+dye^- \rightleftharpoons H^+dye^+ + NH_3(vapour)$$

(red)   (blue)

This sensor is humidity-sensitive and can detect less than 60 ppm NH_3 in ambient air at a relative humidity of 40%. Fibre-optic ammonia sensors based on the changes in absorbance of p-nitrophenol (59) and in fluorescence of several pH indicators (60) have been described. The arrangement of the system in the fluorescence sensor is applicable to both gaseous and liquid ammonia samples.

(vi) *Other sensors.* Optical sensors have been developed for the measurements of several ions in solution, such as cyanide ion (61) based on absorbance measurements, sulphide ion (62) and moisture (63) based on reflectance measurements, and metal ions (64–66) and halide ions (67) based on fluorescence measurements. Fluorescence-based sensors have also been studied for measurement of the widely-used inhalation narcotic halothane (68).

Sensors for metal ions are based on the use of non-fluorescent or weakly fluorescent reagents which form highly fluorescent complexes with the metal ions of interest. For example, the reagent morin (3, 5, 7, 2', 4'-pentahydroxyflavone), immobilized on cellulose by covalent bonding, forms intense fluorescent complexes with aluminium (64) and beryllium (65) which are used to sense these ions. The detection limits for these transducers are reported to be 27 ppb for aluminium and 9 ppb for beryllium.

The sensor for halothane is based on the effect of fluorescence quenching of a highly halothane-sensitive indicator system exposed to the sample (68). Interference by molecular oxygen is taken into account by a second sensor, highly sensitive towards oxygen. The two-sensor combination allows the determination of halothane or oxygen or both with a precision of ± 5%, and is reported to be practically specific for the two analytes. Other gases present in the inhalation gases or blood, including CO, N_2O or fluorans, do not interfere. The response time (15–20 s for halothane and 10–15 s for oxygen for 90% of the final values) is considered short enough to allow analysis of inhaled and exhaled gas. Potential applications of this sensor include the continuous *in-vivo* monitoring of halothane in blood with fibre-optic catheters during operations, and continuous monitoring of anaesthetic gases in the breathing circuit.

An interesting application of an optical-fibre chemical transducer is the measurement of hydrogen, in which the fibre itself is used as the sensing element (69). The optical fibre is coated with palladium and on exposure to hydrogen, the metal expands caused by the formation of palladium hydride (PdH_x) which is dependent on the partial pressure of hydrogen. This expansion stretches the fibre in both axial and radial directions and results in a change in the effective optical pathlength of the fibre. This change can be detected by interferometric techniques, using a Mach–Zehnder interferometer. The effect utilized in this sensor is thermodynamically reversible, and the device is reported to have a good sensitivity with a wide dynamic range (1–30000 ppm) and a temperature coefficient of 2 ppm $H_2 \, °C^{-1}$. The concept used in this sensing device may be applicable to the detection of a large number of other chemical species.

Alkanes have been detected using the formation of alkane-hydrates by interaction of polar solvent molecules with non-polar alkanes and by the technique of multiple total internal optical scattering (70). The alkane hydrates are formed within a layer of water adsorbed on to a glass optical waveguide, and this results in a change in the refractive index of the coating. The instrumentation utilizes LED light sources, and different alkane hydrates are reported to produce different scattering intensities. For methane, its hydrate gas concentrations up to approximately 1000 ppm may be detected using a wavelength of 560 nm at atmospheric pressure and room temperature under saturated air conditions. The minimum concentration of methane required for an explosive reaction in air is about 5000 ppm at standard temperature and pressure.

(vii) *Immunological assay.* Optical techniques for immunoassay are based on the principle of internal reflection spectroscopy or evanescent wave spectroscopy. An evanescent wave of light within a waveguide penetrates into the surrounding medium by a fraction of a wavelength and hence can be used to probe reactions taking place on or near the surface of the guide within a distance of around the wavelength of light. The intensity of an evanescent wave may be increased by plasmon resonance (see later).

A guided mode, represented by a ray of light, will remain within the waveguide if the angle of incidence at the waveguide–liquid interface is greater than the critical angle θ_c (Figure 13.1), which is given by equation (13.1). The electric field E of the evanescent wave, associated with the bound mode, penetrates a given distance z outside the guide (71), where

$$E = E_0 \cdot \exp(-z/d_p) \tag{13.9}$$

E_0 is the electric field at the waveguide–liquid interface and d_p is the penetration depth or the distance at which E_0 falls by $1/e$ of its value, given by

$$d_p = \frac{\lambda}{2\pi n (\sin^2 \theta_i - \sin^2 \theta_c)^{1/2}} \tag{13.10}$$

Thus, for a quartz substrate (72), where $N_1 = 1.46$, and, if immersed in water, $N_2 = 1.33$, and for $\lambda = 300$ nm and $\theta_i = 75°$ (10° larger than $\theta_c = 65.6°$), then the penetration depth can be evaluated as 102 nm. The size of an immunoglobulin (IgG) molecule is approximately 10 nm × 6 nm, which is therefore smaller than the penetration depth.

(viii) *Surface reaction measurement.* In this technique, an antibody is immobilized on the surface of the waveguide and the light probes the antigen–antibody reaction. Several methods of estimating the reaction can be employed, including those involving measurement of light scattered directly from the antigen, or of total absorption or fluorescence if a suitably labelled antigen is used. In this case the fluorescence is also collected and trapped within the waveguide itself. An enhancement effect, as high as 50 times the sideways-emitted light, is due to preferential coupling of fluorescent light with the same electric field intensity as the evanescent exciting wave, and this has been experimentally and theoretically verified (73, 74).

Measurements with evanescent wave spectroscopy have been made, using both glass slides as a waveguide and optical fibres such as polymer-clad silica (PCS) fibre without the cladding.

For fluorescein isothiocyanate-labelled IgG the sensitivity achieved is around 10 nM using scattering light intensity changes and 5 nM using fluorescence measurements (72). The technique eliminates the lengthy separation and washing stages needed in conventional immunoassay, and a dynamic rate monitoring can be used, thus making measurements possible within 5 to 10 minutes.

A large number of different biochemical systems have been studied by optical measurement on continuous surfaces, using ellipsometry, attenuated total reflection, interference techniques and total internal reflection fluorescence (reference 6 lists 45 different experiments). It is therefore likely that evanescent wave spectroscopy will become a widely applied technique in the future.

(ix) *Surface plasmon resonance.* Surface plasmon resonance occurs at a dielectric–metal layer interface where light which is totally reflected within the dielectric induces a collective oscillation in the free-electron plasma at the metal boundary. The condition for this to occur is that the momentum of the photons should match the surface plasmons on the opposite surface of the metal film (75). This occurs at some critical incident angle of light. The resulting effect is to produce a large change in the reflection coefficient at this resonance angle. A typical experimental set-up employs a silver film of 50 nm thickness evaporated on to a glass plate or prism. The sample is placed on the silver layer. The intensity of the electric field of the evanescent wave within the sample layer is enhanced by two orders of magnitude compared to a simple dielectric interface. Changes in the sample properties shift the resonance angle and provide a highly sensitive means of monitoring the surface reaction.

An angular resolution of a typical measurement set-up is better than $5 \times 10^{-2\circ}$ (76).

Gas sensing has been achieved by coating a 43 nM layer of silicon–glycol copolymer on to a silver-coated (56 nm) glass substrate (77). The anaesthetic gas, halothane, for example, swells the film, and a swelling of 0.1% results in a shift of $10^{-2\circ}$. The resulting sensitivity is $3 \times 10^{-5\circ}\text{ppm}^{-1}$ and the response is highly linear up to 0.5% volume with a time response of a few milliseconds.

Immunoassay reactions have also been studied using this technique. A single monolayer (5 nm thick) of IgG produces a shift of 0.6° (75). Injection of antihuman globulin (a-IgG) which binds on to the IgG results in a further shift of 0.9°. From the initial time derivative it is possible to determine the concentration of a-IgG to less than $2\mu\text{g/mL}$.

The film layer thickness of antibody human serum albumin (anti-HSA) bonded to HSA has been calculated and measured by surface plasmon resonance (78). The absorption of HSA on a silver film shifts the resonance peak by 1.1° corresponding to a film thickness of 6 nm; HSA molecular dimensions are $6 \times 6 \times 2$ nm. Reaction with anti-HSA shifts the peak by 3.2° corresponding to 22 nm, or several monolayers.

13.7 Conclusions

The present state of fibre-optic technology has provided miniaturized spectrometers, multiple sensing possibilities, remote measurement capabilities and even miniaturization of optical probes for incorporation into hypodermic needles for clinical and other biomedical measurements. This fast-moving technology offers many exciting possibilities for the future in the area of chemical sensing.

The problems that may be encountered with fibre-optic transducers originate from ambient light interference, which could be eliminated by appropriate modulation of optical signals, the optical fibre coupling employed, and in producing reproducible optical cell designs. Others include limited dynamic ranges compared to other sensors, problems associated with aspects of sensor design involving mass transfer stages due to the reagent and analyte being in different phases, and those due to long-term stability of chemical reagent systems employed in the chemical transduction process. The mass transfer process can limit the response times of the sensing devices. Stability of reagents may be compensated to some extent by employing multiple-wavelength detection or by having a facility to change the reagent phases easily.

In spite of these problems, the technique of chemical sensing based on the use of optical fibres has great potential. Their success not only requires the development of appropriate reagent phases but also the understanding of the principles and mechanism of their operation, and in the investigation of new sensor designs.

References

1. A.L. Harmer, in *Optical Fibre Sensors*, eds. S. Martelucci and A.M. Scheggi, Martinus Nijhoff, The Hague (1987).
2. T. Hirschfeld, J.B. Callis and B.R. Kowalski, *Science* **226** (1984) 312–318.
3. W.R. Seitz, *Anal. Chem.* **56** (1984) 16A–33A.
4. I. Chabbay, *Anal. Chem.* **54** (1982) 1071A–1080A.
5. J.I. Peterson and C.G. Vurek, *Science* **224** (1984) 123–127.
6. J.F. Place, R.M. Sutherland and C. Dahne, *Biosensors* **1** (1985) 321–353.
7. R. Narayanaswamy, *Anal. Proc.* **22** (1985) 204–206.
8. H.G. Hecht, *J.Res. Nat. Bur. Stand.* **80A** (1976) 567–583.
9. H.G. Hecht, *Appl. Spectrosc.* **37** (1983) 348–354.
10. W.W. Wendlandt and H.G. Hecht, *Reflectance Spectroscopy*, John Wiley, New York (1966).
11. H.G. Hecht, *Anal. Chem.* **48** (1976) 1775–1779.
12. H.G. Hecht, *Appl. Spectrosc.* **34** (1980) 161–164.
13. S. Klainer, T. Hirschfeld, H. Bowman, F. Milanovich, D. Perry and D. Johnson, Science Division Annual Report, Lawrence Berkeley Lab., LBL–11981 (1981).
14. G.R. Trott and T.E. Furtak, *Rev. Sci. Instrum.* **51** (1980) 1493–1496.
15. T. Hirschfeld, in *Proc. European Conf. on Industrial Line Spectrographic Analysis*, Rouen, (1985) 1.
16. J.P. Laude, J. Flamand, J.C. Gautherin, D. Lepère, P. Gaçoin, F. Bos and J. Lerner, in *Proc. 9th Eur. Conf. Optical Communications*, Geneva (1983) 417–420.
17. J.P. Laude, J.C. Gautherin, D. Lepère, J. Flamand, P. Gaçoin and F. Bos, *Opto* **19** (1984) 29–31.
18. H.E. Korth, in *Proc. Second Int. Conf. Opt. Fibre Sensors*, Stuttgart (1984) 219–222.
19. H.E. Korth, *Journal de Physique* **44** (C10) (1983) 101–104.
20. P.K. Tien and R.J. Capik, in *Proc. Topical Meeting on Integrated and Guided-Wave Optics*, Nevada (1980) p.TuB 3–1.
21. H. Inaba, T. Kobayashi, M. Hirama and M. Hamza, *Electron. Lett.* **15** (1979) 749–751.
22. A. Hordrik, A. Berg and D. Thingbo, in *Proc. 9th Eur. Conf. on Optical Communications*, Geneva (1983) 317–320.
23. K. Chan, H. Ito and H. Inaba, *J. Lightwave Tech.* **LT–2** (1984) 234–237.
24. K. Chan, H. Ito and H. Inaba, *Appl. Optics* **23** (1984) 3415–3420.
25. S. Stueflotten, T. Christensen, S. Iversen, J.O. Hellvik, K. Almas, T. Wien, A. Graav, in *Proc. Eur. Conf. Industrial Line Spectrographic Analysis*, Rouen (1985) 3.1–3.5
26. K. Chan, T. Furuya, H. Ito and H. Inaba, *Opt. and Quantum Electronics* **17** (1985) 153–155.
27. S. Stueflotten, *Ericsson Review* **F-61** (1984) 24–27.
28. B.G. Gamble, P.G. Hugenholtz, R.G. Monroe, M. Polanyi and A.S. Nadas, *Intracardiac Oximetry Circulation* **31** (1965) 328–343.
29. M.L. Polanyi, in *Dye Curves*, D.A. Bloomfield (ed.), University Park Press, Baltimore (1974) 267–284.
30. S. Klainer, T. Hirschfeld, H. Bowman, F. Milanovich, D. Perry and D. Johnson, Report: Lawrence Berkeley Laboratory, LBL–11981, UC–70 (1981).
31. F.P. Milanovich and T. Hirschfeld, *Advances in Instrumentation, Proc. ISA Int. Conf.* **38** Pt. 1 (1983) 407–418.
32. F.P. Milanovich and T. Hirschfeld, *Intech*, March 1984, 33–36.
33. J.P. Dakin and A.J. King, in *Proc. First Int. Conf. Opt. Fibre Sensors* (IEEE) **221** (1983) 195–199.
34. B.J. Tromberg, J.F. Eastham and M.J. Sepaniak, *Appl. Spectrosc.* **38** (1984) 38–42.
35. M.J. Sepaniak, B.J. Tromberg and J.F. Eastham, *Clin. Chem.* **29** (1983) 1678–1692.
36. D.G. Mitchell, J.S. Garden and K.M. Aldous, *Anal. Chem.* **48** (1976) 2275–2277.
37. R.M. Smith, K.W. Jackson and K.M. Aldous, *Anal. Chem.* **49** (1977) 2051–2053.
38. A.C. Eckbeth, *Appl. Optics* **18** (1979) 3215–3216.
39. R.E. Brenner and R.K. Chang, in *Advances in Research and Development*, B. Bendow and S.S. Mitra (eds.), Plenum Press, New York (1979) 625–640.
40. P. Extance and G.D. Pitt, in *Proc. Int. Conf. on Opt. Tech. in Flow Monitoring and Control*, The Hague (BHRA) (1983) 43–45.
41. D. Snel and G.D. Pitt, in *Proc. Int.Conf. on Optic. Tech in Flow Monitoring and Control*, The Hague (BHRA) (1983) 27–42.

42. J.J. Perez, in *Proc. Eur. Conf. Industrial Line Spectrographic Analysis*, Rouen (1985) 2.1–2.12.
43. L. Papa, E. Piano and C. Pontiggia, *Appl. Opt.* **22** (1983) 375–376.
44. P.C.F. Borsboom and J.J. Ten Bosch, *Paper and Plastic*, SPIE **369** (1983) 417–421.
45. J.I. Peterson, S.R. Goldstein, R.V. Fitzgerald and D.K. Buckhold, *Anal. Chem.* **52** (1980) 864–869.
46. G.F. Kirkbright, R. Narayanaswamy and N.A. Welti, *Analyst* **109** (1984) 1025–1028.
47. L.A. Saari and W.R. Seitz, *Anal. Chem.* **54** (1982) 821–823.
48. Z. Zhujun and W.R. Seitz, *Anal. Chem.* **58** (1986) 220–222.
49. D.W. Lubbers and N. Opitz, *Sensors and Actuators* **4** (1983) 641–645.
50. J.I. Peterson, R.V. Fitzgerald and D.K. Buckhold, *Anal. Chem.* **56** (1984) 62–67.
51. M.E. Cox and B. Dunn, *Appl. Optics* **24** (1985) 2114–2120.
52. S. Mansouri and J.S. Schultz, *Bio/Technology* (Oct. 1984) 885–889.
53. J.S. Schultz, S. Mansouri and I.J. Goldstein, *Diabetes Care* **5** (1982) 245–253.
54. G.G. Vurek, US Pat. Appl. 470920, 22 June 1983, Appl. 01 March 1983, Avail. NTIS Order No. PAT–APPL–6–470920.
55. Z. Zhujun and W.R. Seitz, *Anal. Chim. Acta* **160** (1984) 305–309.
56. D.W. Lubbers and N. Opitz, in *Proc. Int. Meeting on Chemical Sensors*, Fukuoka, Japan, Elsevier, Amsterdam (1983) 609–619.
57. D.L. Smock, T.A. Orofino, G.M. Wooten and W.S. Spencer, *Anal. Chem.* **51** (1979) 505–508.
58. J.F. Giuliani, H. Wohltjen and N.L. Jarvis, *Optics Lett.* **8** (1983) 54–56.
59. M.A. Arnold and T.J. Ostler, *Anal. Chem.* **58** (1986) 1137–1140.
60. O.S. Wolfbeis and H.E. Posch, *Anal. Chim. Acta* **185** (1986) 321–327.
61. E.E. Hardy, D.J. David, N.S. Kapany and F.C. Unterleitner, *Nature* **257** (1975) 666–667.
62. R. Narayanaswamy and F. Sevilla III, *Int. J. Opt. Sensors* **1** (1986) 403–413.
63. A.P. Russell and K.S. Fletcher, *Anal. Chim. Acta* **170** (1985) 209–216.
64. L.A. Saari and W.R. Seitz, *Anal. Chem.* **55** (1983) 667–670.
65. L.A. Saari and W.R. Seitz, *Analyst* **109** (1984) 655–657.
66. Z. Zhujun and W.R. Seitz, *Anal. Chim. Acta* **171** (1985) 251–258.
67. E. Urbano, H. Offenbacher and O.S. Wolfbeis, *Anal. Chem.* **56** (1984) 427–429.
68. O.S. Wolfbeis, H.E. Posch and H. Kroneis, *Anal. Chem.* **57** (1985) 2556–2561.
69. M.A. Butler, *Appl. Phys. Lett.* **45** (1984) 1007–1009.
70. J.F. Giuliani and N.L. Jarvis, *Sensors and Actuators* **6** (1984) 107–112.
71. R.M. Sutherland, C. Dahne, J.F. Place and A.S. Ringrose, *Clin. Chem.* **30/9** (1984) 1533–1538.
72. R.M. Sutherland, C. Dahne, J.F. Place and A.S. Ringrose, *J. Immunol. Meth.* **74** (1984) 253–256.
73. E.H. Lee, R.E. Benner, J.B. Fenn and R.K. Chang, *Appl. Optics* **18** (1979) 862–870.
74. C.K. Carniglia, L. Mandel and H. Drexhage, *J. Opt. Soc. Amer.* **62** (1972).
75. B. Liedberg, C. Nylander and I. Lundstrom, *Sensors and Actuators* **4** (1983) 299–304.
76. I. Pockrand, J.D. Swalen, J.G. Gordon and M.R. Philpott, *Surface Sci.* **74** (1977) 237–244.
77. C. Nylander, B. Liedberg and T. Lind, *Sensors and Actuators* **3** (1982) 79–88.
78. M.T. Flanagan and R.H. Pantell, *Electronics Lett.* **20** (1984) 968–970.

14 Piezoelectric transducers

G.J. BASTIAANS

14.1 Introduction

Piezoelectric devices today play a relatively small role in chemical sensing, but there are indications that their use may expand greatly in the future. Current applications of piezoelectric transducers to chemical analysis and sensing fall into two categories: surface mass detection and strain indication. Examples of strain indicators include strain gauges, accelerometers, and photoacoustic detectors. Piezoelectric mass detectors have been used as humidity gauges, thin film monitors, and gas sensors. However, a great deal of research is being directed towards developing mass detectors into selective chemical sensors, and it is in this area where the potential utility of these devices is greatest.

These present and future applications of piezoelectric devices all hinge upon their ability to transduce physical strains occurring in a solid to electrical potentials and vice versa. Strains are relative displacements of particles within a solid and cause the generation of restoring stress forces. Strains will also result in the establishment of electrical fields within the solid if the material is piezoelectric.

The simplest application of a piezoelectric transducer is its use as a strain indicator, since the strains applied to a piezoelectric element can be quantified simply by measuring the electrical field which is created across the boundaries of that element. In the case of photoacoustic detectors, pressure pulses, created by the absorption of light from a chopped beam, deform a thin plate and are transformed into a series of voltage pulses. In such applications the electrical signals created are either dc or low-frequency, so that the design of the device and interface electronics is straightforward.

A more sophisticated, and potentially very important, application of piezoelectric devices to chemical sensing occurs when they are used to create and conduct acoustic waves. In this form piezoelectric crystals are configured into resonant structures which will transmit acoustic signals in a frequency-dependent manner. The resonant frequencies where signal transmission is optimal are known to be highly dependent upon the physical dimensions and properties of the crystal. One property which causes resonant frequency to change in a quantitative and predictable manner is the mass of foreign material placed on the surface of the piezoelectric crystals. As a result, such devices can be used as mass balances of very high sensitivity. These devices can also be applied as chemical sensors if the deposition of mass upon their

L

surfaces can be achieved in a selective fashion. Recently great progress has been made in obtaining selective surface mass binding under a variety of conditions through the use of chemical surface modification techniques. Thus piezoelectric mass sensors have the potential of being transformed into a wide variety of chemical sensors whose applications are limited only by the surface modification chemistry employed.

In this chapter the basic principles of piezoelectric transducers will first be reviewed. Once we have covered the basics of the propagation of elastic waves through piezoelectric solids, we will turn to the study of how resonant structures respond to surface mass loading. The transformation of mass sensors into chemical sensors will be discussed in terms of existing work and the potential for future developments.

14.2 Acoustic waves and displacements in solids

14.2.1 *Stress and strain*

The physical theory which describes the forces and energies involved in the propagation of acoustic waves through solids and fluids is well established, albeit somewhat complex mathematically. The treatment below is intended to provide the reader with a brief overview in order to allow an understanding of how such properties as elasticity, density, and viscosity affect the operation of piezoelectric transducers and their application to chemical sensing. A detailed description of acoustic waves in solids may be found in several texts (1–3). To simplify notation, vectors and tensors are presented without subscripts, and it is assumed that mathematical operations are summed over the appropriate dimensions.

The element common to the function of all piezoelectric transducers is particle displacement within a solid. Relative particle displacements cause the generation of restoring stress forces and for piezoelectric materials cause the generation of electrical fields as well. Displacement is defined as the vector indicating the difference between equilibrium and perturbed positions of a solid particle. However, this vector is not invariant to translational motion so that strain is used to represent relative displacement. Strain and displacement are related by the equation:

$$S = \nabla s \cdot u \qquad (14.1)$$

where S is strain, u is displacement and ∇s is the symmetric gradient operator (1).

When strains are induced within an elastic solid, restoring forces known as stresses are created. Stresses are internal traction forces defined in terms of force per unit area. Stress is related to strain by the equation:

$$T = cS \qquad (14.2)$$

where T is a second-rank tensor representing stress and c is a matrix of material stiffness constants. The elements of the stiffness matrix are analogous to Hooke's law constants which relate the elongation of a spring and its corresponding return force. There are 81 elements in the stiffness matrix, but mathematical relationships limit the maximum number of independent values to 21. The properties of individual solids usually reduce the number of independent stiffness constants to less than 21. In the case of isotropic solids, for example, there are only two independent stiffness constants which are sometimes represented in the form of the Lamé constants, λ and μ. Fluids, which also support acoustic waves, have only one stiffness constant which relates compressional strains to stress forces.

The inverse relationship between stress and strain is sometimes found useful. In this case strain is equated to stress through the use of a compliance matrix, s, as follows:

$$S = sT \qquad (14.3)$$

The elements of the compliance matrix can be derived from the stiffness constants via relationships dependent upon the symmetry of the solid.

In the case of acoustic waves strains are time-dependent, and equation (14.2) can be expanded to account for any loss of energy caused by damping. For the case of wave propagation in lossy material

$$T = cS + \eta \, dS/dt \qquad (14.4)$$

where η is a matrix of viscosity constants similar to the stiffness constant matrix. In crystalline material, losses due to the viscosity term are quite small at frequencies below 100 MHz and are often neglected.

14.2.2 Piezoelectricity

Additional forces accompany particle displacements in piezoelectric solids due to the fact that macroscopic charge separation (polarization) occurs within the solid. Such charge separation results in the establishment of electrical field gradients in the material. The application of electric fields across the boundaries of a piezoelectric crystal always results in the creation of particle displacement and strains, so the effect is reversible. To account for these additional interactions, strain and its electromagnetic analogue electric displacement, D, can be equated to the stress and electrical forces acting on a solid particle via the following equations:

$$D = \varepsilon E + dT \qquad (14.5)$$

$$S = d'e + sT \qquad (14.6)$$

where E is the electrical field gradient, ε is an electrical permittivity matrix, and d, d' are piezoelectric strain coefficient matrices. It is obvious from equations (14.5) and (14.6) that the magnitude of transduction between electrical and mechanical forces is dependent on the values of the piezoelectric coupling constants.

If a piezoelectric transducer is simply being used as a displacement or strain indicator, then the relationships described above are sufficient to allow the calculation of the electrical signal which will be generated upon perturbation of a device. All that must be known are the stiffness constants (elastic properties), the piezoelectric strain coefficients (piezoelectric properties) of the transducer material, and the strain induced. If however, acoustic waves are being employed for mass sensing applications, then one must consider the dynamics of particle motion and how wave propagation is affected by the finite boundaries of a transducer.

14.2.3 Wave behaviour

The dynamic displacement of particles in an elastic solid can be described by the following equation of motion:

$$\nabla T = \rho d^2 u/dt^2 \tag{14.7}$$

where ρ is mass density. Equation (14.7) is simply the observation that the density of stress force is equal to the mass density multiplied by particle acceleration. For an acoustic plane wave travelling in the x direction, particle displacement is most generally described in terms of time and distance as

$$u = u_0 \cdot \exp[i(\omega t - kx)] \tag{14.8}$$

where u_0 is maximum particle displacement amplitude, ω is angular wave frequency, t is time, and k is the wave vector defined by:

$$k = \omega/V = \frac{2\pi f}{V} = 1/\lambda \tag{14.9}$$

where v is wave phase velocity, f is frequency, and λ is wavelength.

When the equation of motion and the wave equation are combined, the following relationship, known as the Christoffel equation, results:

$$k^2 \Gamma_{ij} \cdot v_j = \rho/\omega^2 v_i \tag{14.10}$$

where Γ_{ij} is the Christoffel matrix, v is particle velocity and the subscripts indicate summation over the dimensions of the coordinate system. For a non-piezoelectric solid, the Christoffel matrix is a function of the stiffness constants and the symmetry of the material with respect to the direction of wave propagation. For piezoelectric solids, the Christoffel matrix also becomes a function of the piezoelectric coupling constants and the electrical permitivity of the material.

When the matrix multiplication represented by the Christoffel equation is performed, a set of simultaneous equations is obtained. These equations are expressions for all allowed modes of plane wave propagation in an unbounded medium (such waves are often referred to as bulk waves). For example, given propagation of acoustic energy in the x direction of an unbounded isotropic solid, the following three independent equations are obtained from the Christoffel equation:

$$k^2 c_{11} V_x = \rho \omega^2 v_x \qquad (14.11a)$$

$$k^2 c_{44} V_y = \rho \omega^2 v_y \qquad (14.11b)$$

$$k^2 c_{44} V_z = \rho \omega^2 v_z \qquad (14.11c)$$

where c_{11} and c_{44} are independent stiffness constants of the solid. These equations define the relationship between the wave vector and frequency, i.e. phase velocity, for substitution in equation (14.8).

Equations (14.11b) and (14.11c) represent acoustic shear waves whose particle displacements are completely orthogonal to the direction of wave propagation. Equation (14.11b) describes polarization along the y axis, and equation (14.11c) polarization along the z axis. These two waves are degenerate because they have the same wave vector. As a consequence, the two waves can be combined arbitrarily to obtain any type of polarization desired.

Equation (14.11a) has a wave vector of different amplitude and represents a compressional or longitudinal wave. In this case particle motion is parallel to the direction of wave propagation. The wave vector has a different value as a consequence of the fact that a different stiffness constant applies to compressive particle motion, and therefore the wave will travel with a different phase velocity. Thus, in isotropic solids three modes of bulk wave propagation exist of which two are degenerate and can only be formally distinguished by their polarization.

In anisotropic solids, three modes of wave propagation can also be derived from the Christoffel equation and their polarizations will all be mutually orthogonal. However the orientation of the polarizations of particle displacement with respect to the direction of wave propagation will vary with the symmetry of the solid and the orientation of the crystallographic axes with respect to wave travel. In cases where symmetry and orientation allow polarizations to be completely orthogonal or parallel to wave propagation, pure acoustic wave modes are said to exist. In other cases polarizations exist which involve particle motion with components both parallel and perpendicular to wave propagation. Such modes are termed quasi-shear or quasi-longitudinal depending upon the relative magnitude of the polarization components.

When anisotropic solids are piezoelectric we gain the ability to create acoustic waves by applying oscillating electrical fields and the ability to detect acoustic waves by measuring electrical fields. This transduction is generally done by plating metal electrodes on the surface of the solid. However, the piezoelectric property of a solid does little to change the manner in which acoustic waves propagate through the material. The most significant effect is often referred to as piezoelectric stiffening, in which case the electrical forces generated by particle displacement cause an effective increase in the stiffness constants of the solid. This perturbation is accounted for in the Christoffel matrix by rewriting it in terms of the piezoelectric coupling constants and electrical permitivity as well as the stiffness constants as mentioned above.

At this point it should be clear that the direction of particle displacement and velocity of an acoustic wave can be varied by using piezoelectric crystals of different symmetries and by generating acoustic waves at different orientations with respect to the crystallographic axes. Theory allows the prediction of all these parameters if the properties of the material are known. Such calculations, however, can be complex and time-consuming, since numerical methods are required in many cases. Of course, the results of many previous calculations may be found in the literature.

Unfortunately the knowledge of how plane acoustic waves behave in an unbounded medium is not sufficient to explain how piezoelectric crystals are used as electrical signal processing devices or, as a consequence, how they can be used as mass sensors. To understand these phenomena one must consider the boundary conditions created by the fact that piezoelectric transducers have finite dimensions. The typical transducer is a thin disk or rectangular device. The existence of flat surfaces and a very limited thickness creates boundary conditions which significantly affect the propagation of acoustic waves in such devices. As a result of these boundary conditions two classes of piezoelectric signal processing devices exist. One class, bulk wave devices, operates by transmitting a bulk wave from one crystal face to the other. The second class, surface acoustic wave (SAW) devices, operates by transmitting waves along a single crystal face from one location to another. These modes of operation induce particle displacement on the device surfaces which in turn allow interactions to occur between the acoustic wave motion and any dissimilar mass deposited upon the surface. Elucidating the mechanisms of these interactions is the key to understanding the operation of piezoelectric mass and chemical sensors. The operation of each class of piezoelectric signal processing device is discussed below.

14.3 Acoustic wave devices

14.3.1 *Bulk wave devices*

Acoustic bulk wave devices have been used as transducers and resonators for many years. In these devices, acoustic waves are excited in a thin slab of piezoelectric material by applying an oscillating electrical field between crystal faces, i.e. along the thickness dimension. Given appropriate crystal symmetry and orientation, either a shear or a longitudinal wave, propagating along the thickness dimension, will be produced. Resonators are usually designed for

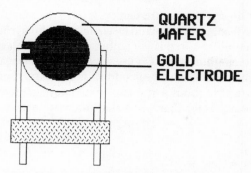

Figure 14.1 A bulk wave resonator.

shear wave operation. Figure 14.1 illustrates the geometry of a typical bulk wave resonator.

The waves generated in this manner differ from bulk waves in unbounded media because certain boundary conditions must exist at each surface. The most important boundary condition is that all stresses normal to the surfaces must be zero. To meet the boundary conditions, standing waves must exist. If it can be assumed for simplification that the lateral boundaries of the crystal are large enough to appear infinite, then the three normal stress forces that exist at each surface can be written as combinations of two opposite travelling waves as follows:

$$T_{xx} = A_1 \exp(ik_1 x) + B_1 \exp - (ik_1 x) \tag{14.12a}$$

$$T_{xy} = A_2 \exp(ik_2 x) + B_2 \exp - (ik_2 x) \tag{14.12b}$$

$$T_{xz} = A_3 \exp(ik_3 x) + B_3 \exp - (ik_3 x) \tag{14.12c}$$

where equation (14.12a) represents a standing longitudinal wave and equations (14.12b) and (14.12c) represent standing shear waves. These waves will only meet the boundary conditions when the wave vector k has the following values:

$$k = \frac{\omega}{V} = \frac{n\pi}{b} \tag{14.13}$$

where b is the thickness of the crystal and n is an integer. Since wave velocity is fixed for a given material, orientation, and wave mode, equation (14.13) defines the resonant frequencies that can exist for a given thickness. Equation (14.13) simply indicates that the thickness of a bulk wave device must be an integral number of half wavelengths for an acoustic wave to be resonant.

Because only acoustic waves at specific frequencies can efficiently propagate between the faces of bulk wave crystals, these devices are electronically used as frequency filters and the frequency-determining elements in oscillator circuits. As filters, bulk wave crystals will only pass electrical signals occurring over narrow frequency ranges centred about each resonant frequency. Additional frequency-selective circuitry is typically used to limit signal transmission to the bandwidth at a single resonant frequency. If this type of crystal filter is placed into the positive feedback path of an electronic amplifier, the signal produced will be an oscillation at a frequency within the bandwidth of the filter. In the practical implementation of bulk wave resonators as electronic components, usually resonant modes having only odd values of n are used. In addition, the size of the bandwidth about a resonant frequency increases with the use of resonances with n greater than 1, so that filter or oscillator performance is degraded.

14.3.2 Bulk wave mass sensors

As explained above the resonant properties of bulk wave devices are dependent upon crystal thickness and boundary conditions at their surfaces. As a consequence, it should not be surprising that the deposition of foreign material on a crystal surface, which alters both the effective thickness and the surface boundary conditions, will result in a change in the resonant frequencies of the device. It is upon this phenomenon that the design of bulk wave mass sensors is based. Any action which results in the addition or subtraction of material from the surface of a bulk wave acoustic resonator will generate a change in its resonant properties. This change is most often determined by placing the resonator in the feedback loop of an oscillator circuit which will undergo a frequency shift in response to the perturbation.

A mathematical relationship between the mass of material placed upon a bulk wave resonator and frequency shift (equation 14.14) was first derived by Sauerbrey (4):

$$\frac{\Delta f}{f_0} = -\frac{\Delta m}{m} \tag{14.14}$$

or

$$\Delta f = -C_1 f_0^2 \Delta m$$

where Δm is surface mass change, m is mass of the crystal, f_0 is resonant frequency, and C_1 is a constant. Equation (14.14) is based upon the following relationships:

$$t = \frac{m}{\rho A}$$

$$f_0 = \frac{V}{2t} \tag{14.15}$$

$$C_1 = \frac{2}{\rho A V} = \frac{2}{(\rho c)^{1/2} A}$$

where A is area of the crystal face and all other variables have been defined earlier. The Sauerbrey equation assumes that the change in resonance due to the uniform deposition of a foreign mass is caused solely by a change in the total thickness of the crystal and that the crystal is operating in a vacuum. The elastic properties of the deposited material and the difference in its density with respect to the crystal are ignored.

Following Sauerbrey's work, more detailed descriptions of the effect of surface mass upon bulk crystal resonance have been developed by many workers (5–13). One can allow for a different density of the deposited layer by using the equation

$$\Delta f = -(2f_0^2 \cdot \rho_2 t_2)/(\rho_1 C)^{1/2} \tag{14.16}$$

where the subscript 1 now refers to the crystal and subscript 2 refers to the deposited material (13). This correction allows for accurate prediction of frequency shift up to mass loadings of two percent of crystal mass (6, 13). If the elastic properties of the deposited mass are known, Sollner has shown that accurate frequency shifts can be predicted for loadings up to 60% (10).

Perhaps the most general mathematical treatment of the surface mass loading effect on bulk shear wave resonators has been presented by Kanazawa (13). In this work, a wave equation was developed for acoustic wave propagation within the deposited layer, assuming the material had both elastic and viscous properties. Boundary conditions between crystal and deposited mass were established by assuming shear forces and particle displacements were equal for both materials at the interface plane. This approach results in a fairly complex mathematical model, but simplified relationships were derived for purely elastic and purely viscous behaviour.

In the case of a purely elastic deposited layer the viscosity coefficient is equated to zero and the resonant angular frequency after mass loading ω, can be found from the expression:

$$\frac{\tan\left[\left(\omega t_2\right)\left(\frac{\rho_2}{c_2}\right)^{1/2}\right]}{\tan\left[\left(\omega t_1\right)\left(\frac{\rho_1}{c_1}\right)^{1/2}\right]} = \left[\frac{(\rho_1 c_1)}{(\rho_2 c_2)}\right]^{1/2} \tag{14.17}$$

This equation applies when a non-viscous solid is deposited on a bulk wave crystal, and can be reduced to the Sauerbrey equation through several simplifying assumptions (13).

If a viscous medium is placed at the surface of a bulk wave resonator, the acoustic wave will extend into the viscous layer but will be damped. The damping constant will be (12, 13):

$$\alpha = \left[\frac{(\omega \rho_2)}{2v_2}\right]^{1/2} \tag{14.18}$$

where v_i is viscosity of medium i.

For a purely viscous medium the frequency shift will be:

$$\Delta f = -f_0^{3/2}\left[\frac{(\rho_2 v_2)}{(\pi \rho_1 c_1)}\right]^{1/2} \tag{14.19}$$

Equation (14.19) is applicable to the case where a thick layer of a liquid is placed atop the resonator surface. Physically, it predicts that only a thin layer of liquid will undergo displacement at the surface the bulk wave device, and device response will be a function of the mass of this layer. Bruckenstein has observed that the response of a resonator which has both an elastic solid deposited layer and liquid atop the surface will be a linear sum of the responses expected for each individual perturbation (12).

An expression for the resonant frequency of a bulk wave device having a viscous and elastic deposited layer has not been derived in any simplified form. The frequency can, however, be obtained by numerical solution of Kanzawa's general viscoelastic solution to the wave equations given below (14):

$$u_y = \{u_+ \exp - [(ik + \alpha)x] + u_- \exp [(ik + \alpha)x]\} \exp (i\omega t) \quad (14.20)$$

where it is assumed that the wave propagates in the x direction and particle displacement is in the y direction. In equation (14.20), u_+ and u_- are the amplitudes of two counter-propagating waves, and the variables k and α are:

$$k = \omega \left(\frac{\rho}{2c}\right)^{1/2} \left\{ \frac{\left[\left[1 + \left(\frac{\omega v}{c}\right)^2\right]^{1/2} + 1\right]^{1/2}}{\left[1 + \left(\frac{\omega v}{c}\right)^2\right]^{1/2}} \right\}$$

$$\alpha = \omega \left(\frac{\rho}{2c}\right)^{1/2} \left\{ \frac{\left[\left[1 + \left(\frac{\omega v}{c}\right)^2\right]^{1/2} - 1\right]^{1/2}}{\left[1 + \left(\frac{\omega v}{c}\right)^2\right]^{1/2}} \right\}$$

Solution of the viscoelastic problem is appropriate when a deposited layer has significant coefficients of both elasticity and viscosity which might occur, for example, in the case of a polymeric coating on the crystal surface.

Unfortunately, much of the theoretical basis for predicting resonator response to different types of mass loading conditions has only recently been developed. As a result many mass sensing and chemical sensing applications were developed within the confines of the expected Sauerbrey equation response. Nevertheless, a broad range of mass and chemical sensing applications has been illustrated using the bulk wave resonator. This work will be discussed in the applications section (14.5) of this chapter.

14.3.3 Surface acoustic wave devices

The phenomenon of surface acoustic waves (SAW) first became of interest in the field of geology because the acoustic energy released by earthquakes moves very efficiently as an SAW on the Earth's crust. Lord Rayleigh (15) described the mathematical basis for a surface wave travelling on the earth's crust, and this type of SAW is known as a Rayleigh wave. As a general class, surface acoustic waves are distinguished by the property that acoustic fields and particle displacements are concentrated at a surface or at an interface and that the fields and displacements decrease with increasing distance from the surface or interface. Individual types of SAWs are distinguished by the directions of particle displacement that occur as the wave propagates, and these displacements are controlled by such factors as physical properties of the media involved, dimensions of the acoustic media, and the method of wave excitation. For example, a Rayleigh wave can exist at the interface between an elastic solid and most fluids and consists of a combination of two waves of equal phase velocity, one wave being longitudinal and the other being a shear wave having displacement polarized perpendicular to the interface plane. Other types of surface waves include Stoneley waves, Love waves, Bluestein–Gulyaev waves, and Lamb waves. In addition, signals can be transmitted through a SAW device via surface-skimming bulk waves which only slowly penetrate into the depth of the device. A complete description of all types of SAWs or a good explanation of their mathematical basis is well beyond the scope of this chapter and the reader must be referred to other texts (1, 2, 3, 16). Here discussion will be limited to those aspects most pertinent to chemical sensing.

Surface acoustic waves became of significant interest to the field of electrical engineering when White developed a simple and efficient method of exciting the waves on piezoelectric solids (17). In this now standard method, metal electrodes are deposited on the surface of a piezoelectric solid in the form of thin overlapping fingers as shown in Figure 14.2. Alternating fingers are connected to two different bus connection bars. When an alternating voltage is applied across the two bus connectors, oscillating electrical fields are induced between each finger pair. The electrical fields created by this interdigital transducer (IDT) induce particle displacement within the solid in directions determined by the piezoelectric coupling coefficients. Given the correct choice of piezoelectric material and orientation, the particle displacements will correspond to an allowed surface wave mode and a wave

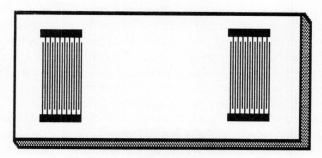

Figure 14.2

will be generated. This wave can later be detected on a surface with a second IDT through the reverse mechanism.

The spacing between the fingers of the IDT determines the operating frequency of a SAW device since efficient transduction requires that adjacent, oppositely polarized, fingers be half an acoustic wavelength apart. Frequency is thus related to finger spacing d by

$$f = \frac{V_s}{2d} \tag{14.21}$$

since

$$\lambda = \frac{V_s}{f}$$

where V_s is the velocity of the surface wave. The transfer function of the IDT is a $(\sin x)/x$ curve having a bandwidth related to the number of finger pairs N, and given by Dieulesaint (18) as

$$\frac{\Delta f}{f} = \frac{1.77}{N-1} \tag{14.22}$$

It can be seen then that an SAW device acts like a bulk wave crystal in the sense that only signals at specific frequencies will be efficiently transmitted. In the case of the bulk wave crystal, frequency is controlled by device thickness while a SAW device's resonant frequency is controlled by the velocity of the SAW and the spacing of the fingers. Modern fabrication techniques allow the creation of very narrow electrode lines and spaces on planar surfaces so that resonant frequencies greater than 1 GHz can be achieved for SAW devices (19). Thus a major advantage of SAW devices over bulk wave devices in electronic applications is the SAW devices' ability to operate at much higher frequencies. Higher resonant frequencies also are an advantage of SAW device chemical sensors since (as will be seen below) sensitivity to mass loading increases with frequency as well.

The electronic applications of SAW devices include use as delay lines, frequency filters, and oscillator components. The velocities of SAWs are approximately five orders of magnitude slower than that of electric signals through conductors. For this reason signals can be transmitted across a SAW device to achieve relatively long time delays. Usage as frequency filters derives from the frequency-selective properties of the IDT and the possibility of operating at very high frequencies is discussed in the paragraph above.

SAW devices can be made the frequency-determining elements in oscillator circuits by using either a delay line or resonator arrangement. In the delay line oscillator arrangement the general bandwidth of oscillator operation is determined by the structure of the IDT. However, the bandwidth is further narrowed by the requirement that the total phase shift around the feedback loop of the oscillator be zero or a multiple of 2π as indicated by

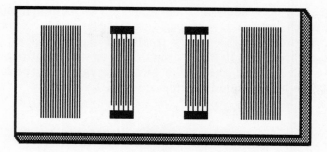

Figure 14.3

$$\phi_1 = \frac{(\omega L)}{V_s} + \phi_a = 2n\pi \qquad (14.23)$$

where L is the centre-to-centre distance between IDTs, ϕ_1 is the total phase shift around the oscillator feedback loop, ϕ_a is the phase shift across the amplifier of the oscillator, and n is 0 or an integer. The effective Q of an oscillator of this type will improve with greater IDT separation distance L, and degrade with increased signal attenuation which also increases with L. An optimal design must balance these two factors.

The resonator-type SAW oscillator operates via surface wave reflection. Reflectors which consist of periodic arrays of surface grooves or thin metal strips are placed on the surface as shown in Figure 14.3. Each groove or strip causes partial wave reflection. When they are spaced half wavelengths apart, they create a standing wave only at the frequency corresponding to the wavelength spacing. Thus efficient signal transmission (and oscillation) is restricted to the frequency of the standing wave in analogy to the operation of a bulk wave resonator. In the case of the SAW, however, very small spacings can be achieved so that very high resonant frequencies are possible. Modern resonator designs generally yield oscillators of superior Q relative to delay line operation, but at the expense of higher fabrication cost and complexity. A more detailed discussion of the signal processing aspects of SAW devices can be found in the literature (2,3,16,20).

For both types of SAW oscillators, the resonant frequency is dependent upon the phase velocity of the surface acoustic wave. In addition, the phase shift across an SAW delay is also dependent upon phase velocity. Any surface perturbation which causes an alteration of the SAW velocity will be detectable either as a frequency or a phase delay shift. The deposition of a foreign mass on an SAW device surface will cause such a velocity change and is the basis for the development of chemical sensors from SAW devices.

14.3.4 *SAW device mass sensors*

The effect of depositing a layer of foreign material upon the surface of an SAW device can be mathematically predicted either by formulating a complete set of equations which describe wave propagation in the layer, wave propagation in the SAW substrate, and boundary conditions, or by modelling through perturbation theory (21). If deposited layers are so thin as to be less than an acoustic wavelength, then perturbation theory is appropriate, and the velocity change due to the deposition of a thin isotropic layer is

$$\frac{\Delta V}{V} = \frac{V' - V}{V} = \left[\frac{(\omega k h \rho')}{4P} \right],$$ (14.24)

$$\left\{ V_s'^2 \left[4 \left(1 - \frac{V_s'^2}{V_1'^2} \right) |v_x|^2 + |v_y|^2 \right] - V^2 [|v_x|^2 + |v_y|^2 + |v_z|^2] \right\}$$

where $'$ refers to the deposited thin film, the subscripts 's' and 'l' refer to shear and longitudinal waves, V is surface wave velocity, and h is the thickness of the thin film (21–23). Equation (14.24) is referenced to a surface wave propagating in the x direction. The quantity $|v_1|^2/4P$ is a normalized particle velocity component which is a function of the physical properties of the SAW substrate and which has been tabulated for many materials (1). In the case of a Rayleigh wave $|v_y|^2 = 0$, and equation (14.24) is simplified somewhat.

An examination of equation (14.24) reveals that the change in velocity is proportional to mass per unit area, $h\rho'$, and this relationship provides the theoretical basis of mass response for SAW devices. It should also be noted that velocity change is a function of the frequency of operation, ω. As a consequence, the sensitivity of mass response of an SAW device increases with frequency. Since SAW devices operating at frequencies higher than 1 GHz are possible, extremely high mass sensitivities can be achieved.

Another important characteristic of equation (14.24) is that the two quantities in square brackets are of nearly equal magnitude. The magnitude and sign of the velocity change is heavily influenced by the relative stiffness constants and densities of the substrate and thin layer materials. It is quite possible to obtain velocity increases (stiffening effect) as well as velocity decreases (loading effect), although decreases are most often observed in sensing applications. Finally it is important to appreciate that equation (14.24) assumes the deposited layer is elastic and ignores viscosity effects. In situations where a viscoelastic material is deposited on a SAW device, an extension of the equation would be required for accurate prediction of response.

The most common approach to electronically detecting wave velocity changes in a SAW sensor is to construct a delay line or resonator oscillator circuit using the device as the frequency control element. In such arrangements relative changes in oscillator frequency are proportional to relative phase velocity changes, so we have

$$\frac{\Delta f}{f} = a \left(\frac{\Delta v}{v} \right) = b \rho' h f$$ (14.25)

and

$$\Delta f = b\rho' h f^2$$

where a and b are proportionality constants. Thus for a given mass per unit area, the absolute frequency shift scales with f^2. This type of relationship indicates again that increases in mass sensitivity for SAW sensors can be obtained by operating at frequencies much higher than those obtainable with bulk wave devices.

While most SAW sensor applications have utilized mass deposition as the means to alter wave phase velocity, it is possible to cause changes via other modes of surface perturbation. Two research groups have reported producing SAW sensors which respond to changes in the conductivity of a thin semiconducting layer on a SAW device (24, 25). Perturbation theory has again been used to predict that the velocity change due to the presence of a conducting surface layer will be (24, 26, 27):

$$\frac{\Delta v}{v} = \left(-\frac{K^2}{2} \right) \left[\frac{\sigma^2}{\sigma^2 + v^2 C_s^2} \right] \qquad (14.26)$$

where C_s is capacitance per unit length of surface, σ is sheet conductivity of the surface film, and K^2 is the substrate's electromechanical coupling constant.

The operation of SAW devices with liquid present at portions of the surface has been less commonly practised. Surface waves which have a vertical shear component (e.g. Rayleigh, Lamb, and Stoneley waves) will be attenuated by the presence of a liquid and are referred to as 'leaky' waves. Such attenuation occurs because compressional waves are launched from the surface into the liquid when the wave velocity in the liquid is less than the surface wave velocity, which is the usual case. Viktorov has provided a theoretical and experimental treatment of the perturbations associated with the interaction of liquids with both Rayleigh and Lamb waves (28). In the case of the Rayleigh wave it was found that thick, 'semi-infinite' layers of liquids will actually increase the wave velocity. One can also use the results of Viktorov to calculate that the attenuation caused by leaky wave behaviour for a combination of water and a typical SAW substrate will be of the order of 4 db $MHz^{-1} cm^{-1}$. Thus a 20 db loss (attenuation by a factor of 10) would occur for a 10 MHz wave travelling over a 5 mm path length. Such a loss would not significantly impair the operation of the SAW device in most sensing applications. As in the case of the bulk wave sensor, it is expected that the individual effects of liquid layers and deposited mass will be additive and distinguishable.

For surface waves which do not have a vertical shear component (e.g. SH, Love, Bleustein–Gulayev waves) the production of compressional waves in a liquid does not occur. Mass loading effects can still be observed but attenuation will only be caused by any viscosity of the liquid at the surface.

14.4 Instrumentation

Bulk wave or SAW crystals cannot individually serve as sensors without additional electronic support circuitry and mechanical fixtures. Currently no widely used commercial supplier for electronics or fixtures exists. However, prototype sensing systems can be assembled from literature designs and commercial components. Some of the more recent piezoelectric sensor system designs are discussed below.

14.4.1 *Bulk wave sensors*

Bulk wave crystals designed for electronic signal processing applications are readily available in a wide variety from a great many vendors. Many of these devices can be converted to sensor usage. Thompson has found the 'keyhole' electrode geometry to be most effective (11).

Many types of oscillator circuits have been used for bulk wave sensors. Guilbault (29) uses a commercial transistor oscillator. For work in the presence of liquids, Thompson (11) reported the design of a system based on the use of an automatic gain control amplifier, while Bruckenstein (12) has used an all-TTL digital logic oscillator design. The jigs and fixtures described in the work cited in this section are also quite typical of what is in common use.

The sensitivity of bulk wave sensor response to mass loading can be predicted from equation (14.14). If an AT cut quartz crystal is being used,

$$\Delta f = -2.26 \times 10^{-6} f^2 \Delta m \qquad (14.27)$$

where in this equation Δm is change in mass per unit area $(g\,cm^{-2})$. If sensitivity can be defined as df/dm, then equation (14.27) yields a sensitivity of $0.226\ Hz\,cm^2\,ng^{-1}$ for a 10 MHz device. This level is typically observed for bulk wave sensors, even when operated in the presence of a liquid (12).

The limit of detection (LOD) of these devices in terms of mass is the mass required to create a response three times greater than the noise level of the measurement system. Assuming a 5 Hz noise level for a 10 MHz crystal, the LOD would be $6.6\ ng\,cm^{-2}$.

Increases in the values of sensitivity can only be achieved by increasing the frequency of oscillation. Unfortunately, operation of bulk wave devices above 10 MHz starts to become impractical because the crystals must be made very thin and so become too fragile to manipulate or expose to sample. However, LOD can be enhanced by reducing the noise level during measurement. One useful approach to reduction of noise is to simultaneously operate a reference oscillator with the sensing oscillator. The difference or ratio of the two frequencies can then be taken. This procedure reduces low-frequency noise (signal drift) and thus improves the LOD.

14.4.2 *SAW sensors*

Fewer commercial components are available for the construction of SAW device mass sensor systems. All SAW devices used for sensing have been custom-fabricated. Design parameters for some of these devices have been given by Wohltjen (30-31) and Venema (25). The same references should be consulted for oscillator design, since no commercial circuits are available.

The mass detection sensitivities and LODs attainable are considerably better for SAW devices because they can operate at much higher frequencies. Using equation (14.24) and assuming an upper frequency limit of 3 GHz, Wohltjen has estimated that an LOD of 300 pg cm^{-2} is achievable. Given an area of 0.0001 cm^2 for such a device, an absolute mass LOD would be 3 femtograms. Similarly, Martin has predicted that a ZnO on Si SAW resonator can resolve a mass change of 50 pg cm^{-2} (32).

The reduction of noise can also be achieved with SAW devices by using a reference oscillator. This procedure should be more effective with these devices, because both sensing and reference oscillators can be placed on the same substrate (e.g. side by side). The two oscillators will then see virtually identical conditions, and compensation for signal drift will be more effective. It should also be possible to construct multisensor arrays by placing many SAW device patterns on a single substrate, since these patterns become small at high frequencies.

14.5 Applications

Piezoelectric transducers have been applied to a wide variety of mass and chemical measurement applications. Bulk wave devices have been used for analytical measurements since the 1960s. Several reviews of their use have been published by Guilbault (33, 34), and a more recent and fairly comprehensive review has been written by Alder and McCallum (35). Applications involving SAW devices first appeared in 1979 (36, 37). Reviews of SAW sensor design have been published by Wohltjen (31) and Venema (25), and the surface chemistry of SAW chemical sensors has been discussed by Nieuwenhuizen (38). Most recently an issue of *IEEE Transactions on Ultrasonics* has been devoted to a review and discussion of acoustic sensor applications (84).

From an examination of literature reports one can classify sensor applications into several categories. The earliest applications of piezoelectric sensors were as simple mass detectors where material was deposited to the device surface in a non-selective manner. In later work, mass sensors were transformed into chemical sensors by developing surface coatings which selectively bound gases from a vapour-phase chemical system. This increase in sophistication represented a major advance in the analytical capabilities of piezoelectric sensors. The manner and selectivity with which surface binding is controlled is the critical element in producing a practical chemical sensor and allows classification of the many types of chemical sensors which now exist.

One may also classify chemical sensors as to whether they operate in the gas or liquid phase. The use of piezoelectric mass sensors was first extended to the liquid phase by dipping or spraying techniques which allowed sensor readings to actually be taken in the vapour phase. Finally, a class of sensors was developed which could operate completely in the presence of a liquid, allowing a great expansion of potential applications. Using this scheme of classifying a sensor as to what its surface binding mechanism is and as to the phase of its sample, a closer examination of the field can be made.

14.5.1 Mass sensors

Bulk wave devices are commonly used as deposited mass or film thickness monitors in vacuum deposition systems. Their use in this context has been described by Czanderna (5) and is the subject of a continuing series of monographs (39). Bulk wave mass detectors have also been applied to the detection and quantitation of aerosols and suspended particles. Surface deposition was achieved by the use of a surface adhesive (40), electrostatic precipitation (41), or inertial impact (42). The mass sensitivity of bulk wave devices appears more than adequate for these applications, and there are no reports in the literature of the development of simple SAW device mass sensors which potentially could yield higher mass sensitivities.

14.5.2 Gas sensors

Gas sensors operate by binding molecules to the device surface via one of several mechanisms, depending how the surface is modified. If a bulk layer of liquid is placed on the surface, absorption of gaseous material can take place. Alternatively, if the surface is covered with a thin molecular layer one or a few monolayers thick, material will be surface-bound by some combination of physical and chemical interactions.

The phenomenon of gas absorption into a layer of liquid coating a surface has been characterized in great detail from the theory and practice of gas–liquid chromatography (GLC). In this theory the thermodynamics of the absorption process are expressed by a partition coefficient which indicates the relative concentrations of a specific molecule in gas and liquid phases. It is well known that the partition coefficients for similar compounds differ by relatively little and that chromatographic separations can only be achieved through repeated partitioning steps. As a consequence in the general case, little selectivity of absorption or sensing can be expected from devices coated with bulk absorbents. Of course exceptions to this trend will occur if the absorbent can be designed to exhibit an especially strong physical or chemical interaction to a specific compound.

Many other gas detectors have utilized thin films of a wide variety of materials to selectively bind mass. If a molecule is bound only by weak van der Waals' forces then physical adsorption to the surface is said to occur. At the opposite extreme, if binding of a molecule involves the making and breaking of chemical bonds, chemisorption occurs. In reality, most binding interactions fall between these two extremes. Examples of such intermediate behaviour include hydrogen bonding and charge transfer complexes. One can collectively refer to this surface binding behaviour as sorption. Nieuwenhuizen (38) has discussed the range of chemical interactions possible and presented data on the use of pthalocyanines as sorbents.

In his paper Nieuwenhuizen properly describes the tradeoffs involved in optimizing the selectivity and reversibility of sorption. If binding energies are high, as in the case of chemisorption, the selectivity of the surface attachment process will be high and sensor response can be limited to one or a few compounds. The cost of high selectivity, however, often will be poor reversibility of the binding process. If surface-bound material cannot be removed, the detection device becomes a dosimeter rather than a sensor. In the less extreme case, the removal of surface-bound molecules may require relatively long times or extreme changes in the physical or chemical conditions at the surface. If poor reversibility is a problem to sensor operation, it can be enhanced by changing the chemistry of the surface layer to reduce the interaction energy between sorbent and sorbate, but this action will reduce binding and detection selectivity. Surface binding capacity and response linearity of the sensor will also influence the need to regenerate a surface.

The influence of the surface layer on response time must also be considered. In a study of the sorption of alcohol vapours by a ZnO surface, Martin observed both a fast and a slow response time which he attributed to surface adsorption and bulk adsorption respectively (32). The fast response had a time constant of less than 100 s and was thought to be due to attachment of molecules to the top of the ZnO layer. A slower response to the alcohols ranging in time constant values from 240 to 750 s was attributed to diffusion of molecules into the bulk of the ZnO followed by adsorption. It was found that smaller molecules exhibited longer response times, which indicated that their smaller size allowed them to diffuse further into the ZnO or into smaller pores. This behaviour is quite analogous to that observed in chromatographic applications. To prevent slow response times, then, sensor surface coatings should be kept thin, and films with small pores should be avoided.

The intelligent design of a piezoelectric chemical sensor must consider all the trends discussed above. It can be seen that optimization of a sensor design is a complex function of analyte, analyte concentration, and sample matrix, as well as the requirements for selectivity, sampling frequency, response time, linearity of response, and sensitivity. No one design will be satisfactory for more than a few applications.

Given the chromatography analogy, it is not surprising to find a number of early gas sensors based upon the use of stationary GLC phases as surface

Table 14.1

Analyte	Reference
(a) Bulk wave sensor applications using liquid-phase absorbents	
Metal salts	48
NH_3	49
SO_2	33 (review)
Aromatic hydrocarbons	50
CO_2	51
Mononitrotoluenes	52
Hydrocarbon, halohydrocarbons	53
HCl	54
Organophosphorous compounds	55
Phosgene	56
Toluene diisocyanate	57
(b) SAW sensor applications using liquid-phase absorbents	
SO_2	59
H_2S	60
NO_2	24, 25, 38
Acetone, methanol	61
Cyclopentadiene	62

coatings. King first proposed the use of a bulk wave device coated with a stationary phase as a detector for a gas chromatograph (43, 44). Karasek studied this detector application (45, 46), and others predicted the kinetics of response using chromatography theory (47). Although positive results were obtained, piezoelectric detectors for gas chromatography were never commercialized, most likely because they could not demonstrate great advantages over existing commercial detectors. Table 14.1 presents a list of other applications involving sensors coated with GLC phases.

Closely related to the use of GLC phases is the coating of piezoelectric devices with other non-volatile liquids to achieve some degree of selectivity. For example, a layer of triethanolamine has been used to absorb and consequently detect SO_2 using bulk wave (58) and SAW (59) devices. Table 14.1 also contains a collection of other applications using liquid layer coatings. In many cases the selectivity of absorption is not reported, or interferences have been observed. Such limited selectivity is a consequence of the relatively small differences bulk layers exhibit in their partitioning behaviour between similar compounds.

Gases can also be surface adsorbed in a partially selective manner by using solid surface coatings. Metal salts have been used to bind methylamines to bulk wave devices (48). More recently, SAW devices have been coated with Pd metal to detect H_2 (63) and WO_3 to detect H_2S (64). As mentioned above, Martin also has used ZnO to adsorb alcohol vapours, but selectivity was not high (32).

Many applications have also appeared where polymer coatings have been placed on piezoelectric sensors (24, 36–38, 61, 62, 65). It is interesting to note that in the case when phthalocyanines are used as the polymer coating, response to adsorbed material occurs through a combination of mass loading and surface conductivity changes because this polymer is a semiconductor (24, 38). It is also significant that while different polymers exhibit widely varying responses to various analytes, selectivity of absorption is still limited. Workers in this area do not claim that a single polymer-coated device will satisfactorily detect a specific gas component in a real mixture. However, it is suggested that the desired selectivity of detection of individual components of a gas mixture can be achieved through the use of multiple sensor arrays.

It has been proposed that sensor arrays can achieve high degrees of detection selectivity by applying pattern recognition techniques to responses obtained from a number of individual piezoelectric sensors, each having a different polymer coating (65–67). Results obtained to date indicate that sensor response can be correlated to solubility properties of the analyte in a given polymer. It also has been demonstrated that pattern recognition techniques were able to classify analytes into subsets on the basis of multiple sensor responses (67). While this approach is still far from being reduced to practice, it does represent a potentially powerful and effective method for enhancing the detection selectivity possible with gas-phase piezoelectric detection.

14.5.3. *Liquid-phase sensors*

It has long been recognized that a much greater degree of selectivity in surface binding interactions can be achieved if the binding process can be performed at the interface between the surface and a liquid phase. Much higher selectivity can readily be achieved by the use of bioactive macromolecules bound to the surface. For example, the binding between enzymes and substrates or antibodies and antigens is known be highly selective. However, until 1979 it was commonly believed that the presence of a liquid at the surface of an active piezoelectric device would severely damp any acoustic wave at the surface and inhibit operation as a sensor. Nevertheless, in 1979 Nomura (68) and Bastiaans (69) published independent accounts of bulk wave sensors operational while one face of the crystal was in contact with liquid. In 1983 Roederer and Bastiaans (70) made the first report of SAW sensor operation in a liquid, and Martin has shown that a surface-skimming bulk wave on a SAW device can be used to make liquid viscosity and mass loading measurements in the presence of a liquid (71).

The advent of piezoelectric sensor operation in liquids was preceded by a number of studies where the advantages of liquid-phase binding reactions were combined with gas-phase operation of the sensor in a two-step procedure. Mieure and Jones (72) demonstrated that mass could be selectively electrodeposited upon the electrode of a bulk wave device and subsequently detected by operating the device in air. More significantly, in the 1970s it was demonstrated that immunoassays, which involve the highly selective binding between antibodies and antigens, could be performed using bulk wave sensors (73–76). In this work antibodies were mixed with a polymer and coated upon a device surface. The sensors were exposed to sample solution, washed, dried, and operated as oscillators in a humidity- and temperature-controlled chamber. Several variations of immunoassay were performed in these studies, and it was shown that one could assay individual antibody subclasses in the presence of other antibodies.

Applications of piezoelectric sensors operating completely in the presence of a liquid have to date fallen into three categories: measurement of the rheological properties of a liquid, electrochemical studies, and bioassays. Kanazawa (13, 14) and Bruckenstein (12), working independently, both developed the theory to predict resonant frequency shifts of a bulk wave device in the presence of a liquid, and both found that responses could be used to measure liquid density and viscosity. In related work, Martin (71, 77) has shown that the attenuation of a surface-skimming bulk wave on an SAW device can be used as an indication of liquid viscosity.

Both Nomura (78, 79) and Bruckenstein (80, 81) have used bulk wave devices oscillating at a liquid interface for electrochemical studies. In one study it was reported that 0.02 monolayer electrode coverage could be detected for a metal of molecular weight 100 (81). These studies indicate that selective detection can occur if the analyte is electroactive such that oxidation or reduction causes a surface mass change and if no other sample matrix component is electroactive at the same electrical potential.

The greatest selectivity of detection, however, has been achieved by covering piezoelectric sensor surfaces with biological antibodies. Roederer and Bastiaans obtained quantitative sensor response to the presence of human IgG dissolved in both buffer and diluted blood serum (70). The sensor was a quartz SAW device which had the antibody anti-human IgG covalently bonded to its surface. Selectivity was exhibited because response to the protein IgG was obtained in the presence of an excess of other blood serum proteins. In later work the limit of detection of an antibody-bonded SAW device was found to be 2 ng of antigen (82).

Immunoassay has also been demonstrated with a bulk wave device in the presence of a liquid (11). In this study anti-human IgG was immobilized to the surface either as a mixture with polyacrylamide or as a covalently bonded layer. In both approaches sensor response to 0.6 to 1.4 $mg\,mL^{-1}$ concentrations of IgG was observed. Changes in sensor response with time after sample exposure were also noted by Thompson. These changes were attributed to increases in surface viscosity due to the formation of an antibody–antigen precipitin. Similar effects were seen by Bruckenstein, who believed they were caused by changes in double layer structure at the surface of a biased electrode (80).

The above studies suggest that while mass loading in the presence of a liquid can be observed with both types of piezoelectric sensors, simultaneous changes in the viscosity of the liquid at the surface or of the material bound to the surface may complicate the response of a bulk wave device to a greater extent. For a SAW device, Martin found that signal attenuation was a more sensitive measure of liquid viscosity than wave velocity (or resonant frequency) change (71, 77). This observation has a theoretical basis, because it is known that acoustic wave attenuation due to viscosity losses has only a second-order effect on the phase velocity of the wave (83).

14.6 Future developments

The major asset of piezoelectric sensors has always been their great mass detection sensitivity. Coupled with the fact that mass is a universal property associated with any potential analyte, these devices could conceivably be applied to a wide variety of analytical problems. A limiting factor in their development, however, has been a lack of methods capable of limiting mass detection to only one or a very few analytes.

Fortunately this limitation on selectivity of detection has been greatly diminished by two recent developments. The relatively new ability to operate piezoelectric sensors in the presence of a liquid not only allows more selective techniques of surface binding to be employed, but also allows a greater range of sample matrices to be examined.

Secondly, the demonstration that biomolecules in the form of antibodies can be used to cause surface binding to specific antigens tremendously enhances the selective detection capability of these devices. Applications are now possible not only in the traditional areas of immunoassay (e.g. clinical analysis) but also to any analytical problem where a biomolecule can be developed to selectively bind to the analyte of interest. Earlier chapters have reviewed the possibilities of biological detection which are now applicable to piezoelectric sensors. Thus one should expect a great deal of growth in the development of these devices as the field matures.

References
1. B.A. Auld, *Acoustic Fields and Waves in Solids*. Wiley–Interscience, New York (1973).
2. E. Dieulesaint and D. Royer, *Elastic Waves in Solids, Applications to Signal Processing*. John Wiley, New York (1980).
3. V.M. Ristic, *Principles of Acoustic Devices*. Wiley–Interscience, New York (1983)
4. G. Sauerbrey, *Z. Phys.* **155** (1959) 206.
5. C.-S. Lu and A.W. Czandenna (eds.), *Applications of Piezoelectric Quartz Crystal Micro-balances*. Elsevier, Amsterdam (1984).
6. C.D. Stockbridge, Vacuum microbalance techniques, in K. Behrndt (ed.), Plenum Press, New York (1966) vol. 5, 193.
7. K.H. Behrndt, *J. Vac. Sci. Technol.* **8** (1971) 622.
8. J.G. Miller and D.I. Bolef, *J. Appl. Phys.* **39** (1968) 4589.
9. C.-S. Lu and O. Lewis, *J. Appl. Phys.* **43** (1972) 4385.

10. E. Sollner *et al. Vacuum* **27** (1977) 367.
11. M. Thompson, C.L. Arthur and G.K. Dhaliwal, *Anal. Chem.* **58** (1986) 1206–1209.
12. S. Bruckenstein and M. Shay, *Electrochim. Acta* **30** (1985) 1295–1300.
13. K.K. Kanazawa and J.G. Gordon, *Anal. Chim. Acta* **175** (1985) 99–105.
14. K.K. Kanazawa, personal communication.
15. Lord Rayleigh, *London Math. Soc. Proc.* **17** (1887) 4.
16. D.P. Morgan, *Surface Wave Devices for Signal Processing.* Elsevier, New York (1985).
17. R.M. White and F.W. Voltmer, *Appl. Phys. Lett.* **7** (1965) 314–316.
18. Ref. 2, p. 334.
19. H.I. Smith, in A.A. Oliner (ed.), *Topics in Applied Physics*, Springer Verlag, Berlin (1978). vol. 24, 305–324.
20. E.A. Ash, in A.A. Oliner (ed.), *Topics in Applied Physics*, Springer Verlag, Berlin (1978) vol. 24, 97–186.
21. Ref. 1, vol. 2, Ch. 12.
22. H.F. Tiersten, *J. Appl. Phys.* **40** (1969) 770.
23. G.W. Farnell, in A.A. Oliner (ed.), *Topics in Applied Physics,* Springer Verlag, Berlin (1978) vol. 24, 13–60.
24. A.J. Ricco, S.J. Martin and T.E. Zipperian, *Sens. Act.* **8** (1985) 319–333.
25. A. Venema, E. Nieuwkoop, M.J. Vellekoop, M.S. Nieuwenhuizen and A.W. Barendsz, *Sens. Act.* **10** (1986) 47–64.
26. K.A. Ingebrigtsen, *J. Appl. Phys.* **44** (1970) 454–459.
27. S. Datta, *Surface Acoustic Waves.* Prentice Hall, Englewood Cliffs (1985).
28. I.A. Viktorov, *Rayleigh and Lamb Waves.* Plenum Press, New York (1967).
29. A. Suleiman and G.G. Guilbault, *Anal. Chim. Acta* **162** (1984) 97–102.
30. H. Wohltjen and R. Dessy, *Anal. Chem.* **51** (1979) 1458–1464.
31. H. Wohltjen, *Sens. Act.* **5** (1984) 307–325.
32. S.J. Martin, K.S. Schweizer, A.J. Ricco and T.E. Zipperian, *Proc. 3rd Int. Conf. Solid State Sensors and Actuators (Transducers 85)*, Philadelphia, PA (1985) 71.
33. J. Hlavey and G.G. Guilbault, *Anal. Chem.* **49** (1977) 1890–1898.
34. G.G. Guilbault, *Ion Sel. Electrode Rev.* **2** (1980) 3.
35. J.F. Alder and J.J. McCallum, *Analyst* **108** (1983) 1169–1189.
36. H. Wohltjen and R. Dessy, *Anal. Chem.* **51** (1979) 1465–1470.
37. H. Wohltjen and R. Dessy, *Anal. Chem.* **51** (1979) 1470.
38. M.S. Nieuwenhuizen, and A.W. Barendsz, *Sens. Act.* **11** (1987) 45–62.
39. K. Behrndt (ed.), *Vacuum Microbalance Techniques.* Plenum Press, New York (1960).
40. R.L. Chuan, *J. Aerosol Sci.* **1** (1970) 111.
41. J.G. Olin and G.J. Sem, *Atmos. Environ.* **5** (1971) 653.
42. T.E. Carpenter and D.L. Brenchley, *Amer. Ind. Hyg. Assoc. J.*, **33** (1973) 503.
43. W.H. King, *Anal. Chem.* **36** (1964) 1735.
44. W.W. Schultz and W.H. King, *J. Chrom. Sci.* **11** (1973) 343.
45. F.W. Karasek and J.M. Tierney, *J. Chrom.* **89** (1974) 31.
46. F.W. Karasek, P. Guy, H.H. Hill and J.M. Tierney, *J. Chrom.* **124** (1976) 179.
47. M. Janghorbani and H. Freund, *Anal. Chem.* **45** (1973) 325.
48. G.G. Guilbault, A. Lopez–Roman and S.M. Billedeau, *Anal. Chim. Acta* **58** (1972) 421–427.
49. S.M. Fraser, T.E. Edmonds and T.S. West, *Analyst* **111** (1986) 1183–1188.
50. K.H. Karmarker and G.G. Guilbault, *Environ. Lett.* **10** (1975) 237.
51. W.W. Fogelman and M.S. Shuman, *Anal. Lett.* **9** (1976) 751.
52. Y. Tomita, M.H. Ho and G.G. Guilbault, *Anal. Chem.* **51** (1979) 1475.
53. A. Kindlund and I. Lundstrom, *Sens. Act.* **3** (1982/83) 63.
54. J. Hlavey and G.G. Guilbault, *Anal. Chem.* **50** (1978) 965.
55. Y. Tomita and G.G. Guilbault, *Anal. Chem.* **52** (1980) 1484–1489.
56. A. Suleiman and G.G. Guilbault, *Anal. Chim. Acta* **162** (1984) 97–102.
57. J.J. McCallum, P.R. Fielden, M. Volkan and J.F. Alder, *Anal. Chim. Acta* **162** (1984) 75–83.
58. K.H. Karmarker and G.G. Guilbault, *Anal. Chim. Acta* **71** (1974) 419.
59. A. Bryant, D.L. Lee and J.F. Vetelino, *Proc. Ultrasonics Symp.* (1981) 171–174.
60. A. Bryant, M. Poirer, G. Riley, D.L. Lee and J.F. Vetelino, *Sen. Act.* **4** (1983) 105.
61. C.T. Chuang, R.M. White and J.J. Bernstein, *Electr. Dev. Lett.* **EDL-3** (1982) 145–148.
62. A. Snow and H. Wohltjen, *Anal. Chem.* **56** (1984) 1411–1416.

63. A. D'Amico, A. Palma and E. Verona, *Sens. Act.* **3** (1982) 31–39.
64. J.F. Vetelino, R. Lade and R.S. Falconer, *Proc. 2nd Int. Meet. Chemical Sensors*, Bordeaux (1986) paper 7–07, 688–691.
65. W.P. Carey and B.R. Kowalski, *Anal. Chem.* **58** (1986) 3077–3084.
66. W.P. Carey, K.R. Beebe, B.R. Kowalski, D. Illman and T. Hirschfeld, *Anal. Chem.* **58** (1986) 149–153.
67. D.S. Ballantine, S.L. Rose, J.W. Grate and H. Wohltjen, *Anal. Chem.* **58** (1986) 3058–3066.
68. T. Nomura and A. Minemura, *Nippon Kagaku* Kaishi: **1980** (1980) 1261.
69. P.L. Konash and G.J. Bastiaans, *Anal. Chem.* **52** (1980) 1929–1931.
70. J.E. Roederer, and G.J. Bastiaans, *Anal. Chem.* **55** (1983) 2333–2336.
71. S.J. Martin, A.J. Ricco and R.C. Hughes, *Proc. 4th Int. Conf. Solid-state Sensors Actuators*, Tokyo 1987, paper B-4.2.
72. J.P. Mieure and J.L. Jones, *Talanta* **16** (1969) 149.
73. A. Shons, F. Dorman and J. Najarian, *J. Biomed. Mater. Res.* **6** (1972) 565–570.
74. T.K. Rice, US Patent 4,236,893.
75. R.J. Oliveira and S.F. Silver, US Patent 4,242,096.
76. T.K. Rice, US Patent 4,314,821.
77. A.J. Ricco and S.J. Martin, *Appl. Phys. Lett.* **50** (1987) 1474.
78. T. Nomura and T. Mimatsu, *Anal. Chim. Acta* **143** (1982) 237–241.
79. T. Nomura and M. Iijima, *Anal. Chim, Acta* **131** (1981) 97.
80. S. Bruckenstein and M. Shay, *J. Electroanal. Chem.* **188** (1985) 131–136.
81. S. Bruckenstein and S. Swathirajan, *Elecrochim. Acta* **30** (1985) 851–855.
82. G.J. Bastiaans, Report to US Army, CRDEC-CR-86040 (1986).
83. Ref. 1, Ch. 3.
84. *IEEE Trans. Ultrason. Ferroelec. Freq. Control* **UFFC–34** (2), (1987) 122–212.

Index